Forest Ecosystems
Concepts and Management

ACADEMIC PRESS, INC.
Harcourt Brace Jovanovich, Publishers
San Diego New York Boston
London Sydney Tokyo Toronto

Forest Ecosystems
Concepts and Management

Richard H. Waring
Department of Forest Science
Oregon State University
Corvallis, Oregon

William H. Schlesinger
Department of Botany
Duke University
Durham, North Carolina

ACADEMIC PRESS, INC
San Diego, California 92101

United Kingdom Edition published by
ACADEMIC PRESS LIMITED
24-28 Oval Road, London NW1 7DX

Library of Congress Cataloging in Publication Data

Waring, Richard H.
Forest ecosystems.

Bibliography: p.
Includes index.
1. Forest ecology. 2. Forest management.
I. Schlesinger, W. H. II. Title.
QH541.5.F6W35 1985 574.5'2642 85-4031
ISBN 0-12-735440-9 (alk. paper)
ISBN 0-12-735441-7 (paperback)

PRINTED IN THE UNITED STATES OF AMERICA
90 91 92 93 94 9 8 7 6 5 4 3

To our teachers and our students

Contents

Preface

Our challenge and reasons for writing this book are to share an emerging insight that there are key linkages between the processes that operate in forests. We emphasize forests in this book because we know them best and because their long life permits us to evaluate the effects of periodic disturbance more readily. Our examples, most often, are drawn from simple cases in which principles are more easily seen and explained. We believe, however, the principles apply widely, as we show in predicting transpiration and other hydrologic properties for a variety of forests in differing climatic settings.

In many cases, scientists cannot accurately predict the effects of acid rain, fire suppression, short-rotation timber harvest, or increasing carbon dioxide levels in the atmosphere on forests or other ecosystems. We believe a diagnostic approach linking a variety of processes is warranted and that with carefully designed experiments the mysteries will unravel.

We have striven to provide a cosmopolitan flavor to the book. Because most experimental work has been focused on rather simple systems, our examples are drawn mainly from temperate and boreal forests. However, the same processes operate in more complex forests, as references denote. The book is written for upper-level students with some background in general ecology, inorganic chemistry, physics, and plant physiology. We hope that specialists will see new implications to their work and be encouraged to develop integrative experiments. Managers of forest resources (wood products, wildlife, and water) will find explanations for some of their observations and predictions of the effects of various management policies.

We owe a debt to earlier studies of ecosystems, particularly those sponsored by the National Science Foundation in the 1970s as part of the International Biological Program, which established a group of five major ecosystem programs in the United States, in addition to earlier work at Hubbard Brook in New Hampshire. For almost a decade, balance sheets were constructed describing how carbon, water, and minerals are stored or transported in a variety of forest, grassland, desert, tundra, and aquatic systems.

Much of the basic information has been published in books and other periodicals. A summary of the international program with listings of data from all woodland sites appeared in 1981, edited by D. E. Reichle.[1] Regional efforts at

[1]Reichle, D. E. (1981). "Dynamic Properties of Forest Ecosystems." Int. Biol. Programme No. 23, Cambridge Univ. Press, London and New York.

synthesis have also been made for the other biomes. These references, as well as the open literature, provide a description of how forest systems operate.

Few of the research programs, however, involved experiments that evaluated linkages between major processes. The influence of fire, erosion, wind storms, and epidemic outbreaks of insects or disease organisms could not be rigorously evaluated until a benchmark for rates of normal processes had been established. The foundation was laid for critical experiments that could test hypotheses involving how and why ecosystems respond to periodic disturbances of various kinds. We propose that integrated experiments based on ecosystem-level insight can provide answers to managers. Whether this is the case, as we emphasize in interpreting the probability of disturbance in forests, awaits future tests.

R. H. W.
W. H. S.

Acknowledgments

In the endeavor of writing this book we received encouragement and support from students and faculty at a number of institutions, including the Swedish University for Agricultural Sciences. To all of the people involved, we are most grateful. For their part in reviewing one or more chapters, we specifically thank Dennis Knight, Pamela Matson, Cindi Jones, John Pastor, Kate Lajtha, Pete Linkins, Steve Running, Dan Livingstone, Paul Jarvis, Sune Linder, John Marshall, Erik Christiansen, Delef Schulze, Douglas Sprugel, Jerry Franklin, Virginia Dale, Erkki Mattson-Djos, Paul Kramer, Adelaida Chaverri, Stig Larsson, Johann Goldammer, Chris Maser, Chris Vaughan, Robert Burgess, Jim Sedell, Fred Swanson, Charles McClaugherty, Tom Hanley, Folke Andersson, Dennis Harr, George Brown III, Robert Peet, John Fox, Wallace Covington, Richard Holmes, Don Knutson, John Melack, and especially Peter Vitousek and Jack Ewel. Doris and Lise Waring provided assistance in translating Swedish. Cindy Romo drafted all of the figures in the book, and the Academic Press editorial staff assisted in many ways besides editing.

R. H. W.
W. H. S.

1

Introduction

Forest vegetation covers about one-third of the Earth's land surface. When we look at forests from a distance, we recognize that large areas can be studied as single experimental units known as ecosystems. A forest ecosystem includes the living organisms of the forest, and extends vertically from the top of the canopy to the lowest soil layers affected by biotic processes. Within a forest ecosystem, the population ecologist may examine the birth and death of species that comprise the forest, while the community ecologist is often interested in the distribution of species and species diversity across the landscape. By considering the entire area of land as an experimental unit, the ecosystem ecologist studies "the circulation, transformation, and accumulation of energy and matter through the medium of living things and their activities" (Evans, 1956). Rather than concentrating on the growth of individual trees, the ecosystem ecologist expresses forest growth as net primary production in the units of kilograms per hectare per year. Ecosystem ecology is less concerned with the diversity of species than with the contribution that species make to the water, energy, and nutrient transfer in large units of landscape.

While the goal of ecosystem ecology is to understand how major landscape units function, the explanation of ecosystem-level phenomena is often found by understanding processes at the physiological, population, and community levels. Ecosystem ecology is an integrative science, which has its roots in the field of system analysis that was first applied to the design of complex systems in electrical engineering (Boulding, 1956). Ecosystem ecologists often build mathematical models of how forests operate. These are always simplifications of nature, but they are useful to evaluate the importance of certain variables and to show the limits of understanding.

Studies of forest ecosystems must consider boundaries in space and time. Often ecosystem ecologists have found it convenient to define the spatial boundaries by watershed catchments (Bormann and Likens, 1967). In other instances, specific stands of trees are appropriate experimental units because vegetation may vary in obvious characteristics within a watershed (Knight *et al.*, 1985). In studies of the movement and role of large animals, the boundary must be less precise, perhaps coinciding with the average home range occupied by these species over a year. Thus, boundaries of forest ecosystems are defined for the specific purposes of study, but once defined, these boundaries are used to identify and quantify the flux of materials and energy entering or leaving the system.

Many ecosystem processes occur continuously. Others, such as daily or seasonal photosynthesis, occur in predictable cycles. Still others, such as forest fires, occur at infrequent intervals, but these events are often somewhat predictable over long periods of time. The duration of ecosystem studies is defined by the questions that are addressed, but most studies require periodic measurements and collections for at least 1 year. In many cases, a decade of research gives an adequate understanding of the variability and long-term effect of processes occurring in the forest. Some questions of ecosystem function can only be addressed when studies span a complete turnover in the population of dominant individuals (Connell and Sousa, 1983). Time and spatial criteria for ecosystem studies are not independent. Processes that are variable from year-to-year on small plots may be more predictable when larger areas are studied. Events that occur infrequently in small plots can be observed in the context of larger landscape units. For example, in many areas, the occurrence of forest fire in a particular stand of trees is random and infrequent, while the occurrence of fire over a forest landscape may be a cyclic process that produces a mosaic of stands of varying age (Van Wagner, 1978).

Forest ecosystems are open systems in the sense that they exchange energy and materials with other systems, including adjacent forests, downstream ecosystems, and the atmosphere. The exchange is essential for the continued persistence of the ecosystem. A forest ecosystem may exist in steady state, if inputs and outputs balance and the storage of materials is not changing rapidly (Johnson, 1971). A forest ecosystem is never in equilibrium, a term appropriate to closed systems in the laboratory. Only the entire globe exists as an equilibrium system, and then only for materials and not for energy.

Ecosystem studies consider not only the flux of energy and materials through a forest, but also the transformations that occur within the forest. These transformations are an index of the role of biota in the behavior of the system. Often it is useful to test ecosystem hypotheses by comparing field observations to what might be expected to occur in systems without a particular biotic component. For example, do present-day forests of the eastern United States capture a different amount of energy in photosynthesis than before the blight eliminated chestnut (*Castanea dentata*) as an important constituent of many forests? Has the absence of chestnut altered the loss of nutrients to streamwater? Although the ecosystem ecologist is not strongly interested in the diversity of species that are present, some species often have more influence on system properties than others. Understory shrubs that fix nitrogen may contribute to the flux of nitrogen through a forest ecosystem far in excess of their contributions to forest biomass or photosynthetic capability. Soil microbial reactions and species that contribute to canopy leaf area are often of great importance to the properties of forest ecosystems. It is often convenient to begin forest ecosystem studies in plantations or in pure natural stands. As important properties are perceived, their generality can be tested in more diverse natural communities.

During the natural process of forest development, the species composition changes in a predictable order known as succession. When a forest develops upon soil that has never supported the growth of vegetation (e.g., volcanic ash), this process is primary succession. Development of vegetation on abandoned farmland or on harvested sites is known as secondary succession. During the successional development of forests, a number of ecosystem properties also change in a predictable manner (Odum, 1969). The important variables that affect forest growth and productivity are often controlled by abiotic factors in early developmental stages and by biotic interactions at later stages. Biomass and canopy leaf area increase to a maximum level that can be maintained at steady state by the resources of the site. During forest growth, the loss of nutrients in streamwater is often much lower than in barren sites. Such changes are due to the combined action of the species that constitute the biotic component of the ecosystem. However, in this book we take the view that there are no emergent properties of ecosystems, such as nutrient conservation, that cannot be predicted from a thorough knowledge of the components of the ecosystem and their interactions.

It is relatively easy to measure the amounts of materials stored in the living biota and dead organic matter in forest ecosystems. A large number of these studies exist, as well as studies of the rate of change in the size of these compartments, made by comparative measurements over a period of years. What *controls* the rate of processes in forest ecosystems is more difficult to study, but an understanding of factors that control forest growth and nutrient cycling is essential for effective forest management and for predicting the response of forest ecosystems to environmental changes. Experimental studies are often needed to establish the important control factors. Studies of some processes, such as soil microbial transformations, can often be performed in the laboratory. In other cases, field experiments must be performed by manipulations of stands or watersheds. The application of unusual conditions to forest ecosystems often allows the ecosystem ecologist to recognize natural processes and the important factors that regulate them.

When forest ecosystems are subjected to unusual conditions caused by environmental fluctuations or experimental manipulations, stress symptoms appear that are useful in understanding how different ecosystem processes are interrelated. For example, fertilization experiments show how nutrient availability controls the amount of carbon that trees allocate to root growth; when there is little root growth in fertile soils, more photosynthate is available for wood growth. In such favorable conditions, trees may also produce compounds that protect against insects and disease. Thus, fertilization experiments have shown the linkage between nutrient availability and internal carbon allocation in forest ecosystems.

Throughout this book we show how a measure of the efficiency of forest wood growth—the amount of wood produced per unit of leaf area—is a useful index of the balance of light, water, and nutrient availability in forest ecosystems. The growth efficiency index was developed from observations of carbon allocation during experimental treatments such as fertilization and thinning, but it is sensitive to changes in a variety of forest conditions during stand development or for comparisons among natural stands. We can expect the efficiency of wood production to decrease as leaf area increases during forest development, due to the mutual shading of leaves (Waring, 1983). This efficiency index declines in conditions of limited water or nutrients and during the exposure of forests to air pollution. Other indices are also available and useful to assess the factors that affect forest growth (Table 1.1). Many forest ecologists are familiar with foliar analysis and leaf water potential measurements as indices of site conditions.

The storage of materials in a forest ecosystem, such as the storage of water in the rooting zone, often determines the relative resistance of the ecosystem to a stress, such as drought. Thus the mass of ecosystem components can provide inertia against environmental perturbations. In contrast, the recovery or resilience of systems that are disturbed is largely determined by the rate of metabolic processes, such as forest productivity. The availability of propagules of fast-growing species speeds the recovery of disturbed ecosystems and the restoration of ecosystem function to predisturbance levels.

Models of forest ecosystems are based on the definition of a system in space and time and recognition of the compartments and transfers within this system. The compartments are known as state variables, and they represent the storage of materials and energy in the ecosystem. The rate at which materials and energy move between compartments or between the compartments and the external environment is described by equations known as state equations. Feedback or control is introduced in models when the rate of one process becomes a factor in the mathematical expression of the rate of a second process, which in turn affects the rate of the first process. Models are empirical if these equations are based on

TABLE 1.1 Forest Ecosystem Stress Indicators

Indication	Examples
Canopy limitation	Maximum leaf area index Duration of leaf display
Production limitation	Diameter growth and cell division Growth efficiency per unit leaf area
Susceptibility to insect and disease attack	Starch content of twigs and large roots Tannin and terpene content of tissues Growth efficiency per unit leaf area
Moisture limitations	Predawn water stress Noon leaf turgor Sapwood relative water content Stomatal closure
Nutrient limitations	Foliar nutrient content Foliar nutrient ratios Nutrient retranslocations Mineralization indices
Physical stress	Sapwood/diameter index Bole taper Symmetry of wood growth Leaf-edge tatter

statistical curve-fitting to field or laboratory observations made over a range of conditions. Models are theoretical if the equations are based on well-known physical properties that determine the movement of materials and energy. If an ecosystem model is based entirely on mathematical equations to describe processes, the model is deterministic. Models that introduce random events, such as fires and storms, are stochastic. Of course, a single model may have both deterministic and stochastic components. Models are introduced in various places in this book to illustrate their use in forming hypotheses and in predicting the response to perturbations. Models are, of course, no better than the knowledge and data that are used to build them, and they are subject to continual validation and improvement.

In this book, we first examine the fixation of solar energy through photosynthesis in forests. We trace the fate of this energy by following the movement of organic compounds in forest ecosystems and the respiration of carbon dioxide to the atmosphere. Then we consider the availability of water for the growth of trees and the movement of water through the forest ecosystem. Movements of water often carry nutrient elements, so our discussion turns to examine the sources, circulation, and losses of nutrient elements in forest ecosystems. Particular emphasis is placed on soil processes as components of the forest ecosystem. Forest ecosystems include a wide variety of animals, ranging from insects to large predatory animals, which affect forest establishment and growth. Several chapters consider the role of animals and the response of forest ecosystems to insects and disease attack. Finally, we consider several ways in which the properties and processes of forest ecosystems affect other ecosystems, ranging from forest streams to the global biosphere.

It is hoped that this approach provides an overview of how forests operate from an ecosystem-level perspective. Theory developed from observations in natural systems and from manipulations in experimental forests should be useful in the improvement of forest management.

2

The Carbon Balance of Trees

INTRODUCTION

Carbon compounds are the currency that plants accumulate, store, and use to build their structure and maintain their physiological processes. The energy captured in molecular bonds of carbon compounds generally represents between 2 and 4% of the radiation absorbed by a tree canopy (Edwards *et al.*, 1981; Kozlowski and Keller, 1966; Monteith, 1972). Stem wood growth often accounts for less than 20% of all the dry matter produced during a year (Linder and Axelsson, 1982; Grier *et al.*, 1981). Even small changes in photosynthesis or in

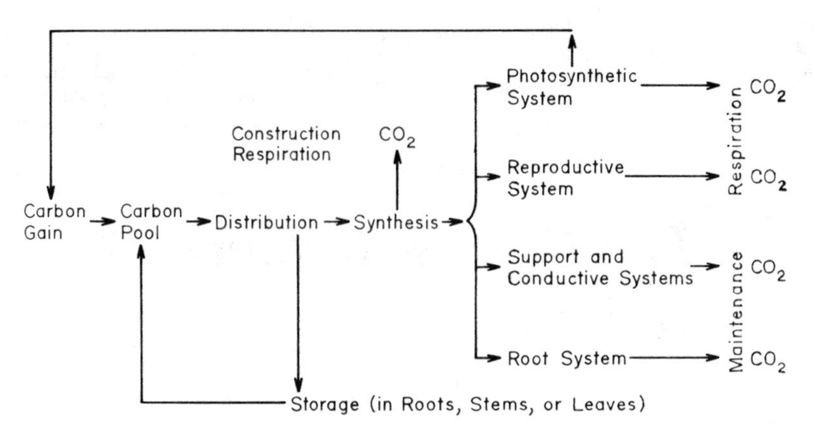

Fig. 2.1. Carbon uptake through the photosynthetic system is made available to a general pool of carbohydrates used in construction and maintenance of various tissues. In metabolic processes carbon dioxide is respired. Carbohydrates may be shifted from one category to another, depending upon the environmental conditions. (From Mooney, 1972.)

the pattern of carbohydrate allocation can profoundly alter the structure of a forest. Such changes affect the forest's value as a habitat and as a source of raw materials.

Just because a tree has a large carbon income does not mean it will produce wood rapidly. Carbon products are distributed to those points within a plant that are most likely to increase the plant's or its progeny's chances of survival. In certain periods growth may proceed entirely from stored reserves. The interactions between carbon gain, storage, growth, and maintenance are summarized diagrammatically in Fig. 2.1.

Because the support and conductive tissues represent a much greater proportion of live biomass in trees than in other forms of life, the carbon balance of trees differs from that of many other kinds of plants. The cost of maintaining these support and conducting tissues is high for trees. In trees reproductive organs may be produced infrequently, but at times profusely. Storage reserves are often large and help permit trees to weather unfavorable periods and to minimize injury to various tissues.

This chapter reviews the major environmental constraints upon carbon uptake. Construction and maintenance costs of various tissues are then assessed with special emphasis on the sapwood, which has a high maintenance cost in large trees. The importance of storage reserves is discussed, and we will review methods for estimating carbon allocation into major structural components. Departures from the normal pattern of allocation are shown to be indicative of stress. By understanding why certain changes in allocation occur, we may discover ways to favor production of stem, roots, secondary products (such as tannins or resins), or seeds.

CARBON UPTAKE

Through the process of photosynthesis, green plants absorb light energy and reduce carbon dioxide (CO_2) to carbohydrates that serve as raw materials for further biochemical synthesis. Both physical and chemical reactions occur during photosynthesis. The chloroplasts, within which photosynthesis takes place, are located mainly in the palisade parenchyma and spongy mesophyll cells of leaves (Fig. 2.2). The principal path of carbon from the atmosphere to the plant is by gaseous diffusion of CO_2 through leaf stomata into the leaf intercellular spaces, and then through the walls of the mesophyll cells to the sites of fixation (carboxylation). The latter transport is in the liquid phase. Carbon dioxide diffusion is dependent on how open the stomata remain and how rapidly the mesophyll can absorb CO_2. Absorption by the mesophyll cells is related to their surface area compared to that of the leaf surface and to the rate of carboxylation (Nobel, 1983).

Chlorophyll and other pigments in the chloroplasts, through photochemical processes, absorb the wavelengths of light that provide energy for the biochemical processes that combine water and carbon dioxide into a simple 3-carbon sugar. Further combinations produce other compounds that are metabolized in the construction and maintenance of plant structures. In respiration, sugars are broken down and CO_2 is released. Because respiration is active at all times, measurements of carbon uptake are net, representing the difference between photosynthesis and leaf respiration. Respiration measured in the dark may

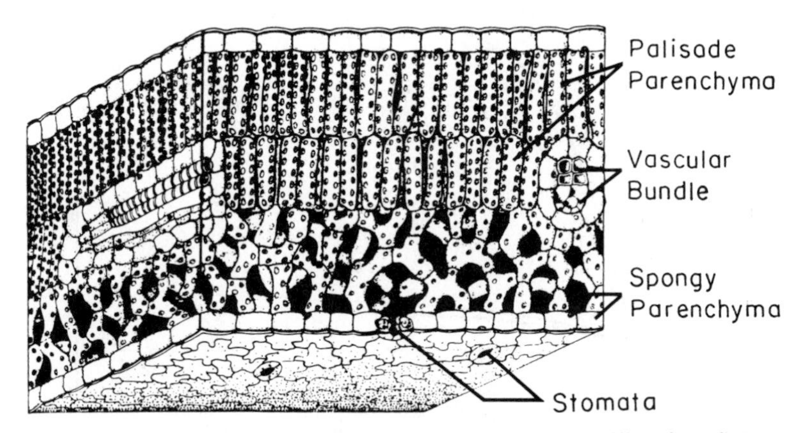

Fig. 2.2. View through a leaf shows how carbon dioxide must diffuse from the atmosphere through stomata in the leaf surface to reach the spongy and palisade layers of parenchyma cells making up the mesophyll where the majority of chloroplasts are concentrated and photosynthesis takes place. The vascular bundle conducts water to the leaf. Sugars move out of the leaves through phloem cells surrounding the vascular tissue. (From Wilson, 1952.)

be only a small fraction of that in light because oxidation of intermediate photosynthetic products may occur in the light (photorespiration) in most forest trees. Certain grasses, desert shrubs, and herbs have photosynthetic systems that lack photorespiration and differ in other ways (Mooney, 1972; Ehleringer, 1979).

In our treatment we consider general relationships between carbon uptake and the environmental variables: (1) light; (2) CO_2 concentration; (3) temperature; (4) water; and (5) nutrient availability. We first evaluate these various factors independently, then integrate them to estimate the uptake of carbon during the growing season or for an entire year.

Environmental Constraints

Photosynthesis uses radiation in the visible spectrum (400–700 nm wavelengths). This photosynthetically active radiation is about 50% of the global radiation (Chapter 5). When individual leaves are exposed to varying irradiance, all show similar patterns of response (Fig. 2.3). Below some irradiance level, carbon uptake is negative, as respiration exceeds photosynthesis. As irradiance increases, a compensation point is eventually reached where uptake of CO_2 through photosynthesis is exactly balanced by losses through respiration. Above

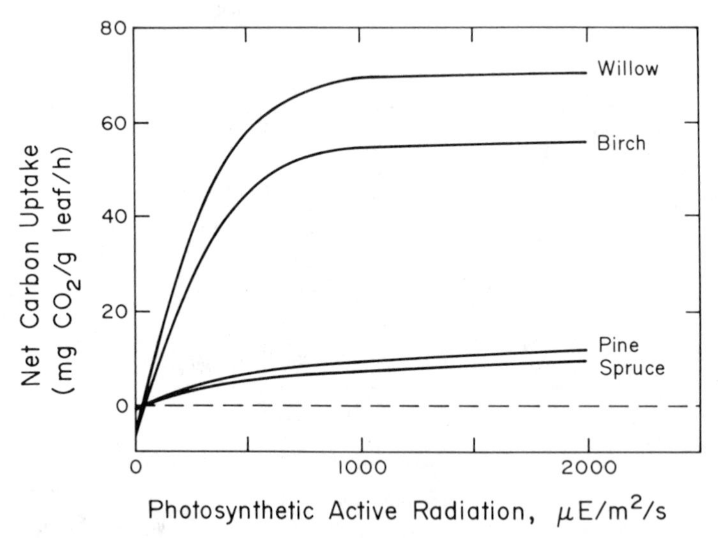

Fig. 2.3. All light-response curves for photosynthesis demonstrate a compensation point where net carbon uptake by the leaf is zero, an intermediate linear response to increasing irradiance, and an upper, light-saturated zone where photosynthesis reaches a plateau. (From Linder, 1981.)

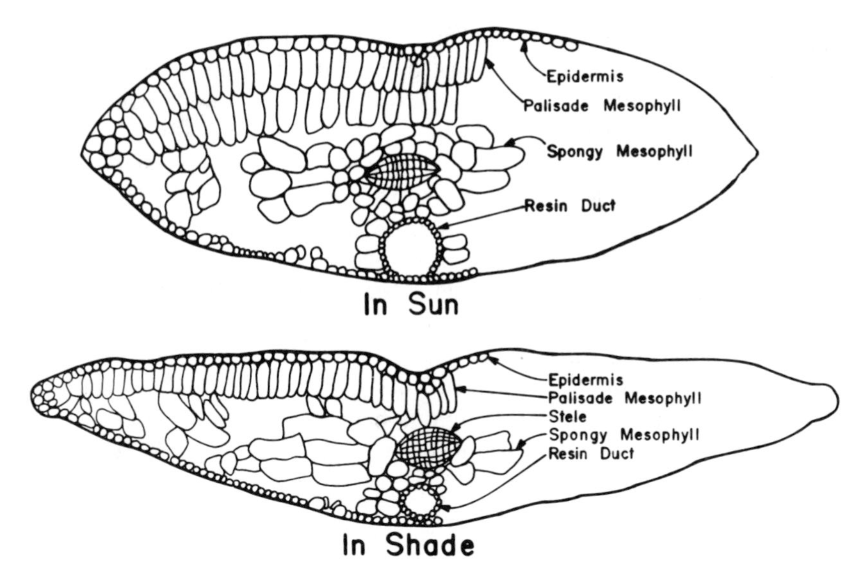

Fig. 2.4. Cross-sectional drawings of western hemlock (*Tsuga heterophylla*) leaves grown in the sun (upper) and in the shade (lower) illustrate that the volume of mesophyll tissue containing chloroplasts is more than twice as large for a comparable surface area in the former than in the latter. (From Tucker and Emmingham, 1977.)

the light compensation point, uptake increases linearly until the amount of carboxylation enzyme or available CO_2 limits the process.

In general, the leaves of deciduous tree species have the potential for higher photosynthetic rates than those of evergreens—particularly when the comparison is made on the basis of leaf dry weight. In terms of carbon uptake per unit of leaf surface area, the difference is less but sometimes still apparent, reflecting inherent differences in the diffusion pathway for carbon dioxide. Even within a single tree, significant differences in light response curves are evident between fully exposed foliage and that positioned in the more shaded part of the canopy. Foliage exposed to high illumination contains more layers of palisade mesophyll cells and has higher concentrations of carboxylation enzyme than more shaded leaves (Mooney, 1972; Berry and Downton, 1982). Exposed foliage is heavier per unit area than shaded foliage as a result of this difference in anatomy and the amount of storage reserves (Nygren and Kellomäki, 1983; Kozlowski and Keller, 1966; Fig. 2.4). In spite of these differences, the shade foliage may contribute as much as 40% to the tree's total carbon uptake (Schulze *et al.*, 1977; Schulze, 1981).

In general, leaf tissue can acclimate to an order of magnitude of variation in irradiance (Berry and Downton, 1982). In trees adapted to full sunlight, it is possible to saturate the light requirements of leaves in the uppermost canopy, but the shade cast by those leaves and others tends to reduce the irradiance that reaches the lower canopy to somewhat below saturation levels. This point is particularly significant when considering carbon uptake by trees maintained in forests with varying canopy densities.

Although complex radiation models have been developed to accommodate the heterogeneity of forest canopies (e.g., Norman and Jarvis, 1975) and to take into account the effects of solar elevation, diffuse and direct radiation, and spectral quality (Monteith, 1972; Hesketh and Baker, 1967), the most common approach is simply to consider the canopy as accumulated layers of foliage through which solar radiation is absorbed exponentially as the amount of leaf area increases. Foliage area is expressed as a leaf area index (LAI) representing all of the upper surfaces of leaves projected downward to a unit area of ground beneath the canopy. The penetration of visible light through leaf layers of different forest canopies approximates the Beer-Lambert law (Fig. 2.5):

$$\ln I_z/I_0 = -k\Sigma\mathrm{LAI}$$

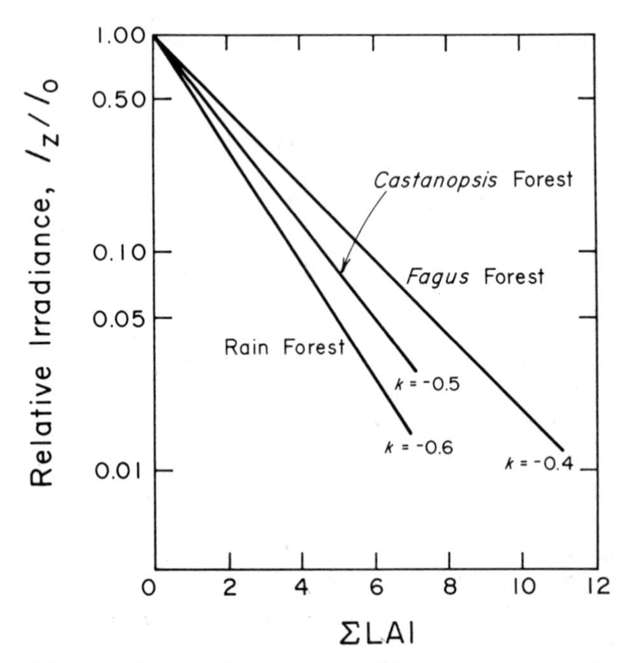

Fig. 2.5. Light penetration through diverse kinds of forest canopies approximates the Beer-Lambert law where the fraction of light transmitted (I_z) to that received above the canopy (I_0) is predicted with knowledge of the light extinction (slope) coefficient and cumulative leaf area index (ΣLAI) above the point of reference. (From Kira *et al.*, 1969.)

where I_z is the intensity of light beneath increasing accumulations of leaf area; I_0, intensity of visible light above the canopy; k, extinction coefficient, representing the slope of the relationship; and ΣLAI, cumulative amount of projected leaf area, in square meters of foliage per square meter of ground surface.

Note that light is absorbed less efficiently by a *Fagus* canopy than in the *Castanopsis* or rain forest (Fig. 2.5). High efficiency, characterized by a more negative extinction coefficient, may indicate low reflective properties of leaves or an overlapping arrangement, such that each leaf tends to shade those below. Foliage displayed on pendant branches casts less shade than foliage with horizontal orientation (Honda and Fisher, 1978). As canopies develop, the amount of branch surface area also increases, adding the equivalent of three or four to leaf area index (Whittaker and Woodwell, 1968; Swank and Schreuder, 1974; Ford, 1982). However, foliage is more efficiently displayed as the canopy deepens (Kira and Shidei, 1967), so extinction coefficients remain fairly constant for a particular type of vegetation.

The presence of chlorophyll pigment beneath the bark of branches, stems, or exposed roots should not be ignored for it indicates the importance of these surfaces for photosynthesis. As stems or branches grow in diameter, the amount of photosynthetic tissue in bark usually decreases. On certain thin-barked species such as *Betula* (birch) and *Populus* (poplar), large trees maintain the ability to fix respired CO_2 throughout the length of the bole, which helps counterbalance respirational losses during the dormant season when the trees lack foliage (Pearson and Lawrence, 1958; Strain and Johnson, 1963; Covington, 1975). Massive bark peeling is typical of many trees with green trunks that grow in arid regions, and bark photosynthesis may be important during the season when leaves are shed (Mooney, 1972).

As the concentration of CO_2 increases in the intercellular space of the leaf and at the chloroplasts, the biochemical reactions in photosynthesis increase linearly in a manner similar to that observed with irradiance. There are two major reasons. First, the amount of carboxylation enzyme is not limiting and, second, the ratio of CO_2 to O_2 is high, which minimizes photorespiration (Powles and Osmond, 1978; Ehleringer, 1979). At very high levels of CO_2, however, insufficient carboxylation enzyme may limit photosynthesis (Farquhar and Sharkey, 1982).

Under relatively stable environmental conditions, the intercellular level of CO_2 equilibrates at some concentration below that of ambient air. Increasing ambient CO_2 experimentally raises the internal CO_2 level (Fig. 2.6). The CO_2 concentration in the atmosphere has increased rapidly in this century (Chapter 11) but is still well below optimal for photosynthesis in most environments and probably not more than one-third to one-half that of 40–100 million yr ago, when the ancestors of today's tree species evolved (Walker, 1977; Arthur, 1982; Berner *et al.*, 1983).

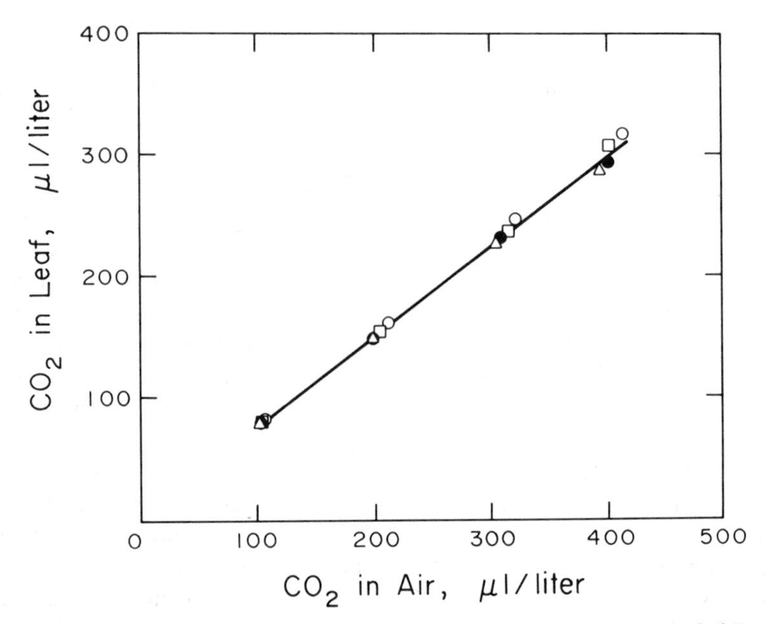

Fig. 2.6. Relationship between ambient CO_2 concentration and that within the leaf of *Eucalyptus pauciflora* at four irradiance levels: △ = 0.25, ● = 0.55, □ = 0.96, and ○ = 2.0 mE/m²/s. The ratio of internal to external CO_2 stays constant only so long as indirect factors do not inhibit diffusion of gases in or out of the leaf. (From Wong *et al.*, 1979, reprinted by permission from *Nature*, **282,** 424–426. Copyright © 1979 Macmillan Journals Limited.)

The optimum temperature range for photosynthesis varies with species but is commonly between 15 and 25°C for temperate trees, with extremes reported between about 10 and 35°C (Stocker, 1960; Kozlowski and Keller, 1966; Mooney, 1972). A shift in temperature optimum may occur if other factors change the intercellular level of CO_2. In general, increased internal CO_2 will shift the temperature optimum upward, as will increased light intensity or improvements in the availability of nitrogen and phosphorus (Cowan and Farquhar, 1977; Hall, 1979; Wong *et al.*, 1978).

Normally, seasonal shifts in the temperature optimum for photosynthesis do not exceed 5 or 10°C (Mooney, 1972). However loblolly pine (*Pinus taeda*) in the southeastern United States adjusts its temperature optimum from 25°C in the summer to 10°C in the winter (Strain *et al.*, 1976). Although photosynthesis may continue to increase with temperature, this leads to a reduction in internal CO_2 levels and favors greatly increased rates of photorespiration. Even when internal CO_2 levels are maintained, gross photosynthesis begins to decrease abruptly at high temperatures as a result of specific changes in chloroplast and enzyme activity (Berry and Downton, 1982). Normal metabolic (dark) respiration increases as well but is not as significant as previously thought (Berry and Bjorkman, 1980). As a result of this combination of factors, the net carbon uptake by leaves in relation to temperature follows a parabola, increasing gradually to the optimum and decreasing gradually thereafter (Fig. 2.7).

Because trees are able to adapt to a wide range in normal seasonal temperatures (Slatyer and Morrow, 1977), the frequency and duration of extremes may be more critical than mean temperatures in limiting photosynthesis. For example, a 1-day exposure of ponderosa pine seedlings to 3°C reduced net carbon uptake rates by 50% when the seedlings were returned to a 23°C environment (Pharis *et al.*, 1972). Nearly a month was required for seedlings to regain their full capacity following exposure to low temperatures for a few weeks (see Neilson *et al.*, 1972).

Exposure to frost may be even more damaging. Frost exposure reduced the carbon uptake rates in *Pinus cembra* and *Picea abies* by as much as 90% for periods of weeks or even months (Polster and Fuchs, 1963; Tranquillini, 1979). Similar results are reported for *Pinus radiata* (Rook, 1969) and *Pinus sylvestris* (Troeng and Linder, 1982). Different mechanisms may operate but increased respiration following exposure to cold temperatures suggests the repair of damaged membranes is at least partly involved (Pharis *et al.*, 1972). High temperatures likewise may be injurious and the effects long-lasting (Levitt, 1972).

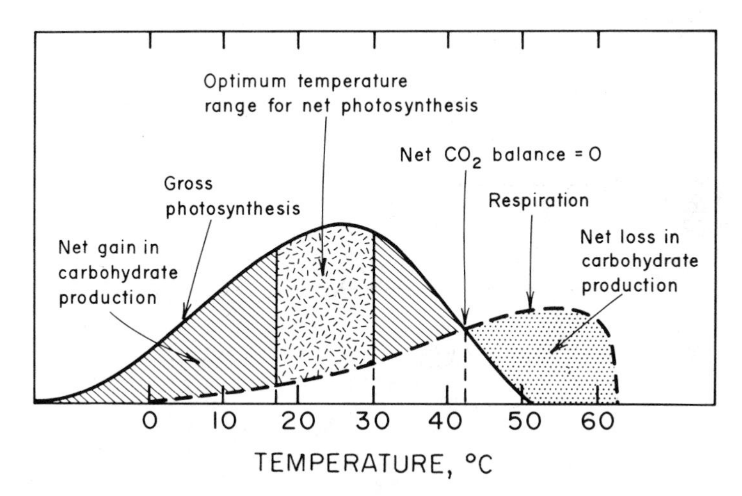

Fig. 2.7. Optimum temperature for net carbon uptake by leaves represents a zone where the difference between gross photosynthesis and losses of CO_2 by respiration are greatest. The optimum temperature varies with species and the season. At high as well as low temperatures, photosynthesis is adversely affected. A large net carbohydrate deficit results at high temperatures because of increasing respiration. Drawing slightly modified to emphasize somewhat greater importance of photosynthesis over respiration at high temperatures. (After Daniel *et al.*, 1979.)

In evaluating the importance of temperature to carbon uptake, we must know something about the normal optimum, possible seasonal adjustments, and exposure to critical extremes.

Water is essential to all living cells so any reduction in its availability might be expected to affect photosynthesis as well as many other processes. In Chapter 4, the water relations of trees is treated in some detail. Here, only the effects of limitations in leaf water supply upon photosynthesis and the related process of photorespiration are discussed.

The availability of water in leaf tissue is not directly dependent on the water content. Changes in cell wall elasticity, in membrane permeability, and in the concentration of solutes in cells counterbalance the effects of decreasing water content (Bradford and Hsiao, 1982). Diurnal variations of 5–10% in leaf water content relative to saturation often have no direct effect on photosynthesis (Hanson and Hitz, 1982). Eventually, if leaf tissue continues to lose water, stomata are forced to close and the rate of CO_2 diffusion into the leaf is reduced. At some point, stomata close completely, which halts all carbon uptake. There is usually a threshold where partial stomatal closure limits diffusion of CO_2 into the leaf and reduces carbon uptake sharply. This threshold is often defined as a particular water stress, measured in water potential (Chapter 4). The water potential threshold where stomatal closure reduces carbon uptake generally ranges from −1.0 to

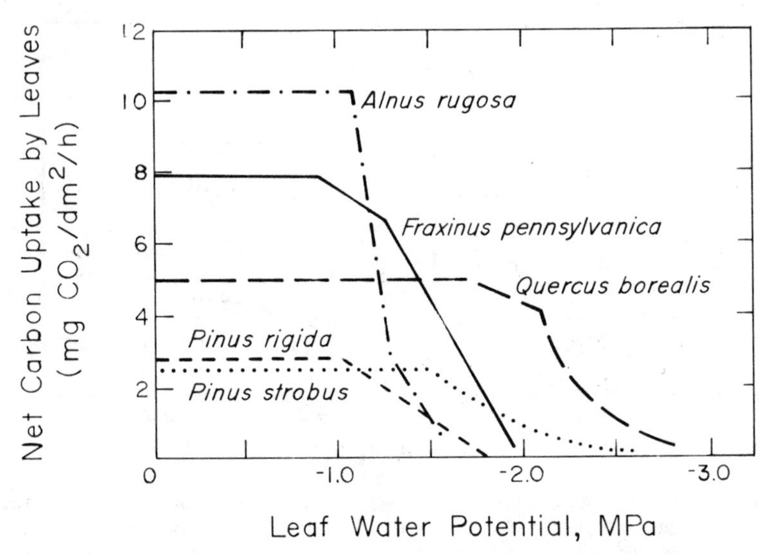

Fig. 2.8. Net carbon uptake by leaves is reduced at critical levels of leaf water potential that initiate stomatal closure and limit gas diffusion. (From Hinckley *et al.*, 1981.)

−2.5 megaPascals (MPa) for trees as illustrated in Fig. 2.8. Stomatal closure may also be initiated by a number of other factors, as discussed in Chapter 4, but the influence upon CO_2 diffusion into the leaf is similar.

As stomata begin to close, the internal CO_2 concentration in the leaf drops, which decreases the ratio of CO_2 to O_2 and favors photorespiration over photosynthesis. Under chronic drought conditions, some carboxylation enzyme will be broken down, reducing the biochemical capacity of the photosynthetic system. Concentrations of chlorophyll and other pigments important in the photochemical reaction may also be reduced (Farquhar and Sharkey, 1982).

Factors that affect diffusion of CO_2 through the stomata but not the photosynthetic capacity of chloroplasts cause the ratio of internal to external CO_2 levels to be reduced below those indicated in Fig. 2.6. From a theoretical mathematical analysis, Cowan and Farquhar (1977) suggest that plants may optimize diffusion through stomata and metabolic processes in the chloroplasts to maintain constant the ratio:

$$\frac{\text{Change in transpiration}}{\text{Change in photosynthesis}}$$

Thus a 20% reduction in transpiration due to drought-induced stomatal closure may be accompanied by a 20% reduction in carboxylation enzyme content and rates of photosynthesis.

Rates of biochemical processes can usually be increased with improved nutritional status of leaves. Nitrogen plays a particularly imporant role through its contribution to photosynthetic enzymes and to light-absorbing pigments. In some deciduous trees, photosynthetic rates can be increased by up to fivefold with higher levels of available nitrogen (Fig. 2.9). Photosynthesis by conifers, on the other hand, is less responsive to variation in leaf nitrogen content, with rates usually increasing by less than 25% upon fertilization (Brix, 1971; Linder and Rook, 1984). Conifers lack the capacity for increasing photosynthetic enzyme levels to the same extent as deciduous trees, and their leaf geometry makes CO_2 diffusion processes more likely to be limiting. Increased nitrogen is more likely to result in production of more foliage. These differences in photosynthetic response to nitrogen are important when interpreting the effects of fertilization upon growth (Chapters 3 and 7).

Of course any deficiency of a critical nutrient may indirectly reduce total carbon uptake by restricting the amount of photosynthetic tissue, by modifying the ability of stomata to open quickly, or by increasing respiration as resources are shifted from one metabolic process to another (Keller and Wehrmann, 1963; Jarvis, 1981a). There is evidence, however, that limitations in phosphate, iron, and even potassium have direct effects that reduce photosynthesis within chloroplasts (Berry and Downton, 1982).

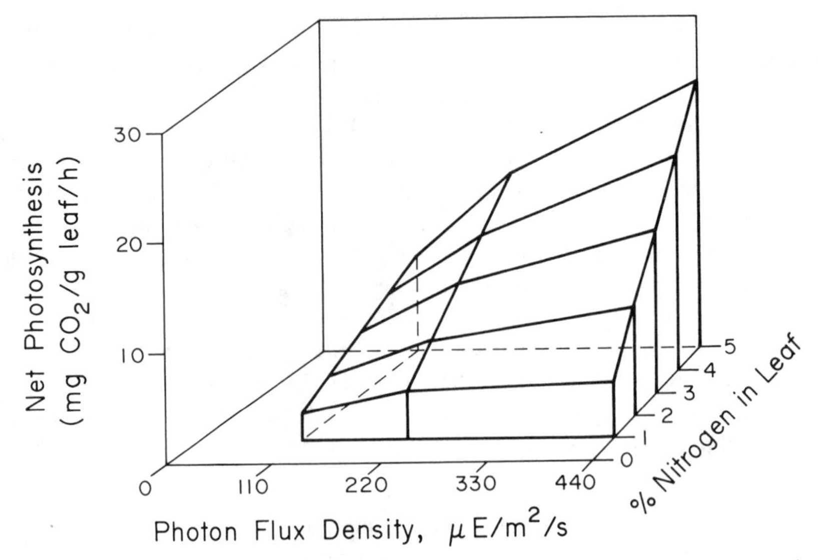

Fig. 2.9. Deciduous trees such as *Betula pendula* increase photosynthetic rates by as much as fivefold as nitrogen content of the foliage is increased. (From Linder *et al.*, 1981.)

Simulation Models

Although a few biochemically based simulation models of photosynthesis have been developed (Hall, 1979; Cowan and Farquhar, 1977; Farquhar *et al.*, 1980), none has been assessed in the field for any tree species. To evaluate interactions among the important environmental variables we must therefore employ an empirical approach, measuring carbon uptake rates over the widest possible range of environmental conditions for a particular tree species. Often the natural range of combinations is extended by providing supplemental light or by varying temperature and humidity within gas exchange cuvettes (Schulze, 1981; Benecke and Nordmeyer, 1982). Then, curves are fitted to the data to describe the various relationships previously discussed, and these are linked together in a computer simulation model.

The major variables required for such a model are solar radiation, temperature, stomatal and mesophyll conductance to CO_2 diffusion (the latter is a measure of biochemical performance), and ambient CO_2 concentration. With this approach, Reed *et al.* (1976) accounted for 90% of the variance in net carbon uptake observed in tulip poplar (*Liriodendron tulipifera*) trees growing in the southeastern United States. The model was applied to Douglas fir (*Pseudotsuga menziesii*) growing in Washington State (Sollins *et al.*, 1981) and to Sitka spruce (*Picea sitchensis*) growing in Scotland (Watts *et al.*, 1976). Applying the simulation model to data from a wide variety of environments in western North America, Emmingham (1982) discovered that many coniferous trees growing in mild winter climates probably accumulate a major portion of their annual carbon uptake during the fall, winter, and spring (Waring and Franklin, 1979). This was particularly true where trees experienced drought-induced stress during the growing season (Fig. 2.10). In simple form, models predict a high potential uptake of carbon, limited only by the amount of light and average temperature. By a comparison of the decrease in photosynthetic rates when variables such as frost or water stress are added to such models, an estimate of the relative importance of those variables is clarified.

These models originally were developed to predict diurnal patterns in net carbon uptake, but they have been adapted to make daily estimates of carbon uptake from seasonal records of solar radiation, temperature, humidity, and predawn plant water potential measurements. The latter two variables are required to predict stomatal conductance without direct measurement (Emmingham and Waring, 1977).

Although it is possible to obtain good agreement with one set of data or possibly several data sets of the same general type, extrapolation of empirical models can be dangerous. This is particularly important if new variables such as toxic pollutants are involved that may affect the photosynthetic process differently, depending on the environment. Nevertheless, some attempt at generalizing

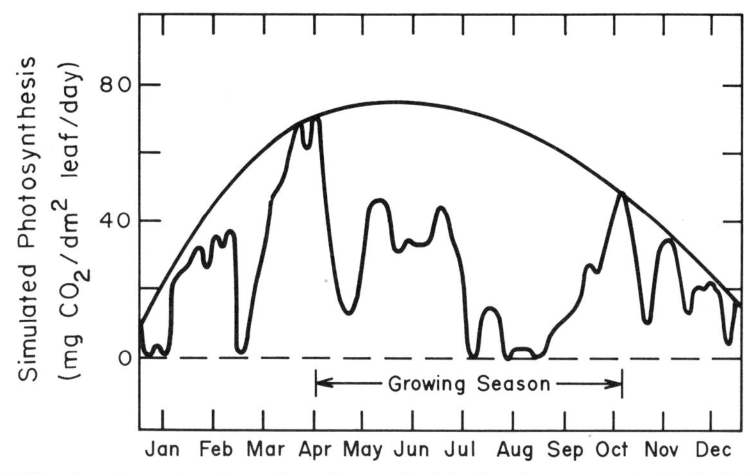

Fig. 2.10. A major portion of annual net photosynthesis by *Pseudotsuga menziesii* in the Pacific Northwest region of the United States is predicted when growth is not occurring. The upper curve represents the effects of seasonal changes in irradiance and mean temperature. The lower curve accounts for additional restrictions upon photosynthesis attributed to water stress and frost. (After Emmingham and Waring, 1977.)

the process of carbon uptake is warranted and simulation models offer an avenue for evaluating the relative importance of major environmental variables in different situations.

RESPIRATION

In addition to the respiration that is associated with photosynthesis, metabolic energy is expended in construction of new tissue and in the maintenance of tissue already synthesized. Respired CO_2 is a final product of such respiratory metabolism. Trees, with such a large amount of conducting and storage tissue associated with their sapwood, require a disproportionate amount of carbohydrates for respiration compared to other plants, in which foliage and seed production are dominant. In this section we discuss growth and maintenance respiration and then evaluate their relative importance as trees grow in different environments.

Construction

The metabolic cost of constructing various compounds or tissue has been assessed by evaluating the biochemical pathways involved in synthesis, assuming glucose is required as a base. For example, the synthesis of 1 g of lipid would require 3.02 g of glucose, whereas 1 g of lignin, protein, or sugar polymer might require, respectively, 1.90, 2.35, and 1.18 g of sucrose (Penning de Vries, 1975). Knowing the biochemical composition of a tissue permits the calculation

TABLE 2.1 Estimated Contruction Costs of Various Organs and Tissues[a]

Type of tree	Organ or Tissue	Construction cost (g sucrose/g dry matter)
Pine	Needles	1.57
	Branches	1.49
	Bark	1.60
	Roots	1.47
Eucalyptus	Phloem	1.45
	Cambium	1.22
	Sapwood	1.36
	Heartwood	1.40

[a] From Chung and Barnes, (1977).

of the cost of synthesis (Table 2.1). The construction cost for a pine shoot is equivalent to 1.57 of the dry weight (Chung and Barnes, 1977). Shoot growth can be monitored throughout a growing season and construction cost allocated proportional to observed elongation.

The seasonal growth of new wood (xylem) can be determined by inserting small pins into the cambial zone of branches, stem, and large roots. Distinct kinds of cells are produced following this treatment, which aid in identifying the number of cells produced between marking periods (Gregory, 1971; Emmingham, 1977). Usually it is adequate to estimate the amount of dry matter produced annually and calculate the synthesis cost.

Maintenance

Once tissue is formed, the living cells still require carbohydrates. Maintenance respiration can be expected to at least double with each increase of 10°C (Amthor, 1984). Tissues with high concentrations of enzymes have higher maintenance cost than tissues that mainly store starch or sugars (Penning de Vries, 1975; Berry and Downton, 1982; Amthor, 1984). Nitrogen content is a crude index of enzyme concentration. Leaf tissue, relatively rich in nitrogen, might respire 2% of its weight daily at 10°C and 4% at 20°C. Wood parenchyma cells that store carbohydrates can respire 1% and 2% of their weight each day at 10° and 20°C, respectively (Penning de Vries, 1975).

Maintenance respiration of nonphotosynthetic tissue can only be accurately determined if no synthesis of cell walls is in progress. When monitored during a growth period, bole respiration cannot be easily separated into its synthesis and maintenance components. However, CO_2 evolved from the lower stem of a pine tree during the dormant season is indicative of maintenance respiration, which can be followed over a range in temperature (Kinerson, 1975). In a dense forest, boles are shielded from direct radiation and remain close to air temperatures. Estimation of exposed branch and bole temperatures, however, is sometimes difficult, particularly in deciduous forests where parts of the stem may be as much as 20°C above ambient during the winter (Sakai, 1966). In stems and branches, some of the respired CO_2 may be refixed by photosynthesis in chloroplasts present under the bark. In such cases, more CO_2 is evolved from a stem or branch in the shade compared to illuminated surfaces (Fig. 2.11).

Assuming the temperature is known or can be predicted from energy budget analyses (Chapter 5), respiration of all living cells in the bole and branches can be predicted. Living cells are concentrated in the cambium and phloem layers that underlie the bark (Fig. 2.12) so in small trees an estimate of plant surface area is an adequate measure of respiring tissue (Whittaker and Woodwell, 1968).

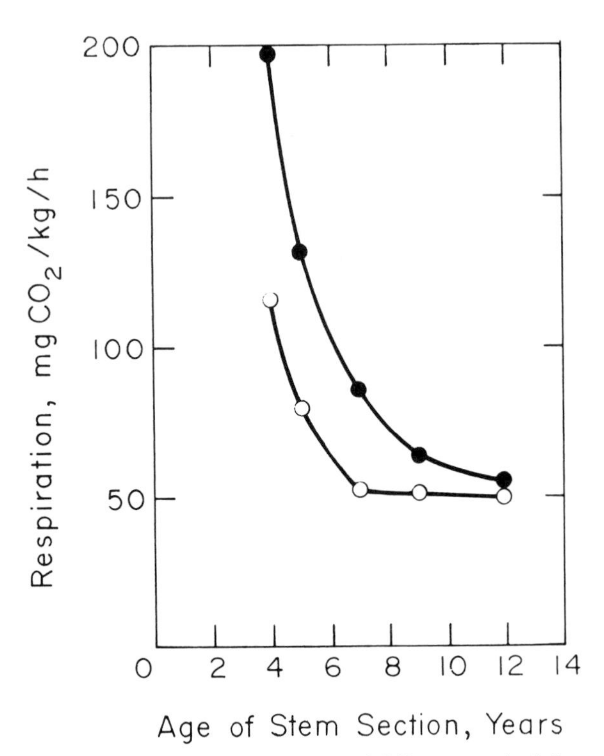

Fig. 2.11. Respiration rates of cut stem sections of different ages (and diameters) from a 15-year-old Scots pine were measured in both the dark (●) and under light (○), when fixation of CO_2 occurred. Temperatures remained constant at 15°C. (After Linder and Troeng, 1981.)

The center of a large stem or branch consists of heartwood that contains no living cells (Panshin *et al.*, 1964), whereas the sapwood, located between the cambium and heartwood (Fig. 2.12), contains a certain volume of living parenchyma cells associated with wood rays (Figs. 2.12 and 2.13). In larger trees, surface area estimates are not adequate because of a large amount of respiration by the living cells in sapwood. The proportion of wood rays in the sapwood is fairly consistent for a particular species, ranging from less than 5% in some conifers to more than 30% in certain oaks (Table 2.2).

Maintenance respiration in nonphotosynthetic cells in the stems, branches, and roots of large trees can represent a large demand for carbohydrates, particularly in warm climates. As trees grow, the percentage of sapwood in the cross section is reduced but the total volume continues to increase. A white oak (*Quercus alba*) of 15 cm diameter has more weight in the living cells of its trunk than in its foliage (from equations in Whittaker and Woodwell, 1968, and data in Table 2.2). For conifers, the amount of foliage accumulated may be much higher than in deciduous trees of equivalent size. Douglas fir is estimated to achieve a

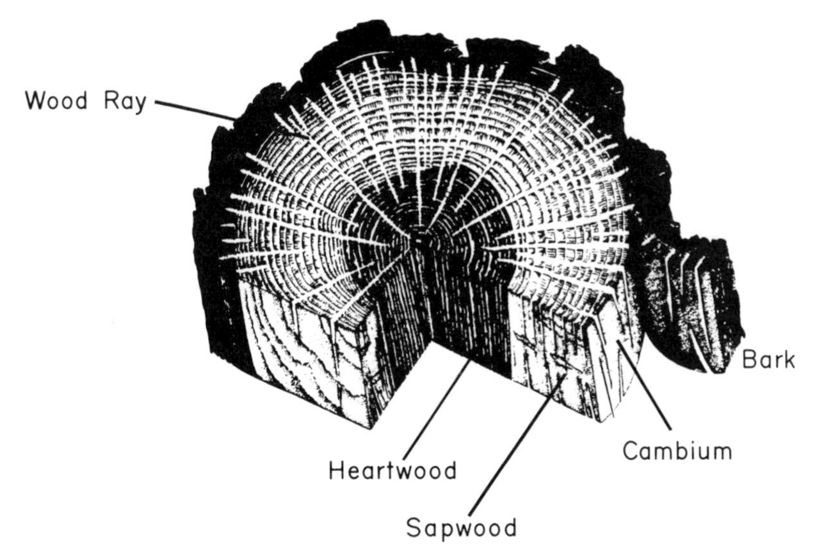

Fig. 2.12. A section of an oak stem shows various components. The outer bark protects the sensitive cambium where tissue formation into wood or phloem is initiated. Water and nutrients are conducted from the roots to the leaves via conducting elements in the sapwood. Large wood rays transverse the sapwood and inner heartwood. The ray cells are alive in the sapwood but no longer function in the heartwood. (From Raven *et al.*, 1981.)

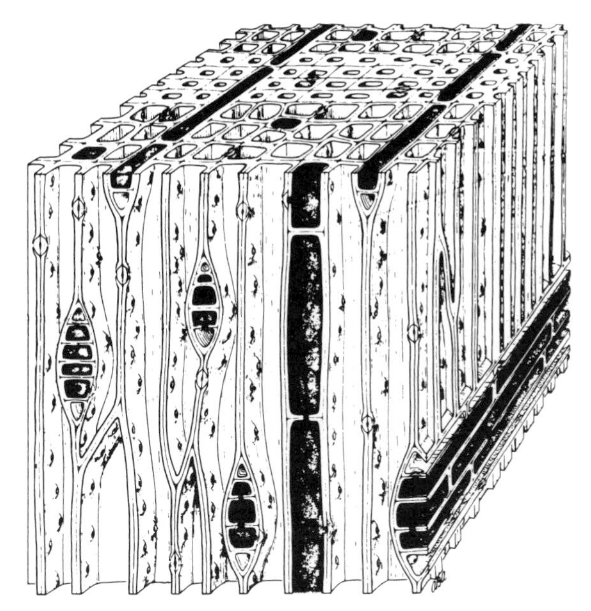

Fig. 2.13. The majority of cells in the sapwood of trees are dead but able to conduct water. Traversing the wood are bands of living parenchyma cells (darkened) that store carbohydrates and perform important functions when trees are under stress. Although the fraction of living cells in sapwood may be low, maintenance cost can be high, depending on the temperature and total amount of sapwood in a tree. Diagram is of redwood (*Sequoia sempervirens*). (From Weier *et al.*, 1974.)

TABLE 2.2 Percentage of Living Parenchyma Cells in the Sapwood of Representative Hardwoods and Conifers Native to the United States[a]

Hardwoods	%	Range	Conifers	%	Range
Populus tremuloides	9.6	4.4	*Pinus taeda*	7.6	1.6
Betula alleghaniensis	10.7	0.9	*Larix occidentalis*	10.0	1.1
Fagus grandifolia	20.4	5.3	*Picea engelmannii*	5.9	2.5
Quercus alba	27.9	—	*Pseudotsuga menziesii*	7.3	2.1
Liriodendron tulipifera	14.2	2.5	*Tsuga canadensis*	5.9	0.7
Robinia pseudoacacia	20.9	3.1	*Abies balsamea*	5.6	2.3
Acer saccharum	17.9	5.2	*Sequoia sempervirens*	7.8	2.5
Tilia americana	6.0	3.8	*Taxodium distichum*	6.6	2.6

[a] After Panshin *et al.* (1964).

balance of foliage and living sapwood cells in trees of 60 cm diameter (calculated from equations by Gholz *et al.*, 1979, and Table 2.2).

Relative Importance of Construction versus Maintenance Respiration

Few studies have attempted to compare metabolic expeditures for construction and maintenance respiration in trees. In Sweden, Linder and Troeng (1981) provide data for a 4.4-m tall Scots pine; in the southern United States, comparable data for an 11.8-m tall loblolly pine are available (Kinerson *et al.*, 1977). In both cases, careful estimates of structural growth were made, and respiration was monitored on various components throughout the entire year. A major difference between the two sites is temperature. Mean annual temperatures averaged 3.8°C in Sweden compared to 15.6°C in the United States. During the growing season, which varied from 145 days in Sweden to 230 days in the United States, the mean temperatures for the two sites were 12.2°C and 20.3°C, respectively (Axelsson and Bräkenhielm, 1980; Kinerson *et al.*, 1977).

Both pines carry relatively small canopies for their size, but growth of various structural components differed significantly. The loblolly pine doubled its foliage in a year whereas the Scots pine added only about 50% above the previous year's canopy. Comparable values for other components were: branches, 69% versus 7%; stems, 15% versus 12%; and roots, 51% versus 288%, for the loblolly and Scots pine, respectively. Assuming a construction respiration of 1.57 times the dry weight of new tissue (or 1.28 kg carbon used for every kilogram of carbon produced), estimates of the carbon respired in synthesis of foliage, branch, stem, and roots were calculated (Table 2.3).

A minimum estimate of maintenance respiration was calculated from measurements of evolved CO_2, less the cost of construction of each structural component. To compare the relative significance of the two metabolic processes, we may evaluate the ratio of construction to maintenance respiration. We find that only for roots was construction respiration the dominant process. The most striking difference between trees was in the respiration ratio for the stems. The loblolly pine had nearly 13 times as much biomass in living tissue as the Scots pine. This, coupled with a nearly 12°C higher annual temperature, resulted in a construction/maintenance respiration ratio of 0.19, whereas the Scots pine's respiration ratio was 0.67, reflecting its smaller size and the cooler climate. When account is taken for differences in stem weight and mean annual temperatures, the respiration ratios for the two pines were identical.

In evaluation of respiration, it is essential to distinguish between construction and maintenance components. Construction costs may be estimated by assessing annual growth and any net change in storage reserves (see next section). Maintenance costs may be assessed by evaluating the fraction of living cells present (and average protein content if variable) in different tissues or organs. Knowledge of temperature throughout the year is essential to estimate maintenance respiration. Fortunately, temperatures throughout a forest can be predicted from an energy budget analysis so continuous measurements of leaf, branch, bole, and root temperatures are not required (Chapter 5).

TABLE 2.3 Comparison of Construction and Maintenance Respiration for Young Pine Trees Growing in the Southeastern United States[a] and in Sweden[b]

Structural components	Carbon content (kg C)	Growth (kg C/yr)	Respiration (kg C/yr)		Ratio C/M
			Construction (C)	Maintenance (M)	
A. *Pinus taeda* (16.7 cm diameter, 11.8 m tall)					
Foliage	1.833	1.833	0.513	4.034	0.13
Branches	3.529	2.450	0.686	1.142	0.60
Stem	27.903	4.242	1.188	6.313	0.19
Roots	7.751	3.917	1.110	—	—
			(3.497)		
B. *Pinus sylvestris* (10.0 cm diameter, 4.4 m tall)					
Foliage	1.091	0.580	0.162	—	—
Branches	1.383	0.097	0.027	0.040	0.67
Stem	2.182	0.256	0.072	0.107	0.67
Roots	0.849	2.563	0.718	0.439	1.64
			(0.979)		

[a] *Pinus taeda* plantation in North Carolina had stocking of 1445 trees per hectare at age 14 (Kinerson *et al.*, 1977).

[b] *Pinus sylvestris* stand in Sweden had stocking of 1500 trees per hectare at age 20 (Linder and Troeng, 1981).

STORAGE CARBOHYDRATES AND SECONDARY PRODUCTS

In addition to structural components and respiration, some photosynthate is allocated to nonstructural compounds that fall into two broad categories: (1) storage compounds such as organic and amino acids, sugars, starch, fats, and proteins; and (2) secondary compounds such as alkaloids, tannins, pigments, and growth regulators. The storage compounds are building blocks for other compounds and a primary source of energy for maintenance. The secondary compounds may also be reconverted into sugars and other compounds useful in construction or maintenance, but their primary function appears to be in protection against predation by animals and microorganisms. This topic is treated in some detail in Chapter 9.

Many secondary compounds are produced through a biochemical pathway based on shikimic acid (Mooney, 1972; Salisbury and Ross, 1978). This same pathway is used for the construction of lignin, a very stable component of cell walls that protects them against attack from microorganisms (Pew, 1967). Lignin may make up 20–30% of the weight of wood cells, so when growth of support and conducting tissue occurs, competition for resources may limit the production of secondary compounds (Mooney, 1972).

Normal Allocation

Competition for resources affects the levels of storage carbohydrates in various tissues. Although concentrations of storage carbohydrates are often high in leaves, twigs, and fine roots, the bulk of stored reserves resides in parenchyma cells scattered throughout the sapwood and inner bark. The preeminence of the lower bole and large diameter roots as storage organs is consistent for all types of trees (Parker and Houston, 1971; McLaughlin *et al.*, 1980; Wargo, 1979; Mooney, 1972; Ford and Deans, 1977).

Not all forms of carbohydrates are equally convertible and transferred quickly from one part of the tree to another. The average velocity of sugar transport in the phloem tissue is only a few centimeters per hour (Kozlowski and Keller, 1966; Zimmermann and Brown, 1971). A tree's ability to withstand a particular kind of stress, therefore, can be assessed by evaluating the soluble and easily mobilized carbohydrate reserves near the points of potential need. Current photosynthate may be reallocated quickly to combat defoliation but is not likely to be a major factor in resistance to herbivory on the stems and roots of tall trees (Chapter 9).

The amount of storage reserves and secondary protective compounds often varies seasonally, depending on growth requirements and photosynthetic efficiency. For example, the evergreen shrub, *Heteromeles* (Christmas-berry), native to the California coastal zone, accumulates few storage carbohydrates or protective compounds during periods of rapid growth (Fig. 2.14). In the dormant season, however, these nonstructural products accumulate even though photosynthetic rates are not particularly high (Mooney and Chu, 1974; Distelbarth *et al.*, 1984).

Starch is a major form of storage carbohydrate, particularly in the spring before shoot growth is initiated (Priestley, 1970). Starch may accumulate in large-diameter roots to support spring root growth (Wargo, 1979) or replacement of foliage following defoliation (Parker and Houston, 1971). Accumulation of starch in the large roots occurs before additional diameter growth is initiated. This is in contrast to stems and branches where some diameter growth precedes the accumulation of reserves. Starch concentration in small-diameter roots (less than 2 mm) appears to be determined at the time of initiation and helps define how long fine roots will live at a given temperature (Marshall and Waring, 1985; see Chapter 3, this volume).

Stress Levels

The maximum level of starch reserves in twigs and stems may serve as an index to the degree of damage a tree can endure and still recover (Fig. 2.15).

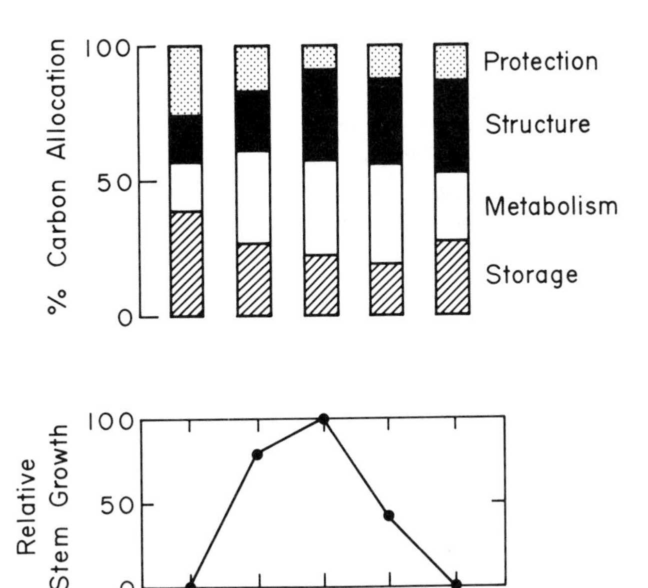

Fig. 2.14. Seasonal allocation of carbohydrates by the evergreen *Heteromeles arbutifolia* shows that structural growth and respiration take relative priority over the production of storage reserves and protective compounds. (From Mooney and Chu, 1974.)

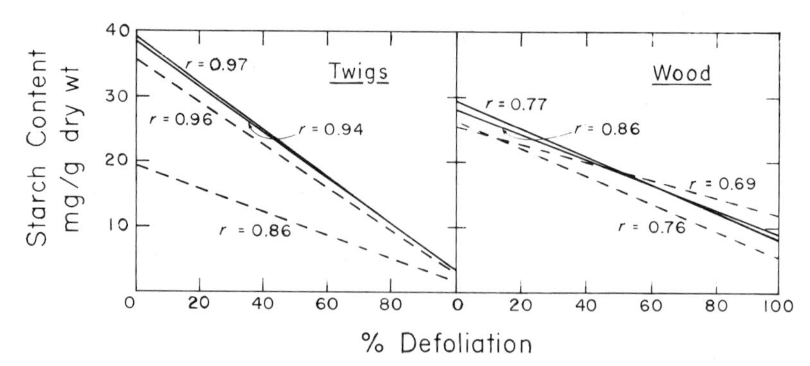

Fig. 2.15. In a dense plantation of 70-yr-old Douglas fir in New Zealand, starch content in the twigs and stem wood decreased linearly in relation to the amount of foliage removed by defoliating insects. The two dotted lines represent trees originally growing under the dense canopy of taller trees. The upper line indicates that starch reserves could increase substantially, at least in twigs, if the overstory trees were removed. Solid lines indicate trees that were in a dominant position initially. Thinning the stand provided no measurable gain in starch reserves for these trees. (After Cranswick, 1979, with permission of the New Zealand Forest Research Institute, Rotorua.)

Webb (1981) reported that Douglas fir and grand fir (*Abies grandis*) could be completely defoliated and still initiate new foliage, as long as twig starch levels remained above a certain minimum. In Australia, Bamber and Humphreys (1965) found that *Eucalyptus* species finally succumbed to repeated insect defoliation only after starch reserves in the sapwood had been completely exhausted. Reserves are particularly important in deciduous trees that must survive much of the year from energy derived from storage. For example, a 410-kg oak may have reserves equivalent to 54 kg of sugar, sufficient to support the replacement of the leaf canopy three times (McLaughlin *et al.*, 1980).

Net changes in the amount of nonstructural carbohydrate can significantly alter growth patterns from one year to the next. Annual variation could also invalidate most attempts to calculate a carbon budget for trees because the assumption is usually made that storage reserves remain constant. A reduction in reserves should result in less foliage and wood production, whereas improvements in reserves should allow for increased elaboration of tissue.

ALLOCATION INTO STRUCTURE

Carbohydrates produced through photosynthesis are used by trees in constructing leaves, branches, stems, roots, and reproductive organs. Except for the latter category, a pattern exists in the way structural tissue is partitioned, which makes it possible to estimate amounts of various structural components from simple correlations with stem diameter at ground level or at breast height, 1.37 m above the ground (Whittaker and Woodwell, 1968; Kira and Ogawa, 1971). Relationships derived from such analyses are known as allometric and are widely applied in forestry and ecology. In a particular region, under comparable conditions, the diameter of the stem is sufficient for estimating branch and foliage biomass as well as that of stem wood (Fig. 2.16). Trees with different forms, of course, must be grouped accordingly. Care should be taken against grouping species with similar form but variable wood densities.

Growth may be estimated by remeasuring stem diameter over a period of time or by coring the tree to determine the width of annual rings and the change in diameter (Whittaker and Woodwell, 1968). The estimates are representative of the environment where sample trees were obtained. Additional sampling is required to estimate growth in other environments or in stand conditions that are different from those originally measured (Benecke and Nordmeyer, 1982; Deans, 1981).

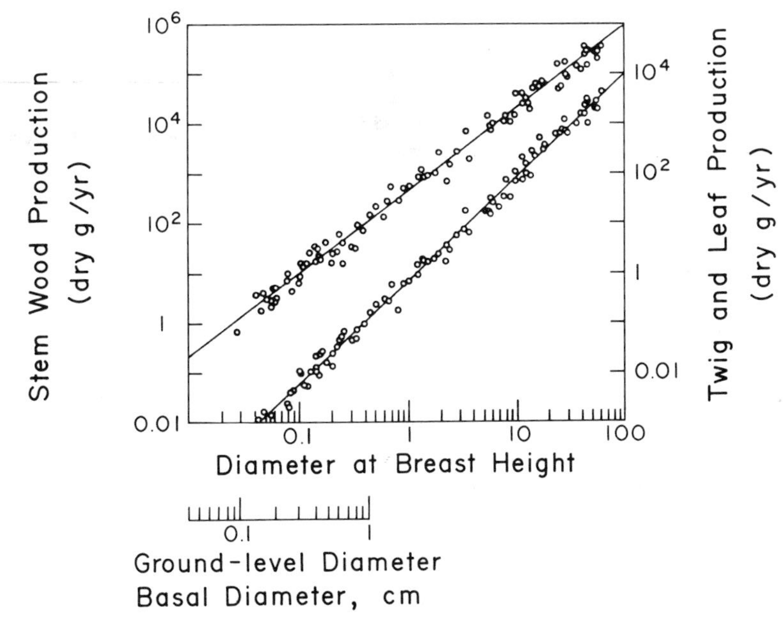

Fig. 2.16. Production of stem wood (lower line) and current twigs and leaves (upper line) for a variety of northern hardwood trees and shrubs can be estimated from logarithmic regressions with stem or basal diameter. (From Whittaker and Woodwell, 1968.)

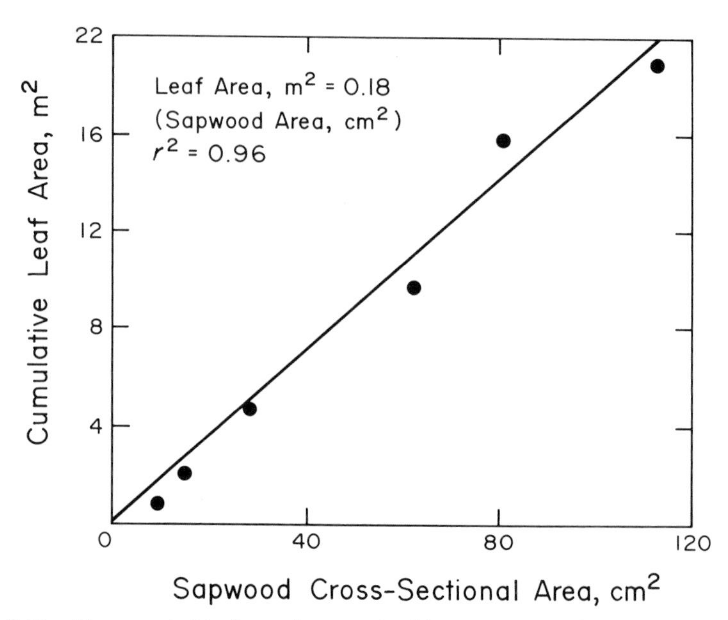

Fig. 2.17. The amount of leaf area above any level in a tree's crown is directly related to the sapwood area at that cross section. Data are from a single 12.4-cm diameter Scots pine showing this relationship from near the top to the bottom of the crown. (Unpublished data from B. Axelsson and R. Waring, Swedish University of Agricultural Sciences, Uppsala, Sweden.)

Sapwood and Leaf Area

Earlier we distinguished the outer functional sapwood in tree stems and branches from the inner heartwood that provides support but has no living cells (Fig. 2.12). The cross-sectional area of sapwood in a tree's stem has proven to be closely correlated with the amount of foliage that the stem supports during the growing season. This relationship appears rather stable for trees of varying sizes growing under similar conditions, which makes sapwood area a better estimator of foliage than stem diameter. The relation is strongest in the foliage-bearing portion of the crown (Fig. 2.17). On large trees, therefore, with a major portion of their bole lacking live branches, sapwood area estimates should be made at the base of the crown (Waring *et al.*, 1982).

Ring-porous hardwoods such as *Quercus* (oaks), which have numerous large-diameter conducting cells (vessels), support more foliage per unit of sapwood area than do trees with diffuse-porous woods with only small-diameter vessels such as *Populus* (Table 2.4). In harsh and exposed environments, more sapwood is apparently required to support a given amount of foliage than in maritime or sheltered locations. Contrast, for example, the subalpine species, *Tsuga mertensiana,* with its more maritime relative, *Tsuga heterophylla,* in Table 2.4. The coastal variety of Douglas fir, *Pseudotsuga menziesii* var *menziesii,* has about twice the ratio of leaf area/sapwood area as the more continental variety, *glauca.* Even within one region where climatic differences are extreme, genetic differences may be expressed, as with *Nothofagus* in montane and subalpine environments in New Zealand (Benecke and Nordmeyer, 1982). Whitehead *et al.* (1984) present evidence that variations in leaf area/sapwood ratios reflect proportional changes in wood permeability (Chapter 4). Carbohydrate storage requirements may also be involved because oak trees use only the outermost vessels of sapwood for water conduction (Rogers and Hinckley, 1979). Future research may clarify the relative importance of sapwood for storage versus conduction of water, nutrients, and carbohydrates.

Because it is relatively easy to estimate the cross-sectional area of sapwood in trees, it is also possible, through correlation, to describe the distribution of foliage in the canopy accurately (Waring *et al.*, 1982). The utility of these leaf area/sapwood ratios is illustrated further in Chapters 3 and 9.

TABLE 2.4 Ratio of Projected Leaf Area to Sapwood Cross-Sectional Area for Selected Tree Species

Leaf area/sapwood area	m^2/cm^2	Reference
A. Conifers		
Abies lasiocarpa	0.75	Kaufmann and Troendle (1981)
Abies procera	0.27	Grier and Waring (1974)
Picea engelmannii	0.35	Waring *et al.* (1982)
Pinus contorta	0.15	Waring *et al.* (1982)
Pinus ponderosa	0.25	Waring *et al.* (1982)
Pinus sylvestris	0.14	Whitehead (1978)
Pseudotsuga menziesii var. *menziesii*	0.54	Waring *et al.* (1982)
Pseudotsuga menziesii var. *glauca*	0.25	Snell and Brown (1978)
Tsuga heterophylla	0.46	Waring *et al.* (1982)
Tsuga mertensiana	0.16	Waring *et al.* (1982)
B. Hardwoods		
Acer macrophyllum	0.21	Waring *et al.* (1977)
Castanopsis chrysophylla	0.46	Waring *et al.* (1977)
Nothofagus solandri (montane)	0.12	Benecke and Nordmeyer (1982)
Nothofagus solandri (subalpine)	0.07	Benecke and Nordmeyer (1982)
Populus tremuloides	0.10	Kaufmann and Troendle (1981)
Quercus alba	0.40	Rogers and Hinckley (1979)
Tectona grandis	0.65	Whitehead *et al.* (1981)

Allocation under Stress

In Scots pine, about 20% of the annual photosynthesis may go into branch and foliage production, leaving about 80% for export to other parts of the tree or for maintenance (Linder and Axelsson, 1982). Although branch growth is correlated with stem wood growth, considerable variation can be observed from the top to the lowermost branch on a tree. Linder and Axelsson (1982) found that the current branch growth of each whorl was directly proportional ($r^2 = 0.95$) to the net annual uptake of carbon by each of 15 separate whorls of Scots pine branches. In most cases, carbon uptake of the lower branches was so limited that little photosynthate was available for export. Branch growth or shoot elongation can provide an index of annual photosynthesis by different whorls and a basis for interpreting the contributions from various parts of a tree. This correlation may be particularly useful in assessing the results of various silvicultural practices such as thinning and pruning. It may also have application in calculating total tree carbon balances.

The normal pattern of structural growth may be predictable, but changes in the relative availability of a critical resource may cause deviations. Some examples of how certain factors might stimulate the growth of some tissues over others are provided in Table 2.5. A variety of physiological adjustments result in these shifts in allocation, but the reasons for the adjustments are not fully understood.

Drought is rather complicated because it affects allocation processes in a variety of ways. Under mild drought, carbon uptake is not much reduced but shoot growth is inhibited. Carbohydrates are transported down the phloem toward the roots. Roots, being nearest the source of water, are less water stressed than cells in the stem and branches and are able, with an improved carbohydrate supply, to grow actively (Hsiao, 1973; Hanson and Hitz, 1982). With additional root growth, nitrogen and other minerals taken up by the plant may concentrate

TABLE 2.5 Departures from Normal Pattern of Carbon Allocation in Trees

Component	Conditions favoring development
Reproductive organs	Moderate water stress, limited root growth, or restricted phloem transport
Foliage	Shade, abundant nitrogen, and water
Stem wood	Wind or snow load on canopy or optimum environment
Fine roots	Water supply inadequate in upper rooting horizons, limited nitrogen, or unstable soils
Protective compounds and other defensive reactions	Injury, sometimes specific response to the kind of injury

in the foliage. Changes in the relative availability of various growth regulators and receptors may initiate flower buds (Trewavas, 1982). Carbohydrates that normally would be exported to other places are available for production of flowers or strobili. These observations help explain why moderate drought can, in some cases, trigger development of a good seed crop. Limited root growth or restricted phloem transport may also stimulate seed production. Once seed production is initiated, the process may require additional resources beyond those available through current photosynthesis, resulting in the use of storage reserves (Gifford and Evans, 1981).

Inadequate light affects allocation differently. Shaded branches export little photosynthate and have limited carbohydrates available for shoot or branch growth. If the entire canopy of a tree is shaded, lower branches will die and an umbrella-like canopy results (Steingraeber *et al.*, 1979; Kohyama, 1980). Root growth and seed production are greatly reduced under shade; new foliage and terminal elongation are favored instead.

Wind or heavy snow loads on the canopy or against the stems of trees exert strong physical forces that stimulate a disproportional allocation of carbohydrates to the lower bole and result in increasing taper in tree trunks. Leaning trees sometimes develop unusual asymmetrical forms for similar reasons. The ratio of crown weight to lower stem diameter changes during growth to prevent the tree from buckling under its own weight (King, 1981) or the weight of a snow load (Petty and Worrell, 1981). When the canopy of a forest normally exposed to wind or snow is opened, damage may occur to the remaining trees before they have time to adjust to the greater exposure.

Nearly all nutrient uptake by trees is via their roots and associated symbiotic bacteria or fungi (Chapter 7). When a certain nutrient such as nitrogen is inadequate, construction of photosynthetic enzymes in new foliage is restricted, canopy development slows, and an increasing proportion of carbohydrate moves toward the roots. Additional root growth may increase the uptake of nitrogen and permit shoot growth to continue, but at a slower rate than if nitrogen were supplied in abundance (Ericsson, 1981).

Recently, field studies with young forests of Scots pine have shown that nutritionally stressed trees allocate more than 60% of their photosynthate belowground, whereas those receiving nutrient supplements throughout each growing season allocate less than 40% of their carbohydrate resources in this manner (Linder and Axelsson, 1982). In Fig. 2.18, a comparison of carbon allocation to shoot (foliage, branches, and stem) versus small-diameter roots (less than 2 mm) is presented from experiments involving not only fertilization but also irrigation and a combination of the two treatments. Each of the treatments progressively increased shoot/fine root production. The combination treatment stimulated the most striking response with a ratio of shoot/fine root production nearly five times greater than in the untreated trees. Larger-diameter roots showed increments

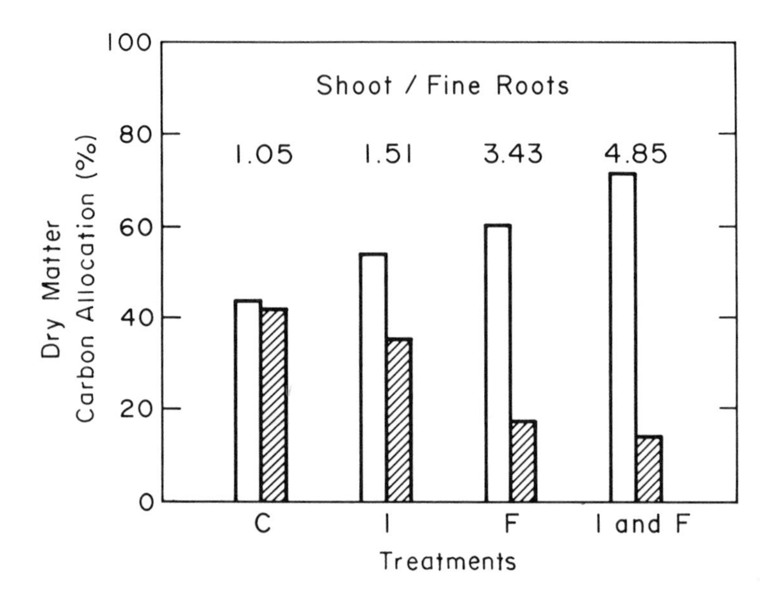

Fig. 2.18. The percentage of carbohydrates allocated into shoot (□) versus fine root production (▨) by Scots pine differs depending on the availability of soil water and nutrients as demonstrated by an experiment in Sweden where trees were irrigated (I), fertilized (F), or received both treatments (F and I). Each treatment provided an increase in shoot/fine root production compared to untreated (C) stands. (After Axelsson, 1981.)

paralleling growth in stem diameter (Axelsson, 1981). Deciduous hardwoods may be expected to show similar responses to fertilization but, in addition, may greatly improve their photosynthesis rates per unit of leaf area. Chapter 3 discusses the implications of improved nutrition and water relations upon stand growth.

Growth Efficiency Index

Much is still to be learned about the rules that govern carbon allocation under various kinds of stress. We may, however, gain considerable insight by comparing departures from the normal allocation pattern. The canopy represents the photosynthetic tissue and generally has high priority for carbohydrates. Except when trees are in danger of buckling under their own weight, stem wood growth is of relatively low priority. Many secondary compounds, such as protective chemicals, are even less essential than diameter growth because new foliage

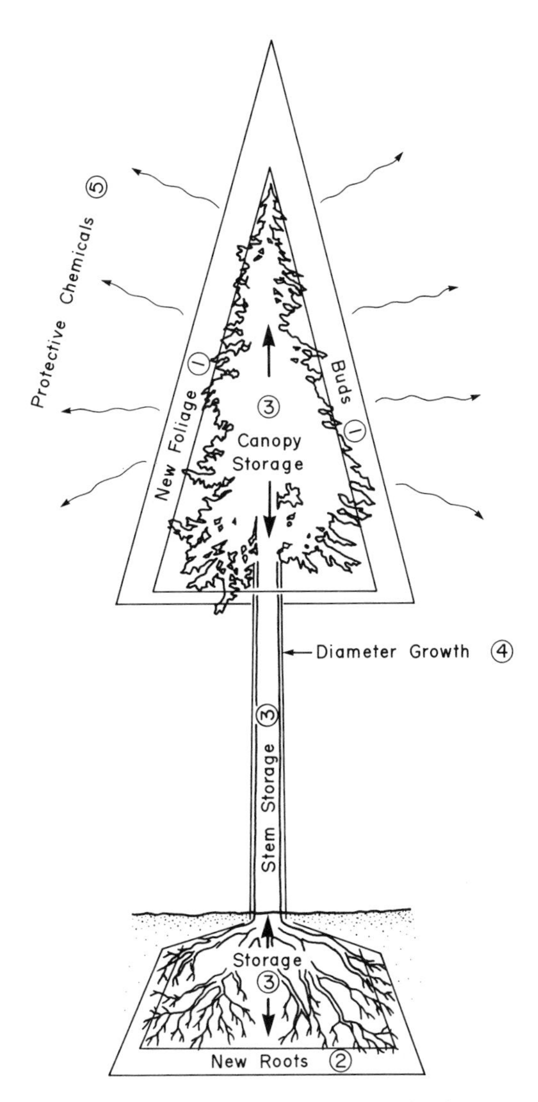

Fig. 2.19. A postulated hierachy for normal carbon allocation in a tree suggests that photosynthetic tissue represented by buds and new foliage (1) and new roots (2) have high priority. If additional carbohydrates are available, they are likely to go progressively into storage reserves (3), diameter growth (4), and protective chemicals (5). Trees do not produce reproductive organs annually but once seed production is initiated it may draw down carbohydrate reserves and limit growth of other components. Under stress, allocation patterns can be expected to vary from those suggested here. (From "Modifying lodgepole pine stands to change susceptibility to mountain pine beetle attack" by R. H. Waring and G. B. Pitman, *Ecology,* 1985, **66,** 889–897. Copyright © 1985 by the Ecological Society of America. Reprinted by permission.)

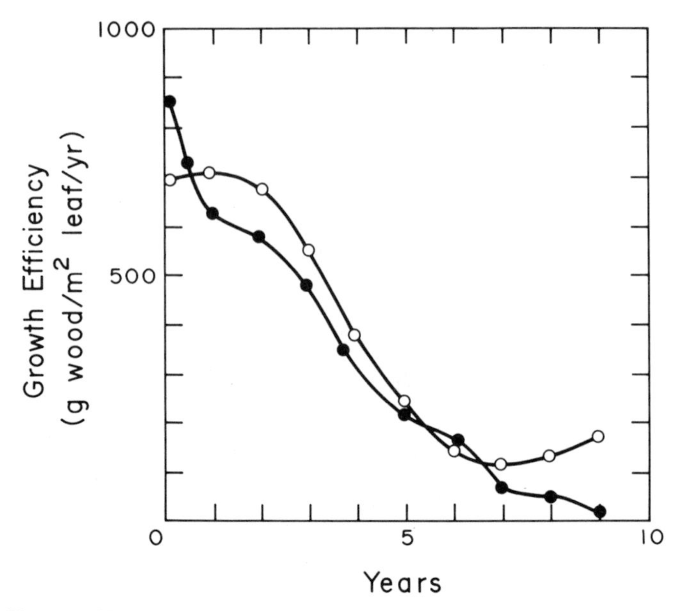

Fig. 2.20. Two 20-yr-old Scots pine with approximately similar dimensions showed more than a fivefold reduction in their growth efficiency in less than a decade. During the last 3 yr, canopy closure occurred and one tree (●) was overtopped. The other tree (○) achieved a position of dominance and increased its growth efficiency. Analyses were made from wood cores (trees were all sapwood) extracted from the two trees. Allometric equations developed from trees in the same stand were used to estimate growth from changes in stem diameter. Leaf area was calculated from the sapwood area present each year, assuming 1 cm^2 of sapwood supports 0.18 m^2 of projected leaf area. (From R. H. Waring, B. Axelsson, and S. Linder, unpublished data. Swedish Univ. of Agricultural Sciences, Uppsala, Sweden.)

must have new sapwood. Because carbon allocation to stem wood has low priority (Fig. 2.19), the annual growth of stem wood per unit of foliage may serve as a "growth efficiency" index. High values indicate a reasonable balance in the distribution of carbohydrates whereas low values would suggest limitations in storage reserves and in the ability of the tree to produce various protective chemicals.

The index is akin to "net assimilation rate," the ratio of nonphotosynthetic tissue produced per mass of foliage (Briggs *et al.*, 1920) but focuses on the main harvestable portion of the stem and expresses foliage in unit area rather than mass. Net assimilation rate has been widely applied in agriculture (see review by Williams, 1946) and was first introduced into forestry by Burger (1929).

Growth can be estimated applying allometric equations to changes in stem diameter, and leaf area is predictable from knowledge of the cross-sectional area of sapwood. Growth of both stem wood and sapwood area can be determined by extracting a wood core from the stem at breast height or at the base of the crown. This index is sensitive to changes in the competitive position of trees in the forest canopy (Fig. 2.20). Growth efficiency can serve as an index to competition and to the general vigor or disease resistance (Chapter 9). Under specified conditions this index may also aid in interpreting the relative constraints of various environmental factors upon growth, as discussed in Chapter 3.

SUMMARY

In this chapter we have identified five key environmental variables that affect photosynthesis: light, CO_2, temperature, water, and nutrients. We have also provided some explanation as to how these environmental variables interact. Knowledge of these environmental interactions permits net carbon uptake by a tree canopy to be predicted fairly accurately, as demonstrated from empirical models. Some of the carbohydrate produced through photosynthesis is respired as CO_2 during the construction of tissue. More is used in the maintenance of living cells. These costs vary with the kind of tissue and environment. In trees, sapwood in the trunk deserves special consideration because it represents a large amount of living and respiring tissue and is a major storage organ.

Normally, a balance exists between the construction of various tissues: foliage, support and conducting tissue, and roots. Because of this, there are correlations between stem diameter and foliage, branch, and supporting root biomass. Departures from the normal allocation pattern occur, however, in response to limitations of a critical resource or to various stresses such as wind, heavy snow, or insect attack. By recognizing how such allocation differs from normal, we can often identify the cause. Structural and biochemical indices are available to assess whether a tree is inadequately supplied with carbohydrates. Both kinds of indices are used in other chapters of this book to assess a number of important questions concerning the functioning and management of forest ecosystems.

3

Forest Productivity and Succession

INTRODUCTION

The net primary productivity of forests consists of the accumulation of wood in standing trees as well as the growth of other tissues that are usually short lived. Net primary production supports all life in the forest, from large animals to microbes. When we discuss forest productivity we must consider the accumulation of photosynthate by the canopy and its allocation into tissue, respiration losses by plants, and the consumption and respiration by animals and microbes (Fig. 3.1). Residues that are not metabolized increase the organic content of the soil and forest floor.

Only in the last few decades has the forestry literature begun to report produc-

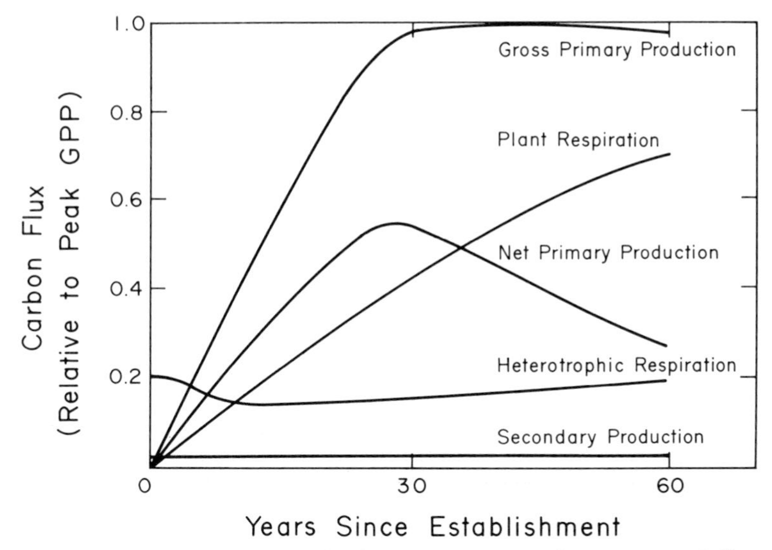

Fig. 3.1. Generalized relationships showing how components of ecosystem metabolism might change over the course of development of a forest from establishment to maturity. Gross primary production is shown to peak after about 30 yr corresponding to canopy closure. Plant respiration continues to increase as living tissue accumulates. Net primary production peaks with canopy closure and then decreases as a result of continuing increases in plant respiration. Heterotrophic respiration of microbes and other nonphotosynthetic organisms is initially high, following removal of a previous forest. With canopy closure, respiration decreases slightly. Later, as gaps in the canopy appear, respiration of heterotrophs again may increase. Secondary production of nonphotosynthetic organisms is always a small component of GPP.

tion values that include more than bole wood (Pardé, 1980). This reflects a growing interest in the use of all components of the forest as well as an ecosystem perspective. Unfortunately, the majority of studies still report only aboveground production or unsubstantiated estimates of root production. Belowground processes are an essential component of ecosystem studies. For example, Keyes and Grier (1981) examined two forests of *Pseudotsuga menziesii*. Both forests had similar net primary production, but one showed much higher timber yields; the difference was mainly in the proportion of carbohydrates allocated to roots.

Forest managers are now interested in predicting the total accumulation of potential fuels that might support fires or serve as a material or energy resource for man. Thus, there is interest in the amount of residue left following forest disturbance and in the amount of plant tissue that dies each year. The latter is closely related to nutrient cycling and to the accumulation or loss of organic matter and minerals from soils (Chapters 8 and 10). Large standing dead trees or fallen boles are now recognized as critical habitats for certain groups of terrestrial and aquatic organisms (Chapters 9 and 10). Decayed boles may serve as sites for establishment of certain tree species (Harmon, 1985; Denslow, 1980).

Although net primary production may be of special interest, it is ultimately dependent upon gross photosynthesis and respiration. From the standpoint of an ecosystem, respiration includes both plants and animals. Gross primary production not lost to respiration (or otherwise removed from the system) results in accumulation of biomass and soil organic matter. In this chapter we present a brief discussion on how various components of ecosystem production are estimated and then compare values in some widely different forest systems. The emphasis, however, remains on primary production.

The fraction of production allocated to foliage, bole wood, and roots changes as forests grow or as the microenvironment is modified. We review some studies that contrast changes in canopy development, biomass accumulation, and growth rates as forest age or change in composition. We suggest that general underlying principles explain the observed variation. Differences in canopy leaf area, in the proportion of carbohydrates allocated to stem wood, and in the maintenance respiration of living cells are used to interpret why forests differ in aboveground net primary production and biomass accumulation. The chapter ends by reviewing some recent forest succession models that incorporate ecosystem principles and have promise of wide general application. In predicting changes in ecosystem properties during succession, the frequency and scale of disturbance are shown to be particularly significant.

COMPONENTS OF ECOSYSTEM PRODUCTIVITY

The small fraction of radiant energy that is transformed by photosynthesis into organic compounds is termed gross primary production (GPP). The actual amount of organic matter created by green plants, net primary production (NPP), is less than GPP because of losses through plant (autotrophic) respiration (R_A). Net primary production includes all increments in the biomass of stems, leaves, reproductive organs, and roots, and the amount of plant tissue that is consumed by animals or that become detritus (dead material) over a fixed time, usually a year. Units are in grams per square meter per year or metric tonnes (t) per hectare per year of dry matter produced, or its carbon (approximately 50% of dry matter content) or energy equivalent (averaging about 4.25 kcal or 4.2 J/dry g). Only the net primary productivity is available for harvest by man or other organisms.

The total biomass produced during a year by all heterotrophic organisms—animals and saprobes—is termed secondary production (SP). The energy content of animal tissue is higher than that of plants, averaging 5.0 kcal/dry g. Although production of heterotrophic organisms is usually a very small fraction of NPP in forests, heterotrophs eventually consume most of the primary production. Their activity, reflected by heterotrophic respiration (R_H), is a major component of ecosystem metabolism (Fig. 3.1).

The total respiration by an ecosystem during a year (R_E) represents the sum of autotrophic (R_A) and heterotrophic (R_H) activity. Net ecosystem production (NEP) is the difference between GPP and R_E, assuming no other losses of organic matter through harvesting, fire, erosion, or export from the system. Often the structure of forests is characterized by the total amount of organic matter (or carbon or energy equivalent) in various components. The sum of all organic matter in living vegetation, which includes nonliving tissue in trees, is termed total standing crop (TSC). In forests this may be an extremely large value since organic matter may accumulate for centuries.

Estimating Primary Production

Gross primary production, representing carbon compounds acquired through photosynthesis, can not be measured easily because leaf respiration produces CO_2 while photosynthesis occurs. Estimates of respiration can be made in the dark, but these ignore the process of photorespiration which may be up to four times that of dark respiration (Chapter 2). Thus, in calculating GPP, net photosynthesis or NPP is measured and estimates of plant respiration are added to compute daily, seasonal, or annual GPP.

Quantification of plant respiration (R_A) is logistically difficult because plant surfaces may be occupied by other respiring organisms, including canopy and stem epiphytes as well as symbiotic root bacteria and fungi. Moreover, as trees grow larger they usually acquire a massive inner core of nonrespiring heartwood. Conventionally, organisms that occupy plant surfaces aboveground are separated from the tree, but the belowground activity of symbiotic organisms is included as a part of plant respiration, although this metabolism is really a form of consumption. Also, because growth of root symbionts is difficult to estimate separately, it is usually included as part of NPP.

Total plant respiration is derived by summing estimates for various plant organs. Rates differ depending on the amount and kind of tissue produced each year. Leaf tissue is costly to construct and to maintain whereas a similar amount of large root biomass makes a small drain upon photosynthate. Respiration measurements are generally taken seasonally and over the full length of a day, because growth processes may be more active at night in both stems and roots (Edwards and McLaughlin, 1978; Edwards and Sollins, 1973). Often respiration rates are calculated on the basis of surface area rather than dry weight. The surface area of leaves, stems, and branches is relatively accurately estimated using allometric relationships with stem diameter or sapwood area (Whittaker and Woodwell, 1967; Waring *et al.*, 1982). Estimating respiration of large roots follows procedures developed for stems and branches, but direct estimation of fine root respiration is most difficult because the standing crop of live roots varies seasonally and any attempt to measure *in situ* respiration includes heterotrophic activity in the soil (Schlesinger, 1977; Chapter 8). Changes in fine root biomass have been estimated by periodically sampling the soil with a coring device (Edwards and Harris, 1977). Root respiration is assumed proportional to soil temperature and the amount of live roots present. This and alternative approaches are described in more detail later.

Because much of the biomass in large trees consists of dead heartwood, the proportion of heartwood in stems, branches, and roots should be estimated before calculating respiration. In Chapter 2, the amount of living tissue in large stems and branches was shown to increase linearly with diameter and to be closely related to the cross-sectional area of sapwood (Hari *et al.*, 1985). This suggests that the volume of living cells might be approximated by estimating the amount of sapwood in proportion to a tree's total weight or volume. If we assume that after a certain age, the canopy leaf area and photosynthetic capacity of a forest stabilize, the cross-sectional area of sapwood should also stabilize (Chapter 2, Table 2.4). As trees continue to grow in height, however, the total volume of sapwood and related living tissue continues to increase. As a result, annual stem growth and total NPP must decrease.

Increases in bole sapwood volume are directly correlated with a 50% reduction in growth in Norwegian Scots pine forests, as illustrated in Fig. 3.2. Wood production and the rate of decline following canopy closure differ depending on the site. On the best sites, growth rates are reduced most rapidly, reflecting the fact that these environments are warmer and found at lower elevations and more southern latitudes. When a similar analysis was performed for Norway spruce, *Picea abies* (Braastad, 1975), growth rates on comparable sites were found to decrease more slowly at equivalent volumes of sapwood (Chapter 2, Table 2.4). Spruce has nearly twice the canopy and presumably about twice the photosynthetic capacity of pine, so the differences in growth response are not surprising.

The biomass accumulation ratio (biomass/NPP) increases during stand development. As the amount of total biomass increases, the mass of living tissue will increase too, although not proportionally. As forests accumulate biomass, an increasing amount of GPP is used in respiration. Although respiration rates of tissue vary among species, some generalizations are possible. The fraction of GPP allocated to plant respiration will vary with temperature. For example, plant respiration is about 15% higher in warm lowland tropical forests as compared to cooler temperate forests with similar biomass accumulation ratios (Fig. 3.3).

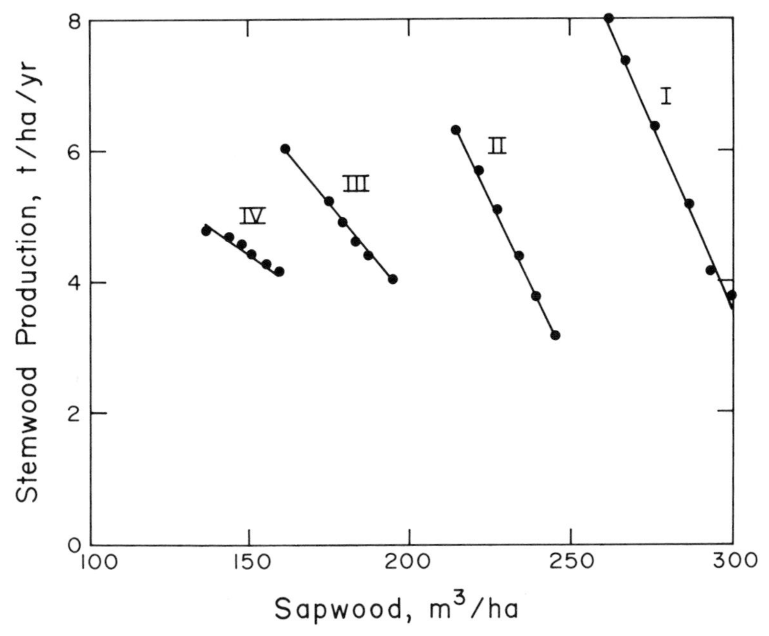

Fig. 3.2. As the volume of sapwood in the stems of unthinned Scots pine forests increases between the age of 60 to 120 yr, the production of wood decreases proportionally. Relationships are shown for a range of site production classes, I–IV, reflecting a range in tree height from 26 to less than 19 m at 100 yr. Comparisons begin when canopy closure occurs at 60 to 90 yr, depending on the site, and continue for 30 yr, at 5-yr intervals. After canopy closure, sapwood cross-sectional area per hectare was assumed constant on each site but the sapwood volume was assumed to increase in proportion to the ratio of sapwood/total basal area of stems. (Data from yield tables by Brantseg, 1969.)

Because the amount of *living* biomass provides the basis for accurately estimating plant respiration, we strongly recommend distinguishing sapwood from heartwood in trees.

Theoretically, NPP could be calculated from knowledge of photosynthesis and plant respiration. As we have seen, the difficulties involved in accurate measurement of these two variables are immense. For this reason, NPP is normally estimated by measuring changes in the biomass of various tissues over a period of a year, accounting for any biomass that may have been consumed by animals or transferred to detritus through mortality of individual trees, leaves, roots, and other parts. The equation is:

$$\text{NPP} = \Delta B + D + C$$

where ΔB is the change in biomass over a period of a year; D, detritus produced during the year; and C, consumption of biomass by animals during the year.

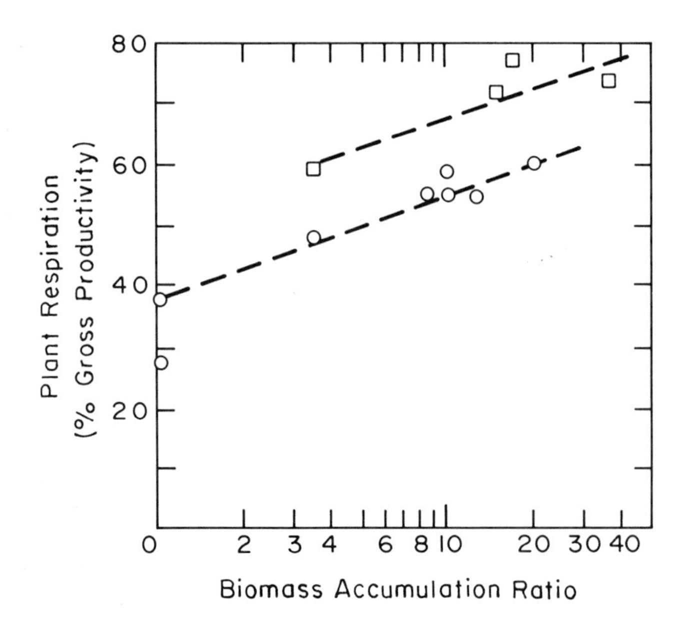

Fig. 3.3. Plant community respiration in relationship to biomass accumulation ratio (TSC/aboveground NPP). The circles (○) represent temperate and the squares (□) lowland tropical communities. The trend lines show increasing fractions of GPP used for plant respiration from agricultural fields on the left to forests (with biomass accumulation ratios from 10 to 40) on the right. About 15% more gross production is expended in respiration under the higher temperatures of the tropics. (After Whittaker, 1975.)

Increments in biomass, as reported for stem wood, bark, branches, leaves, fruits and flowers, and large-diameter roots are obtained using logarithmic regressions with stem diameter at breast height or other dimensions of interest (Whittaker and Woodwell, 1967; Gholz *et al.*, 1979; Deans, 1981; Chapter 2, Fig. 2.16). The regressions are used to compute forest biomass. Production is determined by periodic remeasurement or by extracting wood cores and measuring annual ring widths. Biomass and production of shrubs and herbs in forest undergrowth are estimated separately. Clipping of the current growth of herbs and small shrubs may be required to obtain growth estimates or correlations with cover.

Biomass increment is calculated for a unit area of sample forest by measuring all trees within a known area or by using plotless sampling based on the dimensions of trees intercepted by a selected angle. The second approach, called variable-plot survey, wedge prism cruising, Bitterlich method, or Relaskop surveys in the forestry literature, is generally much more efficient because only a few trees require measurement at each sampling point (Sukwong *et al.*, 1971).

Detritus production by plants includes all organic matter that died during the year, whether or not any residue remains at the end of the sampling period. In the tropics, leaf litter must be gathered quickly because decomposition is rapid. In all regions a similar concern is required in estimating fine root death, because the standing pool of roots less than 5 mm in diameter may turn over a number of times during a year (McGinty, 1976; Harris *et al.*, 1975; Persson, 1978). Most estimates of detritus production have ignored fine root turnover although the latter may range from 3.5 to more than 11.0 t/ha/yr in conifer forests (Persson, 1978; Grier *et al.*, 1981) and from 4.0 to 7.5 t/ha/yr in typical temperate hardwood forests (McGinty, 1976; Harris *et al.*, 1975; McClaugherty *et al.*, 1982). These are more than twice the values obtained from simply assuming that fine root production is proportional to large root growth or to foliage production, as done in many studies. In fact, root production may represent the major part of NPP in some forests (Linder and Axelsson, 1982; Grier *et al.*, 1981).

Where both litterfall and fine root turnover have been estimated, the latter dominates. For example, forests of Douglas fir (Keyes and Grier, 1981), *Liriodendron* (Cox *et al.*, 1978), and *Abies amabilis* (Grier *et al.*, 1981) have ratios of litterfall/root detritus production ranging from 0.21 to 0.27. Only on a very fertile site, where fine root production was small, did Keyes and Grier (1981) find that litterfall approached the detritus produced from fine roots.

Not only is the amount of fine root production high compared to aboveground NPP, but it may be variable from year to year and at different stages in forest development. Estimates of root production and turnover have been obtained by laboriously sampling with coring devices at monthly intervals when growth is occurring and then calculating the net change in any month (Fig. 3.4). A large number of samples is required to reduce the standard error of estimate to less than 10% of the mean. Marshall (1984) provides a physiological approach to estimating root growth and turnover by observing that fine roots in conifers live on starch reserves deposited when roots are first formed. Roots live as long as they have some starch content [also noted by McClaugherty *et al.* (1982) on cut roots of deciduous hardwoods]. Marshall estimated the amount of turnover as a function of soil temperature and estimated changes in biomass by comparing net differences in root starch content over a given time interval. His estimates, made only from an initial knowledge of fine root biomass and seasonal changes in soil temperature, corresponded very closely with periodic measurements of the standing crop of living and dead roots in three Douglas fir sites (Santantonio, 1982).

The importance of fine root production for mineral and water uptake has long been recognized, but the significance of fine roots in the carbon budget of plants and ecosystems is only beginning to be fully appreciated. Because the amount of carbon allocated to roots may be modified through management, this area has special implications that are discussed in more detail in later sections.

Usually consumption by animals is a small fraction of total NPP in forests

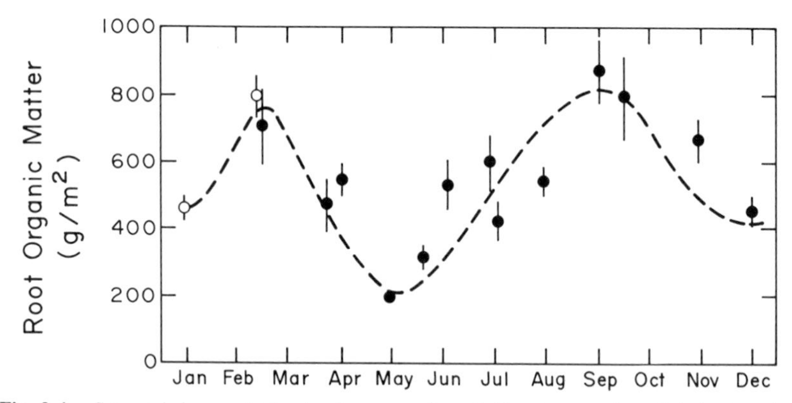

Fig. 3.4. Seasonal changes in the standing crop of small-diameter roots in a *Liriodendron* forest were obtained by core samples collected throughout the year. Estimates of production and turnover rates are obtained from differences in peak and nadir biomass. Organic matter production in this case was about 1460 $g/m^2/yr$ with an average standing crop of 680 g/m^2. (After "Carbon cycling in a mixed deciduous forest floor" by N. T. Edwards and W. F. Harris, *Ecology,* 1977, **58,** 431–437. Copyright © 1977 by the Ecological Society of America. Reprinted with permission.)

unless there is an outbreak of defoliating insects. Even if 10% of the entire foliage is consumed, this represents less than 3% of total NPP in most deciduous forests and less than 2% in most evergreen communities. Estimates of foliage consumption are made by deducting the area of sample leaves that has been consumed (Reichle *et al.*, 1973), comparing foliage on twigs that developed normally with those that were partly defoliated, or by measuring insect frass in litterfall traps and calculating the weight of tissue consumed to produce the amount of frass. Other animals may eat roots or fruits but these are ignored in most calculations of NPP.

Heterotrophic Activity

Secondary production by animals is much more difficult to estimate than primary production. Only a small amount of the plant material consumed is actually digested and a large fraction of what is assimilated is expended on respiration. For mature animals, nearly all of the energy obtained from food goes for maintenance; net production is seen only in reproduction. Populations of animals usually include growing and reproducing individuals so populations generally have some measurable secondary production over a year.

Most estimates of secondary production require knowledge of population densities, average weight of individuals, assimilation efficiencies (fraction of food eaten that is not passed through the body), and respiration rates, plus a survey of the animal's diet and its energy content (Holmes and Sturges, 1973; Burton and Likens, 1975). If an animal that ate 1 g of leaves had an assimilation efficiency of 40% and a respiration expediture of 90% of that assimilated, its secondary productivity in terms of weight gain from the leaves eaten would be 4% (1 g ingested = 0.4 assimilated − 0.36 respired = 0.04 SP). Assimilation efficiencies for terrestrial animals are mostly between 20 and 60% for herbivores, 50 and 90% for carnivores. The production efficiency in terms of net gain per unit of assimilate, however, can be expected to decrease by five- to tenfold progressing up food chains or trophic levels, primarily as a result of decreasing efficiencies in obtaining food (Edwards *et al.*, 1981).

In forests with few large grazing animals and little insect defoliation, the activity of heterotrophic organisms is concentrated in the forest floor and upper soil horizons. Reichle *et al.* (1973) demonstrated that over 95% of all heterotrophic metabolism in forests could be attributed to decomposers. Microflora dominate the metabolism (90%), but they in turn provide a substrate for invertebrate decomposers. Decomposition is treated in detail in Chapter 8.

General estimates of heterotrophic respiration in forests range from about 6 to 14 t/ha/yr of organic matter, which represents 20 to 30% of total ecosystem respiration (Edwards *et al.*, 1981). In massive coniferous forests, R_H may represent less than 5% of total ecosystem respiration (Grier and Logan, 1977).

Ecosystem Metabolism

With some knowledge of the various components of ecosystem production and how they are measured, we are now ready to compare a variety of forest systems for which data are available. To facilitate comparisons, ratios of various components are often employed. Production efficiency (GPP/R_A) is related exclusively to primary production. In Table 3.1, R_A varies from about 10 to 150 t/ha/yr and GPP ranges from about 20 to nearly 160 t/ha/yr. However, the ratio of GPP/R_A ranges only from 1.08 to 2.33 because GPP and R_A both increase with temperature. The amount of respiring tissue increases substantially as forests grow older and larger so plant respiration (R_A) may approach GPP (note Douglas fir forests in Table 3.1.) Forest ecosystems, unlike those dominated by plants of small stature, expend most of GPP in plant rather than heterotrophic respiration (Edwards *et al.*, 1981).

Ecosystem productivity (NEP/GPP) is obtained by subtracting R_H from NPP to obtain an estimate of NEP and adding NPP to R_A to estimate GPP. High values of ecosystem efficiency are associated with rapidly growing forests with relatively small amounts of respiring plant biomass (southeastern pine forest in

TABLE 3.1 Production and Carbon Allocation in Different Forest Ecosystems[a]

Variable	Mixed hardwoods, U.S.[b]	Pine, U.S.[c]	Subalpine conifers, Japan[b]	Beech, Denmark[b]	Rain forest, Africa[b]	Douglas fir, U.S.[d]
Total standing crop (TSC)	175.2	141.2	318.1	—	—	870.4
Gross primary production (GPP)	43.2	77.3	38.2	23.5	53.5	160.9
Net primary production (NPP)	14.5	36.0	10.0	13.5	13.5	10.9
Plant respiration (R_A)	28.7	41.3	27.5	10.0	40.0	150.0
Heterotrophic respiration (R_H)	13.4	13.9	6.6	13.5	13.5	7.6
Ecosystem respiration (R_E)	42.0	55.2	34.1	23.5	53.5	157.6
Net ecosystem production (NEP)	1.1	22.1	4.1	0.0	0.0	3.3
GPP/R_A	1.51	1.89	1.38	2.33	1.33	1.08
NEP/GPP	0.03	0.29	0.11	—	—	0.02

[a] Values other than ratios are in tonnes per hectare per year or tonnes per hectare (TSC).
[b] Edwards *et al.*, 1981.
[c] Kinerson *et al.*, 1977.
[d] Grier and Logan, 1977.

Table 3.1 or young Sitka spruce plantations in Scotland) (Ford, 1982). In cooler climates, the efficiency of organic matter accumulation may be similar to that in warmer climates; however, the amount of live standing crop and accumulated detritus often reaches higher values in cooler environments, e.g., redwood forests and boreal peat bogs.

REGIONAL PRODUCTIVITY OF FORESTS

The ability to predict NPP in various forest systems is of immense practical and theoretical value. Net primary production is the energy base for all secondary production and represents a large amount of potentially harvestable material or organic matter for long-term storage in soils and the forest floor. The range in forest production (NPP) is from less than 1 to nearly 40 t/ha/yr with an average about 15 t/ha/yr (Table 3.1; also refer to Chapter 11, Table 11.1).

Climatic Factors

In a general way, terrestrial NPP is limited, at least at the extremes, by cold temperatures and inadequate precipitation. Leith (1975) suggested that NPP for various ecosystems increased positively but curvilinearly with increasing precipitation and mean annual temperature. Recent studies of forests indicate that these simple relationships are very imprecise (Fig. 3.5). In part, the poor correlation reflects the variable amount of respiring tissue in forests of different ages. Also, because deep-rooted trees generally can obtain sufficient water (Chapter 4), there is no simple relationship between increasing temperature and restrictions on water use that would affect NPP (Webb *et al.*, 1978). Periods of wintertime photosynthesis by evergreen forests (Chapter 2) weaken a general correlation with evapotranspiration suggested in earlier studies by Rosenzweig (1968) that included forests as well as other vegetation. Climatic variables alone should not be expected to adequately predict aboveground NPP of forests. Even GPP cannot be estimated realistically without knowledge of seasonal changes in the amount of forest canopy displayed. Nevertheless, there are clear restrictions upon potential forest production related to climate, as indicated by topographic relationships (Fig. 3.6). Even in a particular area, such as the Great Smoky Mountains, local differences in soil depth, slope, and aspect may cause restrictions in available water and differences in NPP. For example, in Fig. 3.6 production rates by drought-adapted pine are substantially lower than for other forest types growing at comparable elevations.

In the Pacific Northwest region of the United States, much steeper gradients in climate exist than in the eastern part of the country as a result of higher mountain ranges and a strong maritime influence by the Pacific Ocean. Transects from the coast across the mountain range begin with coastal *Picea sitchensis* (Sitka spruce) forests receiving an annual precipitation of 2600 mm and extend to *Juniperus* woodland and *Artemisia* (sagebrush) with one-tenth as much precipitation (Franklin and Dyrness, 1973). The distance involved is less than 200 km.

At eight representative sites along a transect at 44° 30′ N latitude, Gholz (1982) measured soil water use, precipitation, and evaporation to compute a site water balance (Grier and Running, 1977) encompassing the period from May 15 to October 15, during which less than 10% of annual precipitation falls. Aboveground NPP of relatively mature forest was correlated with water availability (Fig. 3.7a) except in the subalpine zone, where winter minimum tem-

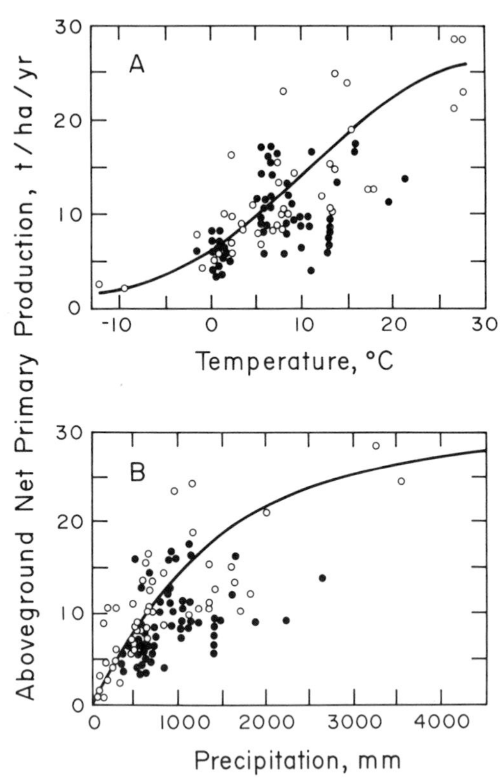

Fig. 3.5. Relationship proposed by Lieth (1975) between NPP and temperature (A) or precipitation (B) fitted to his original data (○), with supplemental data (●) from the International Biological Program. (From O'Neill and DeAngelis, 1981.)

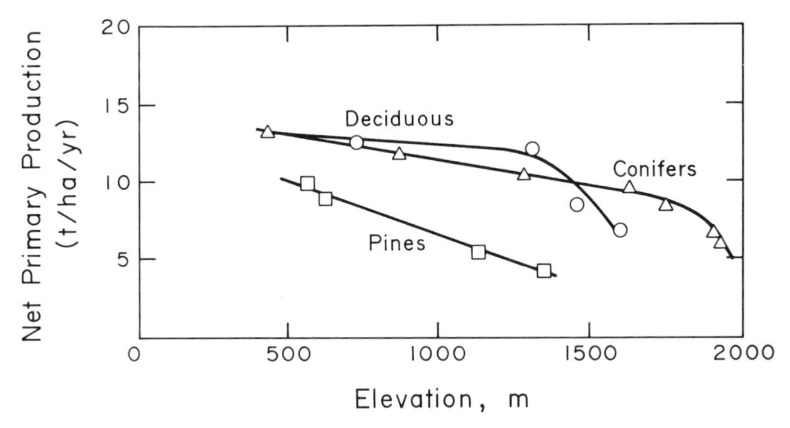

Fig. 3.6. Net primary production along an elevational gradient in the Great Smoky Mountains, southeastern United States, for broad-leaved deciduous forests of moist environments; evergreen coniferous forests (*Tsuga*, *Picea*, and *Abies*) of moist environments; and pine forests of dry environments. (After Whittaker, 1975.)

peratures appeared more important (Fig. 3.7b). January minimum temperatures served as a good index for predicting NPP in all vegetation zones, probably because nongrowing season photosynthesis plays an increasingly important role in lower elevation forests west of the Cascade crest (Emmingham and Waring, 1977; Waring and Franklin, 1979).

Structural Indices of Productivity

Webb *et al.* (1983) found that NPP was well correlated ($r^2 = 0.8$) with the maximum foliage present in a broad survey including coniferous and deciduous forests, grasslands, and desert ecosystems. Gholz (1982) estimated leaf area index (LAI; projected canopy area/unit of ground) for the vegetation types sampled along his Oregon transect and found that NPP appeared to correlate better with maximum (sustainable) leaf area than with any of the environmental gradients alone (Fig. 3.7c). A correlation beween production and maximum LAI was also found along a similar transect 2° further north in latitude (Schroeder *et al.*, 1982). In Europe where deciduous stands of beech (*Fagus sylvatica*) are widely planted, maximum LAI decreases from 6.9 to 3.2, progressing from France north to the limits of the species in southern Sweden. Net primary production followed a similar trend (DeAngelis *et al.*, 1981).

The photosynthetic season is important to define in such comparisons. For example, GPP in a hardwood forest of northeastern United States was only about

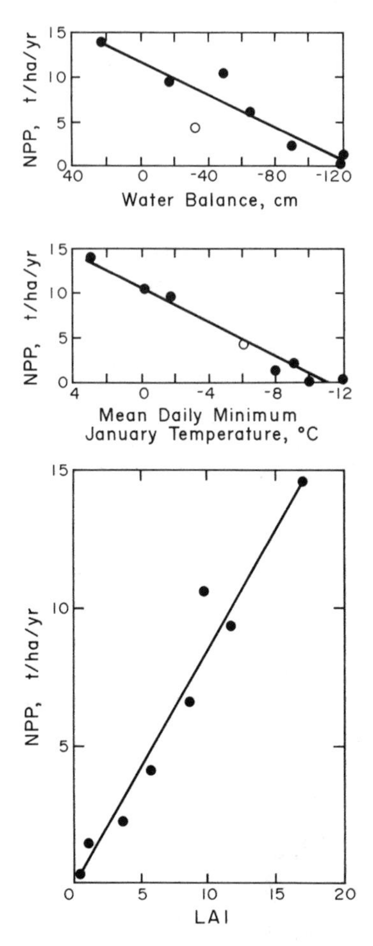

Fig. 3.7. Aboveground NPP decreases progressively across a 200-km transect at 44°30′ N latitude from Pacific coastal forests of *Picea sitchensis* to *Juniperus* woodland. The relationship between NPP and a water balance index for the period from May 15 through October 15 is shown in (a). In (b) NPP values are plotted in relation to January temperatures. This improved the prediction for a subalpine forest, denoted by the open circle. Maximum LAI showed close agreement with NPP at the represented sites (c). (After "Environmental limits on aboveground net primary production, leaf area, and biomass in vegetation zones of the Pacific Northwest" by H. L. Gholz, *Ecology*, 1982, **63**, 469–481. Copyright © 1982 by the Ecological Society of America. Reprinted by permission.)

half that reported for a forest with similar canopy LAI located in the southeast, where the growing season was nearly twice as long (110 versus 180 days) (Whittaker *et al.*, 1974; Harris *et al.*, 1975). Thus, a leaf area duration (LAD) index, representing the product of LAI and months in which the canopy may conduct photosynthesis, provides a more general estimator of GPP (Fig. 3.8). In the eastern United States where deciduous forests dominate, the maximum LAI may not vary much; the length of the growing season, however, varies from less than 100 days in northern New England to more than 300 days in some parts of Florida. Net primary production tends to reflect these differences (Fig. 3.9).

In the western half of the United States, NPP follows a different trend with highest production rates along the Pacific Coast and a steep decrease progressing inland (Fig. 3.9). The general dominance of evergreens in the West makes an analysis of yearlong photosynthesis essential. In such an analysis, climatic differences between the western and eastern parts of the country become obvious (Waring and Franklin, 1979). In the eastern United States, the humidity is generally high during the growing season and precipitation is relatively evenly distributed. Summer night temperatures are warm, except at higher elevations or in the most northern latitudes. These conditions favor trees with few, but efficient, broad leaves. In the West, summer is a time of relative drought with extremely low humidity. Thus for equivalent temperatures, the potential evaporation may be at least a third higher than in the eastern United States (Waring and Franklin, 1979). This favors needle-bearing species that can limit transpiration without at the same time increasing leaf temperatures to near lethal levels (Chapter 4). Moreover, in the West, the major forests grow in mountains where

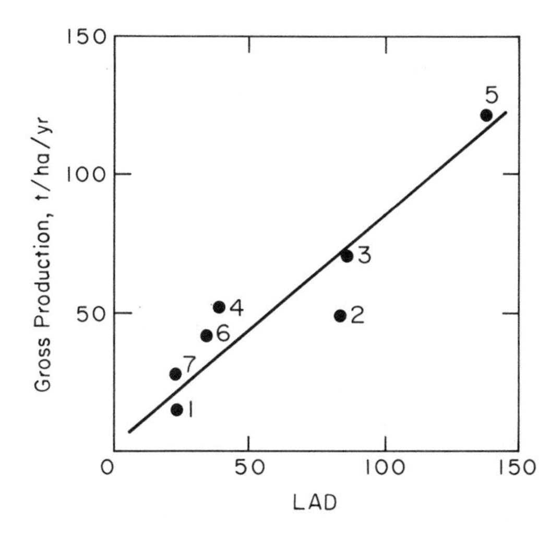

Fig. 3.8. Gross primary production in a wide range of broad-leaf forests increases as the product of LAI and number of months in the growing season increases. This index is termed leaf area duration (LAD). The numbered points refer to (1) *Fagus* forest in Japan; (2) *Castanopsis* forest in Japan; (3) broad-leafed forests in Japan; (4) tropical humid forests of the Ivory Coast in Africa; and (5) tropical forests of southern Thailand (Kira and Shidei, 1967). Point number (6) is a *Liriodendron* forest in the southeastern United States (Harris *et al.*, 1975) and (7) is a mixed hardwood forest from the northeastern United States (Whittaker *et al.*, 1974). (Additional information in DeAngelis *et al.*, 1981.)

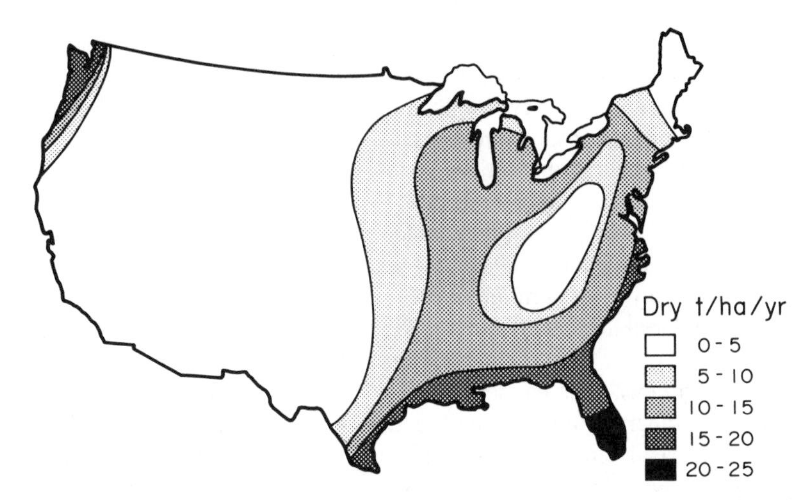

Fig. 3.9. Estimated aboveground production across the contiguous United States shows both maritime effects and latitudinal relationships in the hardwood-dominated eastern half of the country. In the western part, where evergreen vegetation dominates, the maritime influence is enhanced by the mountainous topography. (After Ranney and Cushman, 1982; Oak Ridge National Laboratory, operated by Martin Marietta Energy Systems, Inc.)

cold air often passes downslope in the evening. Cool night temperatures reduce respiration expenditures by the massive accumulation of living cells in mature conifers. Other factors are also involved, such as the mild winters along the Pacific Coast which permit photosynthesis to continue in that season (Emmingham and Waring, 1977). At higher elevations in the Rocky Mountains, minimum temperatures alone limit the presence of hardwoods to a few boreal species (see Chapter 4; Burke *et al.*, 1976).

Although the maximum LAI or LAD are merely structural indices, they link directly to physiological processes controlling GPP. The more inhospitable an environment, the more restricted trees are in accumulating and maintaining a high leaf area. Either of these two indices might serve well as long-term integrators of the influence of environmental change. Factors such as air pollution should, when damaging, reduce both the amount of foliage and its duration of display (Mann *et al.*, 1980).

STAND PRODUCTIVITY

The amount of canopy affects not only the efficiency of leaf photosynthesis but also the relative allocation of NPP into various structural components. In this section, we first review canopy dynamics and related changes as forests develop and then evaluate how carbon allocation to stem wood, the major harvest product, is affected by changes in the relative availability of light, water, and nutrients.

Canopy Dynamics

Initially, following destruction of a forest and before establishment of other vegetation, the canopy leaf area is near zero. The speed of vegetation recovery depends on the availability of propagules and harshness of the environment. After removal of tropical rain forests, the canopy LAI may return to that typical of mature forests within a period of months if soil fertility is not drastically altered (Ewel, 1977). In temperate forests of the northeastern United States, shrub cover may reestablish a canopy leaf area equal to that of a mature deciduous forest within a period of 3 yr after harvest (Marks, 1974). In other areas, particularly those dominated by long-lived conifers, the process of canopy closure may be very slow, extending to a century or more (Peet, 1981).

Even before LAI in a forest approaches a maximum, competition among trees results in some mortality. For example, in a pure forest of *Abies* growing in Japan, the maximum LAI may require 50 yr to develop, yet by that time, stocking has been reduced by nearly 90% through natural mortality (Tadaki *et al.*, 1977; Chapter 9, Fig. 9.1). LAI actually peaks and then decreases slightly, maintaining a plateau for half a century before further decreasing as the stand dies (Fig. 3.10). The maximum sustained LAI is often 10–20% less than the peak LAI observed in many even-aged forests (Forrest and Ovington, 1970; Kira and Shidei, 1967; Ford, 1982). Some have suggested that the peak in LAI just precedes the period of most intense competition for light and nutrients when mortality increases dramatically (Ford, 1975; Mohler *et al.*, 1978).

Forests of mixed composition sometimes show a temporary drop in LAI following loss of dominants from earlier successional stages (Fig. 3.10). The death and fall of a large number of even-aged trees create gaps that release species already established in the understory. There is a lag, however, that may account for a temporary decrease in LAI before full canopy is reestablished (Peet, 1981). Of course, if gap or patch regeneration continues, even at a slower rate, the LAI may never reach the level obtained when a single even-aged forest dominated (Bormann and Likens, 1979a).

In some boreal forests, leaf area first increases and then decreases following a transition from deciduous to coniferous species. Under evergreen cover, the level of permafrost rises, restricting root growth and the availability of nitrogen and phosphorus. Without fire or other disturbance the site may eventually deteriorate

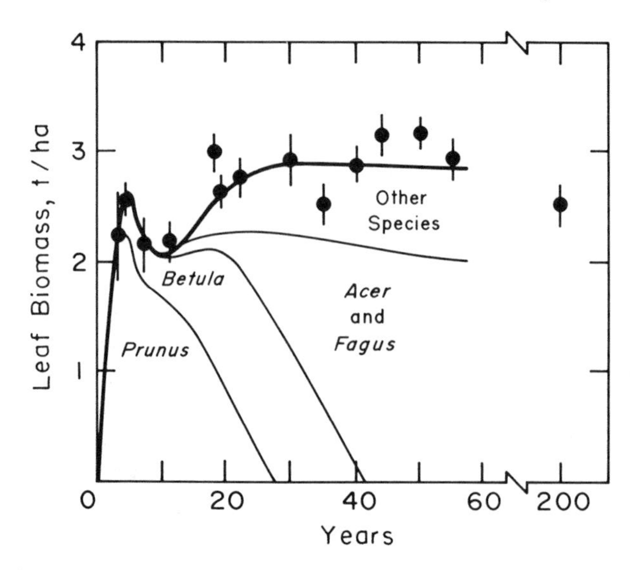

Fig. 3.10. Leaf biomass for a secondary successional sequence in northern hardwoods. Data are means and 95% confidence intervals. (After Bormann and Likens, 1979a.)

to muskeg with very sparse canopy cover (Reiners *et al.*, 1971; Van Cleve and Viereck, 1981; Van Cleve *et al.*, 1983).

Productivity and Biomass Accumulation

When secondary forests are established following removal of more advanced stages, they often accumulate biomass more rapidly and to a greater extent than their predecessors. This is true of bamboo in tropical forests, cottonwood (*Populus deltoidea*) along floodplains in the midwestern United States, pine forests in the southeastern coastal plain and northern lake states, and *Liriodendron* forests in the southeast (U. S. Department of Agriculture, 1965; Peet, 1981). Foresters often select early successional species because of rapid growth under intensive management.

Why some advanced stages of forest succession are less productive than earlier ones is still a matter of some speculation. Horn (1971, 1974) argued that the canopy geometry of early successional trees is more efficient, e.g., has lower light extinction coefficients than the canopy of species appearing later, but Peet (1981) cites exceptions. Inherently higher rates of photosynthesis are generally associated with species adapted to well-illuminated environments (Chapter 2;

Bazzaz and Pickett, 1980; Stephens and Waggoner, 1970). In addition, even-aged forests that establish following a disturbance may have more access to nutrients. The cause of reduced production observed in some advanced stages of forest succession could be determined by some rather simple experiments where established saplings are allowed to grow following complete removal of the overstory. Alternatively, fertilizer could be added or drainage improved to discern how these factors might improve growth. Unfortunately, these kinds of experiments have been restricted, almost exclusively, to even-aged forests. We can, nevertheless, benefit by reviewing the results.

At the beginning of a stand's development, a disproportionate amount of carbohydrate is expended upon foliage production. As a stand develops, the relative proportion of biomass in stems increases and that in foliage and branches decreases (Fig. 3.11). About 50% of aboveground production goes to stem and branch growth over an extended period when the canopy LAI has stabilized (Fig. 3.12). In older stands a reduction in stem and branch growth reflects the increasing respirational demands associated with long-lived nonphotosynthetic tissue in the branches, bole, and large roots. Changes in the relationship of foliage production to stem and branch wood growth are a sensitive measure of whether allocation patterns are being altered throughout the tree.

Environmental Effects

In forests at the same stage in development, partitioning of NPP varies with the environment. In Chapter 2 we introduced a growth efficiency index (E) based

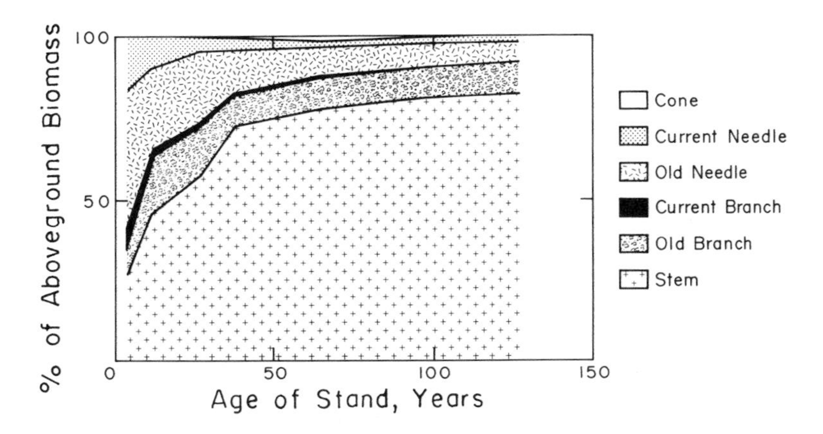

Fig. 3.11. Relative biomass distribution in an *Abies* stand changes with age. (After Tadaki *et al.*, 1977.)

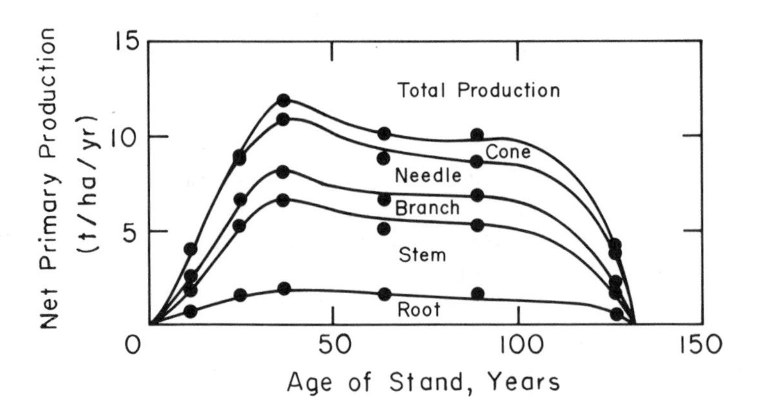

Fig. 3.12. Change in NPP and allocation into various structural components as a stand of *Abies* ages. (After Tadaki *et al.*, 1977.)

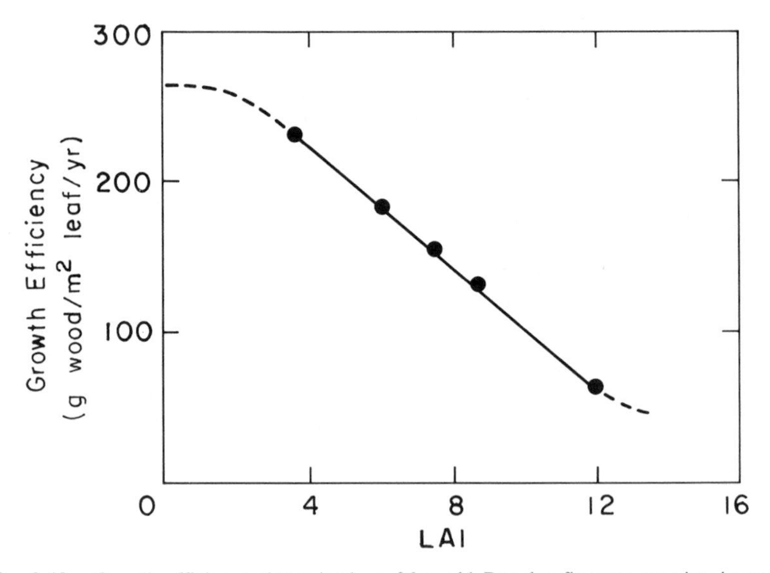

Fig. 3.13. Growth efficiency determined on 36-yr-old Douglas fir trees growing in stands thinned to different stocking levels showed a fourfold decrease over a range in LAI from 4 to 12. Points represent averages of measurements made on 30 randomly selected trees. (After Waring *et al.*, 1981.)

on the ratio of stem wood production to leaf area. The index E was shown to reflect patterns of carbohydrate allocation and to be easily estimated from current annual ring growth and sapwood thickness. It is instructive to see how this index changes following a reduction in canopy leaf area.

In a series of thinning plots in a 36-yr-old Douglas fir forest growing on well-watered, nutrient-rich soils, trees ranged in stocking from less than 200 to nearly 2000 stems/ha. Growth efficiency decreased linearly by fourfold with increasing LAI over the range from 4 to 12 (Fig. 3.13). At lower levels of LAI the ratio of wood production to leaf area eventually stabilized, and at very high stocking densities the relationship became curvilinear as all foliage became confined to a narrow zone near the very top of the trees (Ford, 1975). If maintained at such high densities, the LAI may actually decrease (Bormann and Gordon, 1984).

Although growth efficiency of the forest decreases with increasing LAI, total aboveground NPP increases up to a constant level (Fig. 3.14). In forests unable to support canopy LAI above about 6.0, no plateau is reached (Madgwick and Olson, 1974; Waring, 1983). Production per unit of ground area is the product of growth efficiency and total canopy LAI, but a sparse canopy with high efficiency

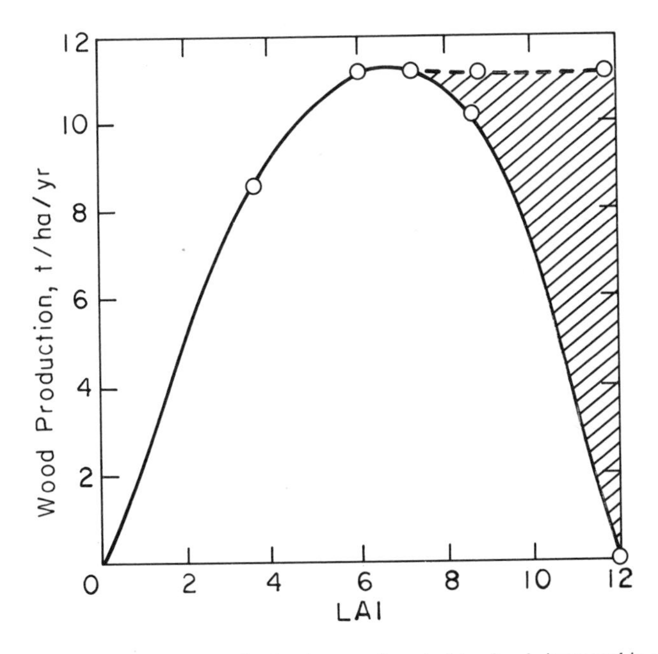

Fig. 3.14. Aboveground wood production increased as stocking levels increased in a 36-yr-old Douglas fir forest until the canopy LAI reached about 6. Deducting the biomass lost in annual tree mortality (shaded area) indicated the rates of net harvestable wood growth decreased when stands were maintained near maximum LAI. (After Waring *et al.*, 1981a.)

cannot produce as much wood as a dense canopy with lower efficiency. The annual production of merchantable stem wood follows a parabolic curve with increasing LAI since at the higher levels of competition there is considerable mortality (Fig. 3.14).

In a more arid region where forests of *Pinus ponderosa* dominate, a long-term experiment was conducted to assess whether removal of undergrowth shrubs might provide more available water to pines thinned to various canopy densities and stocking levels (Barrett, 1970). Measurements of soil water confirmed that heavy thinning and shrub removal increased the available water substantially (Barrett and Youngberg, 1965). At the highest stocking level, the available water was used exclusively by pine trees since all shrubs were shaded out (Fig. 3.15). The lower the stocking or LAI, the greater the influence of removing the shrubs upon growth efficiency. For example, at a pine LAI of 1.0, production increased by more than 50% following shrub removal, whereas at a LAI of 2.0, growth efficiency increased by only 30%. Because the maximum LAI obtained was still less than 3.0, total wood (or biomass) production per hectare continued to increase with stocking. The economic value of wood products, however, peaked at intermediate values of LAI.

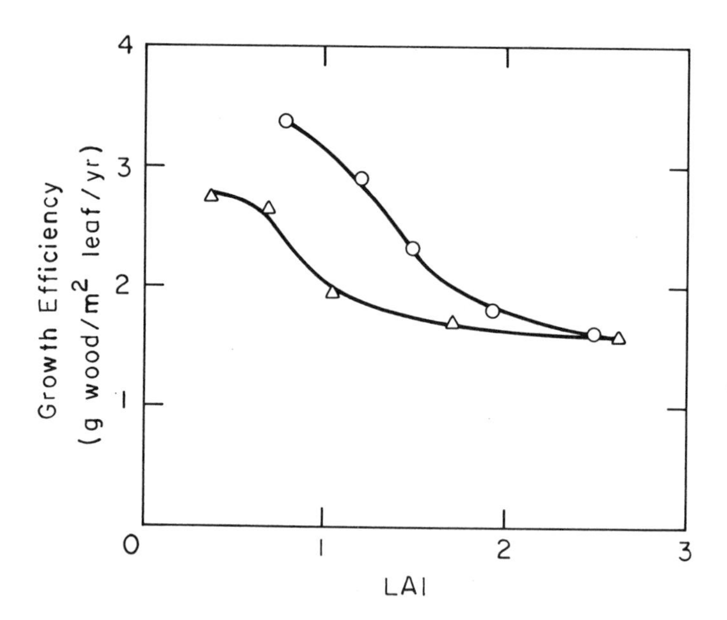

Fig. 3.15. When shrub cover was removed (○) from beneath an arid zone ponderosa pine forest, the growth efficiency of trees significantly increased over that observed with shrub cover present (△). Above an LAI of 2, however, the differences between treatments disappeared. (Calculated from Barrett, 1970; after Waring, 1983.)

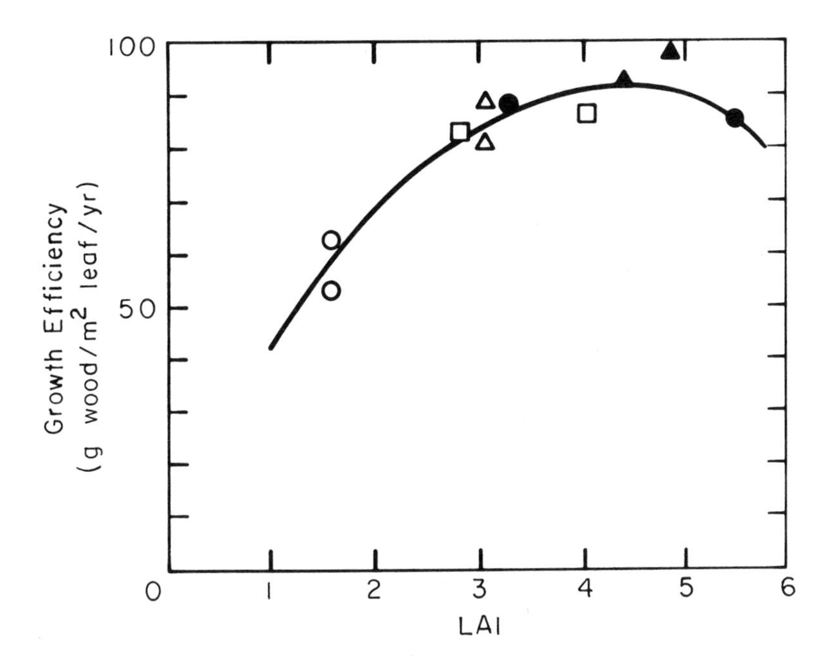

Fig. 3.16. In a Norway spruce plantation where increasing levels of nitrogen and phosphorus fertilizer were applied annually (○, N_0P_0; □, N_1P_1; △, N_2P_0; ▲, N_2P_1; ●, N_3P_1), growth efficiency cy increased and was maintained as LAI increased from 1 to 5.5. At higher values of LAI, growth efficiency begins to decrease abruptly. (After Albrektson *et al.*, 1977, and unpublished data of C. O. Tamm, Swedish University of Agricultural Sciences, Uppsala, Sweden; figure from Waring, 1983.)

Rarely have fertilizer experiments in forestry been designed to take into account the effects of changing leaf area upon growth (Miller, 1981). Usually fertilizer is applied only once or at infrequent intervals, all stands are thinned to the same density, and growth is compared after a given interval. In Sweden, a group headed by C. O. Tamm has conducted long-term experiments on *Picea* and *Pinus* forests that involved annual additions of various amounts of nitrogen and phosphorus designed to maintain different internal levels of these nutrients in the tree foliage (Albrektson *et al.*, 1977). Stocking was similar in all treatments. Increasing amounts of nutrients increased the proportion of carbon allocated to wood compared to foliage, maintaining a high growth efficiency even as LAI continued to expand (Fig. 3.16). In later years, as LAI on some treatments exceeded 8.0, growth efficiency fell (C. O. Tamm, Swedish Agricultural University, Uppsala, Sweden, personal communication).

In a recent experiment in Sweden, stands of 10-yr-old Scots pine were: (1) irrigated each day during the growing season; (2) fertilized annually; (3) fertilized and irrigated; or (4) left untreated (Aronsson *et al.*, 1977). Leaf area index and growth efficiency were calculated by analyzing changes in stem growth and sapwood area of sample trees (Fig. 3.17). Initially, growth efficiency decreased rapidly as the LAI on each plot began to increase. The actual rates of decline were quite similar among treatments as were the ultimate levels approached. The maximum or near maximum LAI ranged from 1.0 on the untreated plots to 3.6 on the two treatments receiving fertilizer. The latter values are more than 50% higher than that estimated for an adjacent 150-yr-old forest (Lindroth and Perttu, 1981). If we compare growth efficiency at similar LAI, the choice is limited to values below 1.0, where the irrigated, fertilized, and combination treatments averaged 1.9, 2.8, and 3.4 times greater growth efficiency than the untreated stand, respectively. These differences reflect a shift in NPP from roots to aboveground production of foliage and wood (Linder and Axelsson, 1982; Chapter 2). At near maximum values of LAI, no real difference in growth efficiency appeared.

In this experiment, production rates per hectare were directly related to maximum canopy LAI (Fig. 3.18) because intense competition and mortality were experienced. This experimental evidence supports the view that forests continue to develop their canopies dependent on the availability of water, nutrients, and other resources until a maximum is attained. This maximum may be very low on dry, cold, or infertile sites. Once reached, however, the majority of trees may

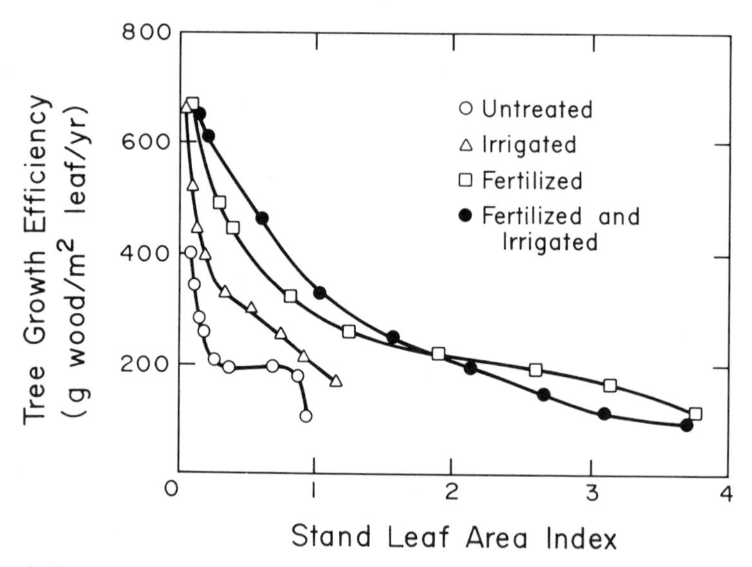

Fig. 3.17. Saplings of Scots pine grown over a decade under differing nutrient and water regimes showed large differences in growth efficiency at a comparable LAI of 1.0. As the stands approached their maximum leaf area after 10 yr, however, growth efficiencies approached similar values. At that time LAI differed more than threefold. (After Waring, 1985.)

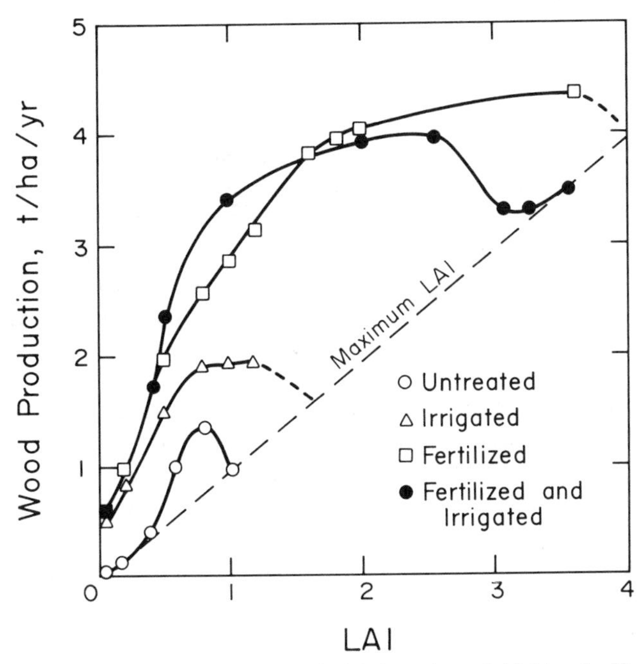

Fig. 3.18. As Scots pine plantations approached their maximum LAI determined by experimental conditions, stem wood production per hectare becomes closely related to maximum LAI. (From unpublished data by B. Axelsson, A. Aronsson, and R. H. Waring, Swedish University of Agricultural Sciences, Uppsala, Sweden.)

exhibit a similar balance in the way carbohydrates are allocated between foliage and wood production. With this insight, we would expect that forests of mixed composition in the same region would exhibit a close correlation between production and maximum LAI (Schroeder *et al.*, 1982; Gholz, 1982). In Chapters 2 and 9, the foundation and implication of these principles are considered, respectively.

FOREST SUCCESSION MODELS

The ecological literature is rich with descriptions of how the vegetation in many regions varies in sequence over time (succession) following a major disturbance (Braun-Blanquet, 1951; Daubenmire, 1968; Golley, 1977; McIntosh, 1978; West *et al.*, 1981). Few studies have attempted to evaluate the underlying mechanisms, and almost none have done so experimentally. Often, the projected patterns of change do not occur because succession is interrupted by additional disturbances. Frequent disturbance tends to decrease LAI, allowing more light to penetrate to the forest floor. This selectively stimulates the growth of certain species. Although factors other than light and seed supply are known to vary during succession, they have not received equal attention. In particular, the introduction or loss of competitive plant species, herbivores, or disease organisms may invalidate theories of succession based on historical evidence. Similarly, forest practices that involve shorter crop rotation cycles, complete tree utilization, plowing, and herbicide or pesticide applications provide situations requiring new insight as to how various factors interact. Unfortunately, a historical analysis of succession, although helpful, is no longer adequate to predict the future.

With the introduction of computer simulation techniques, the structural and physiological information important for determining growth and reproduction of different species can be incorporated into models (Botkin *et al.*, 1972). The general validity of these models can be evaluated where known gradients in vegetation are described or historical changes in forest composition are well documented (Shugart and West, 1977). Although there are now more than 100 computer simulation models available in the United States alone, many of these are restricted to use in pure stands of one species, and often with trees of similar size and spacing (Shugart and West, 1980).

Most models applicable for projecting changes in forests of mixed composition over centuries have a similar basic structure. At their heart is an interactive accounting system (Fig. 3.19) that keeps a record of the size distribution of trees and other vegetation on imaginary plots (Shugart and West, 1980). The more advanced models also record what happens when trees die, e.g., whether they remain standing, fall to the ground, or are removed from the site by fire or logging. Each year, a reassessment is made and trees are allowed to (1) become established; (2) grow; (3) die; or (4) decay.

Whether the initial site is bare, burned, flooded, or covered with debris is important to determine the establishment of various species. Information about the reproductive behavior of species must be stored in the model—which species produce seeds or sprout from roots, how far and by what agents may seeds be dispersed, and how long propagules remain viable? When conditions are deemed suitable for establishment of certain groups of species, their presence upon the imaginary plots is a matter of statistical probability.

Growth of the established plants is then predicted, making rough allowances for how photosynthesis is affected by the amount of light penetrating through a defined amount of canopy leaf area. Some reduction in growth is also made as trees increase in size, reflecting changes in the amount of nonphotosynthetic tissue. Plants are grouped with regard to their relative sensitivity to temperature, light, moisture, browsing, pollution, etc. These groupings are often very broad;

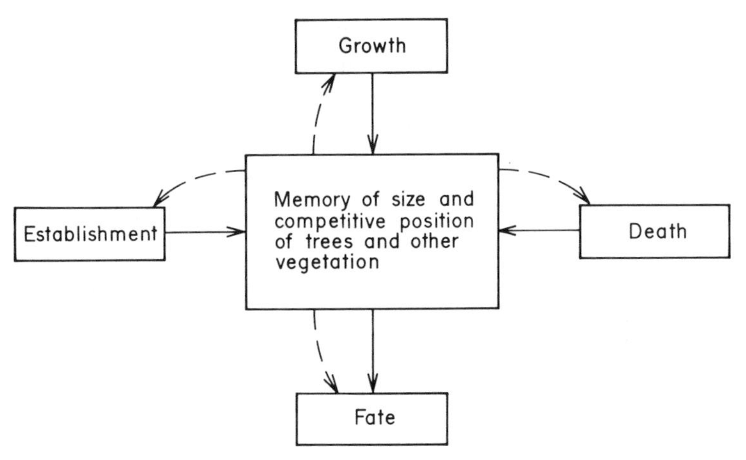

Fig. 3.19. The basic structure of interactive simulation models consists of a memory of size and competitive position of trees and other vegetation on imaginary plots and four interactive processes (transfer of information denoted by - - -) that determine the establishment of trees, their growth, death, and ultimate fate (changes in number of trees, size, and state result denoted by ———).

for example, most species are considered to be either shade tolerant or intolerant. Sensitivity to temperature is often determined from the present distribution patterns in latitude and elevation. In spite of these simplifications, the resulting predictions, when restricted to particular regions, are amazingly well supported by field data and historical surveys.

Depending on the general environment, some groups of trees are predicted to grow rapidly while others drop behind. Once trees are overtopped, light becomes limiting to growth and the probability of death is allowed to increase markedly. Death of a tree may be a simple function of age, sensitivity to fire, wind, snow breakage, insects, diseases, or the selective activity of man. This last point is important. These models differ from most previous successional schemes by incorporating both natural and human effects.

When a tree dies, its fate is often critical for assessing the future direction and rate of succession. If left on the site, a tree may serve as shade, as a nesting site, as a substrate for germinating seedlings, and eventually as a component of soil humus as it progresses from a standing snag to a fallen and decaying bole. On the other hand, when trees are harvested, mineral and organic resources are lost from the ecosystem. Whether foliage and branches are removed is often more critical to the assessment than the amount of bole wood because of the difference in nutrient content and resistance to decay (Chapters 6 and 8). Only recently have these added structural refinements and the ability to account for nutrients been incorporated into forest succession models (Aber and Melillo, 1982b; Aber *et al.*, 1979; Kimmins *et al.*, 1981). These models also permit us to predict the effect of additions of inorganic fertilizers, and they represent the first quantitative link between primary production and decomposition processes in determining the course and rate of forest succession.

The major computer simulation models of forest succession have been partly validated in three ways using independent data. First, they have predicted changes in composition and growth in particular areas. For example, Doyle (1981) predicted the diameter and frequency distribution of different species in the rain forests of Puerto Rico. Shugart and West (1977) simulated the known changes in forests in the uplands of Arkansas, and Botkin *et al.* (1972) did the same for elevational gradients in the northeastern United States. Second, the models have been validated by predicting internal relations not required to run the model. The total live biomass, litter biomass, rates of production by the whole community, and light absorbed by the canopy are examples of the kinds of variables predicted in this kind of test (Aber and Melillo, 1982b). The third kind of validation involves prediction of changes following documented disturbances. The responses of *Eucalyptus* forests to different intensities of fire (Shugart and Noble, 1981), changes following the introduction of chestnut blight to the southern Appalachian forests (Shugart and West, 1977), and the effects of different intensities of flooding in Arkansas wetlands (Phipps, 1979) represent this kind of validation.

In addition to predicting changes in species composition and growth rates following natural or man-induced disturbances, succession models have the potential of assessing the behavior of vegetation under environmental conditions different from those encountered today. We might ask, for example, what are the effects of fire frequency upon forest composition and biomass accumulation in a particular region (Shugart and West, 1981)? Similarly, the effects of flood control upon cypress (*Taxodium*) forests can be assessed (Tharp, 1979). With a growing concern for endangered species of wildlife, we can test various management options designed to provide critical habitat such as dead trees (Shugart and West, 1981).

The succession models described have provided special insight when used to evaluate the effects of chronic air pollution (West *et al.*, 1980) and increasing levels of carbon dioxide (Botkin *et al.*, 1973). We have learned that a species' relative tolerance to a particular kind of stress is insufficient by itself to predict the success of that species in competitive associations. Also, the time when stress is imposed is shown to make a difference in how the forest responds. Some species react very quickly when small in stature, e.g., to grazing or drought. Later, when a forest is composed of larger trees, there may be no effect until regeneration of the species is involved (West *et al.*, 1980).

Predictions of change in response to stress are probably conservative. None of the models yet incorporates insects or disease outbreaks or the contribution of wildlife in disseminating propagules from refugia (Chapter 9). With chronic stress we might expect more epidemic populations of insects and diseases, hastening death of slow-growing trees. Reduction in wildlife or refugia could reduce the availability of seeds.

Earlier we discussed a variety of possible explanations for observed changes in forest productivity and accumulation of biomass based on observations made almost exclusively on small plots. Experience gained with computer simulation models indicates that successional trends can only be evaluated when interpreted for large areas. For example, the living biomass on a small plot may change drastically as dominant canopy trees die and are slowly replaced (Fig. 3.20). A similar response is expected over thousands of hectares of forest land following a fire or other destructive agent, if most of the species exhibit similar growth rates and longevities.

In the northeastern United States where wind and human activity are the major disturbing agents, the initial forests are relatively even-aged and develop maximum biomass. Later, an uneven-aged forest of relatively few hardwood species dominates and the total biomass is predicted to equilibrate at a lower level (Bormann and Likens, 1979a). If forest succession is interrupted periodically, or an extreme stress is experienced, or if the tree species are more diverse in their ecological behavior, then the biomass peak is lost and the maximum occurs during more mature phases, as in the tropical rain forests of Puerto Rico (Fig. 3.21). On the other hand, if there are a few species that can colonize disturbed

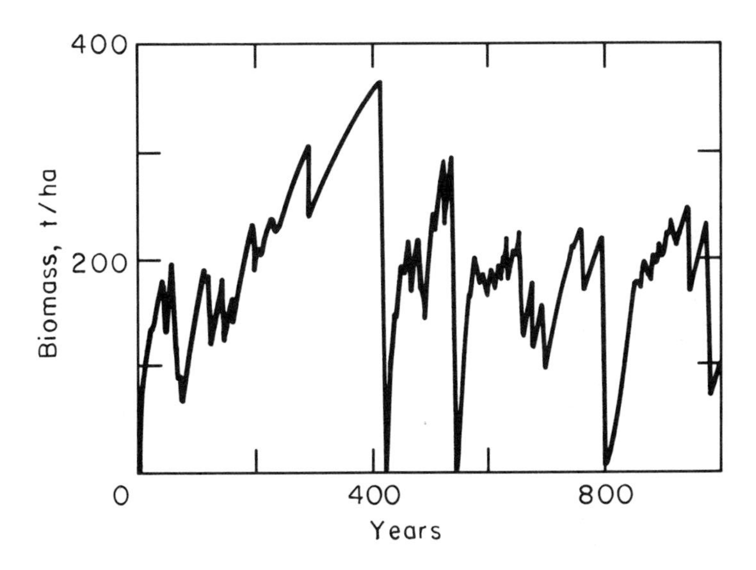

Fig. 3.20. Biomass accumulation on a typical small patch ($\frac{1}{12}$ ha) reflects the development of dominant trees that die periodically as the initial forest ages. (From Shugart and West, 1981.)

areas and attain a large size, then a biomass peak will appear early in the successional sequence (Shugart and West, 1981).

In evaluating the succession models, a very large number of imaginary plots are needed to even out variation attributed to the death of individual large trees. Plot size must be increased in proportion to the height of trees and the area affected by their death. At least 50 plots of $\frac{1}{12}$ ha each are required to obtain steady state at the landscape level in the eastern United States. More plots of even larger size would be required in areas with larger trees and more mountainous landscape (Hemstrom and Adams, 1982; Reed, 1982). In Fig. 3.22, the area required for forests to approach steady state is shown to vary from small watersheds of 10^4 to 10^5 m^2 when disturbances involve the fall of individual trees to areas between 10^8 and 10^{12} m^2 for large wildfires or hurricanes (Lugo *et al.*, 1983). Thus, many areas are too small to ever attain a real successional steady state. In Australia, the amount of forest burned annually exceeds the entire range of many species of *Eucalyptus* (Luke and McArthur, 1978). Likewise, in the Caribbean where hurricanes are frequent, only the larger islands such as Cuba might possibly have a landscape in steady-state (Doyle, 1981).

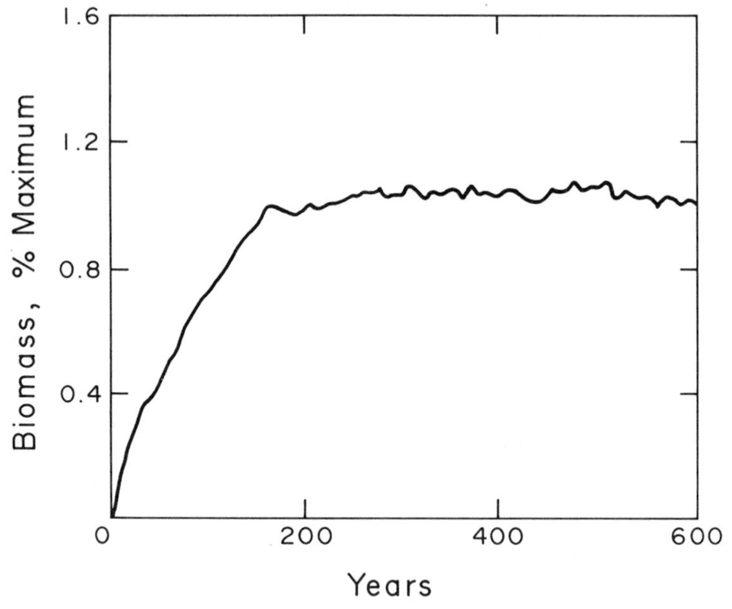

Fig. 3.21. Where there are many species with differing ecological requirements and none attain strikingly different sizes, the accumulation of biomass tends to peak and be maintained in later stages of forest succession, as shown for tropical rain forests of Puerto Rico. (From Doyle, 1981.) Biomass scale is relative to values predicted for forests at 600 yr.

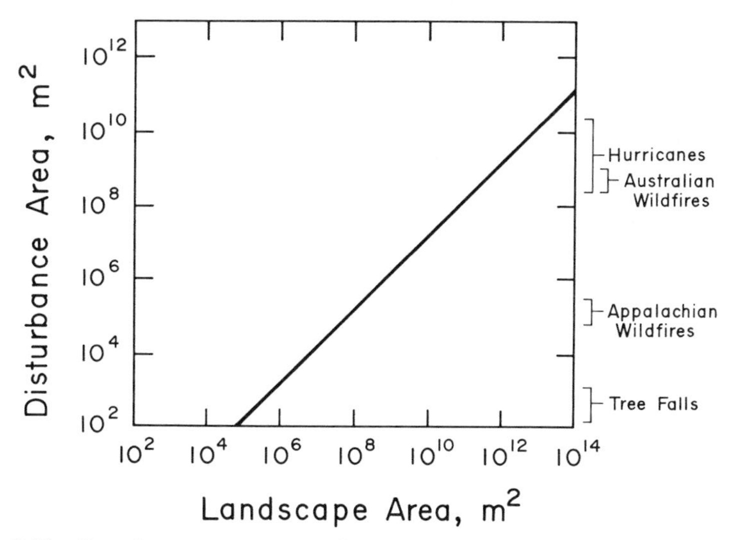

Fig. 3.22. Some forest areas are too small to reach steady state because they are subjected to large-scale disturbance. Thus, a small watershed is relatively unaffected by the fall of a tree but drastically altered by a forest fire. (After Shugart and West, 1981.)

SUMMARY

In this chapter we learned the importance of knowing the maximum canopy LAI that a forest can maintain and its seasonal duration or activity. Once a forest has reached maximum LAI, its primary production can be expected to decrease in proportion to an increasing biomass of sapwood and associated respiring, nonphotosynthetic tissue. Although the maximum LAI or its duration of display (LAD) vary with environment, the growth efficiency of trees may be quite similar, explaining why production is often linearly related to the two canopy indices. Aboveground biomass increment and wood production may peak or reach a plateau when LAI equals 5 or 6 because of a corresponding peak in growth efficiency.

Growth efficiency is strongly influenced by the competing canopy LAI but is also affected by the availability of water and essential nutrients. To assess the relative importance of these and other environmental variables, comparisons of wood production or other attributes of NPP should be made at equivalent levels of canopy LAI.

Although even-aged forests often produce the greatest biomass and are relative easy to manage, the ability to predict changes in forest composition and growth in mixed-aged forests is equally important. Computer simulation models are beginning to incorporate basic principles that affect both primary production and decomposition processes. These models at present employ very primitive environmental indices and incorporate only the most rudimentary knowledge of species tolerances. Still, they show the necessity of evaluating competition among species to assess management options or the effects of environmental stress. Furthermore, they indicate that large areas are required for successional patterns to reach steady state, depending on the scale and frequency of disturbance.

4

Tree–Water Relations

INTRODUCTION

All functional living cells in a tree must remain turgid with relative water content above 75% (Bradford and Hsiao, 1982). Turgidity is threatened as water evaporates and diffuses out through the stomata of leaf surfaces in the process called transpiration. Turgidity is maintained by water absorbed by roots. The rate of absorption is controlled to a large extent by tension that develops in the stream of sap that connects leaves to roots. If the flow of water is halted, stomata respond by closing and photosynthesis stops.

Maintenance of turgor is essential for plant growth and the availability of water often determines the kinds of trees present in a forest. In this chapter we compare how trees have adapted to water stress by discussing (1) how roots extract water from the soil; (2) how sapwood both stores and conducts water; and (3) how stomatal behavior is affected by what happens along the conducting pathway as well as in the environment around the leaves. Finally, information concerning flow through various components will be incorporated in models that calculate water uptake, tissue water status, and transpiration. These factors have significance at the ecosystem level through their influence upon hydrology (Chapter 5), primary production (Chapters 2 and 3), and mineral cycling (Chapter 6).

The degree of attraction among water molecules provides a common basis for comparing the state of water because it reflects the thermodynamic capacity of water to do work (Slatyer, 1967; Milburn, 1979). If pure water at atmospheric pressure and 20°C is taken as a reference having zero potential, increasing pressures provide positive potentials with increased attraction among molecules, whereas decreasing pressures result in negative values with less attraction. As we shall see, water in the soil usually has less negative potential or cohesive tension than that in the foliage. Water always flows in the direction of more negative potential. Standard international units are megaPascals (MPa) with dimensions of m^{-1} kg s^{-1}. One MPa is equivalent to 10^7 dyne/cm^2, 10 bars, or 10.13 atms.

Plant water potential consists of osmotic, matric, and pressure components. Within an individual cell, the vacuole, which contains both ions and other molecules in solution, has a negative solute or osmotic potential (ψ_s). Water diffuses into the solute-enriched (0.5–1.0 *M*) vacuole through its semipermeable membrane. Water is not absorbed indefinitely, however, because counterbalancing forces are exerted by the cell wall upon the swelling membrane, like a tire upon an inflated tube. When the cell is turgid, the wall pressure (ψ_p) is positive. Such positive turgor pressures permit expansion and growth. In addition to turgor and osmotic potentials, living cells may exhibit slight matric potentials associated with collodial surfaces or interfaces so that:

$$\psi = \psi_p - \psi_s - \psi_m$$

In soils, where water is held by capillary forces in crevices between irregularly shaped particles and in films on the surface of the particles, matric potential is the dominant component of water potential. In cells, however, matrix potential is so small that it can usually be ignored and, in any event, it is in equilibrium with the osmotic potential.

The actual water potential in a wide variety of forest trees and shrubs rarely drops below −2.0 MPa irrespective of the size or amount of canopy (Scholander *et al.*, 1965; Jarvis, 1975). The water potential gradient between adjacent cells is usually small. From the soil to the atmosphere surrounding a leaf, however, the gradient in water potential may exceed 100 MPa on clear days. Water in the leaf mesophyll cells differs chemically from the almost pure water flowing through the sapwood. When comparing water potentials at a given place along the conducting pathway from soil to atmosphere, it is necessary to know: (1) the soil water potential around roots actively extracting water; (2) the gravitational potential, equivalent to about −0.01 MPa/m; and (3) the frictional potential associated with moving water along the pathway at different rates. Although the pathway of water flow through roots and the stem is tortuous, the frictional potential usually remains in the range from −0.01 to −0.02 MPa/m even during times of maximum transpiration (Nobel, 1983; Zimmermann and Brown, 1971).

WATER UPTAKE BY ROOTS

Availability of Soil Water

Generally, roots must grow through the soil to obtain water because in soils drier than field capacity, water movement is relatively slow (refer to Chapter 5). The water potential of the soil is determined by the relative amount of water held in pore spaces. Usually about 50% of the soil volume contains pores with diameters between 0.2 and 50 μm, the range from which roots may extract water (Ulrich *et al.,* 1981). Very compacted soils and soils with a substantial amount of roots have low pore volume and water-holding capacities. The amount of water available to tree roots varies depending on the soil texture. A sandy soil with a bulk density of 1 g/cm^3 might hold 25 mm of water to a depth of 50 cm, whereas a loam soil of similar depth might hold 75 mm (Fig. 4.1).

As the water content of the soil decreases, so does the water potential (Fig. 4.1). Tree roots do not grow at water potentials below about −0.7 MPa (Leshem, 1970; Day and MacGillivray, 1975; Larson, 1980) but with the aid of mycorrhizal fungi, whose mycelia are one-hundredth of the diameter of tree roots (Duddridge *et al.,* 1980), trees may extract water from the soil until water potentials reach −1.5 to −2.0 MPa. Below these values, absorption is so slow that trees are unable to maintain positive cell turgor.

When soil water is unavailable, predawn water potentials in trees fall until irreversible damage occurs. For most temperate forest species, water potentials much below −5.0 MPa are lethal (Hinckley *et al.,* 1981, 1983). However, trees adapted to arid or saline environments, such as *Acacia* and *Eucalyptus,* may endure potentials as low as −10.0 MPa (Doley, 1981).

Even when roots completely colonize the soil, water uptake is restricted almost entirely to the upper soil horizons until the potential drops below −0.2 MPa (Patric *et al.,* 1965; Krygier, 1971; Hinckley *et al.,* 1981; Woods and O'Neal, 1965). Water is then extracted from progressively lower horizons, until maximum rooting is reached (Patric *et al.,* 1965; Waggoner and Turner, 1971; Nnyamah and Black, 1977). Often 75% of the available water (refer to Chapter

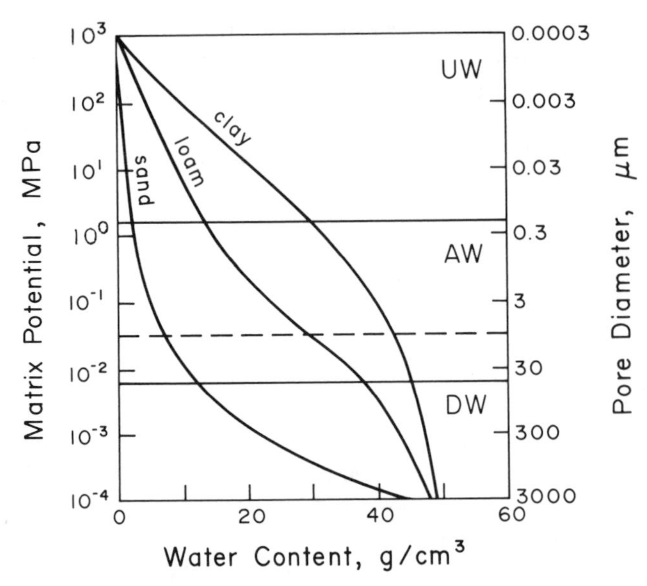

Fig. 4.1. Water storage capacity of soils varies with texture, reflecting pore size distribution. Water in pores below 0.2 μm diameter is unavailable to roots (UW), being held at soil matrix potentials at or below −1.5 MPa. Between this value and that held in pores at a matrix potential of about −0.01 MPa, the field capacity to which soils drain after 24 h, is water that is available to plants (AW zone in drawing). Soils of similar density (here about 1.2 g/cm^3) varied in the volume of available water from 5 to 15%, depending on their texture. More water may be held temporarily in nearly saturated soils, but as pore size increases above 50 μm, capillary forces are unable to hold water and it drains rapidly (DW zone in drawing). (After Ulrich *et al.*, 1981.)

4, Fig. 4.1) can be extracted before trees show decreasing water potentials under nontranspiring (predawn) conditions (Fahey and Young, 1984; Fig. 4.2).

Different species exploit the soil to differing extents. Shallowly rooted trees or shrubs exhibit more negative water potentials than more deeply rooted associates under droughty conditions (Waring and Cleary, 1967; Hinckley *et al.*, 1978, 1983). Some arid zone forests of *Acacia* extend roots to depths of 15 m (Doley, 1981) and only rarely experience water stress. In areas where precipitation is abundant, trees often develop rather shallow root systems. In such ecosystems a brief period without normal precipitation may stress the vegetation and reduce the rate of transpiration to a greater extent than where drought is frequent.

Constraints on Water Absorption by Roots

Water uptake depends on the absorbing surface area, its permeability, and the driving force, i.e., $\psi_{root} - \psi_{soil}$ (Landsberg and Fowkes, 1978). Any factor that inhibits root growth, reduces root permeability, or hastens root mortality reduces the water uptake capacity. Even with a well-established root system, water absorption may be restricted by low temperatures, poor aeration, hardpans, or high salt concentrations in the soil.

Low root temperatures may be particularly important because the permeability of cell membranes decreases at low temperatures. Various species show differential sensitivity. Monterey pine, *Pinus radiata,* a native of the California coastal region at 35°N latitude, is sensitive to soil temperatures below 15°C. Reduction in root temperature to 6.5°C results in a 66% reduction in water uptake by seedlings for an equivalent water potential gradient between roots and leaves (Kaufmann, 1977). At the other extreme, species with more northern distributions such as Sitka spruce (*Picea sitchensis*) and Scots pine (*Pinus sylvestris*) show no restrictions in water absorption by roots until temperatures fall below 3°C (Turner and Jarvis, 1975; Linder, 1973).

In the coniferous forests of the Pacific Northwest, subalpine species such as Noble fir (*Abies procera*) and silver fir (*Abies amabilis*) demonstrate higher root permeability in cold soils than more temperate trees such as western hemlock (*Tsuga heterophylla*) and Douglas fir (*Pseudotsuga menziesii*) (Teskey, 1982). It is difficult to establish trees in soils that are much colder than those in their native habitat. Also trees growing in regions where snow packs persist during the spring

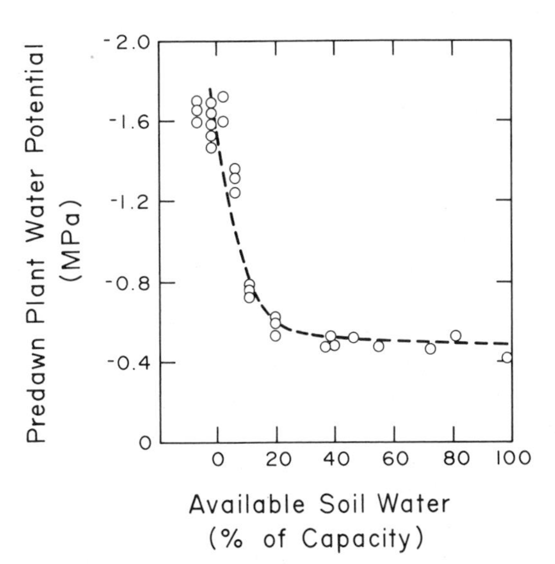

Fig. 4.2. Until more than three-quarters of the available water is depleted from the soil by *Pinus resinosa,* plantation trees recovered to the same predawn water potential. Thereafter, an abrupt decrease began until all available water was used. (After "Water potential in red pine: Soil moisture, evapotranspiration, crown position" by E. Sucoff, *Ecology,* 1972, **53,** 681–686. Copyright © 1972 by the Ecological Society of America. Reprinted by permission.)

while air temperatures rise well above freezing may suffer damage associated with restricted water uptake by roots (Anderson and McNaughton, 1973; Whitehead and Jarvis, 1981; Tranquillini, 1979).

In flooded soil the air is displaced and oxygen in the water is rapidly depleted by respiration of roots and microorganisms. Some species are resistant to flooding, such as *Fraxinus pennsylvanica, Nyssa aquatica,* and *Taxodium distichum,* and continue to grow roots when flooded (Hook *et al.*, 1970; Dickson and Broyer, 1972; Keeley, 1979). The ability to grow under such conditions indicates the passage of oxygen from above the water level. Upland plants subjected to flooding are rarely observed to exhibit water stress, probably because their stomata remain closed (Pereira and Kozlowski, 1977; Kozlowski and Pallardy, 1979) as a result of altered hormone balance (Davies *et al.*, 1981).

Compacted soils may become a physical barrier to root growth. Root extension into compacted soil decreases proportionally as the bulk density of the soil increases (Heilman, 1981). Eventually all growth may be halted at bulk densities between about 1.5 and 1.8 g/cm^3, depending on the species and type of soil (Youngberg, 1959; Forristall and Gessel, 1955; Minore *et al.*, 1969; Heilman, 1981; Tworkoski *et al.*, 1983). Some studies indicate soil densities of 1.8 g/cm^3 may not be completely limiting to root growth for a variety of conifers (Zisa *et al.*, 1980). The ultimate limitation to penetration, of course, is when all pores in a compacted zone have diameters less than the smallest possible root diameter, about 60 μm (cf. p. 4, Kramer, 1983). There is a high probability, however, that nutrients or oxygen may also become limiting as soil density increases, particularly in fine textured soils. To compare different textured soils, resistance to penetration by a metal probe is recommended over soil density measurements (Zisa *et al.*, 1980; Sands *et al.*, 1979). Better seedling establishment and increased growth are sometimes possible by breaking through hardpans beneath the point where seedlings are planted (Hatchell, 1981).

Deficiencies or imbalances in various nutrients or abnormally high concentrations of certain elements may restrict root growth and water absorption (Proctor and Woodell, 1975; Reich and Hinckley, 1980). Generalizations about how specific nutrients or other dissolved compounds affect root growth are difficult to make. Even the exudates from some roots and leaf litter may be toxic to the growth of plants (MacClaren, 1983). This subject is discussed in more detail in Chapters 8 and 9.

Where high concentrations of salts accumulate, they provide the equivalent of dry soils, because a 1 *M* solution is equal to a soil water potential of less than −2.0 MPa. Those plants adapted to soils high in salts, such as *Acacia harpophylla* in Australia or *Juniperus occidentalis* in the semiarid woodlands of western North America, are able to adjust the osmotic potential in their tissue accordingly (Doley, 1981; Scholander *et al.*, 1965). Plants unable to maintain high osmotic concentrations are excluded from such areas.

We may summarize this section by saying that when tree seedlings are first becoming established they may experience water stress, even on deep, moist soils. Once tree roots are well established, however, they normally provide sufficient water to maintain transpiration at relatively constant rates until most of the available water is extracted from the root zone or the supply becomes inaccessible owing to cold soil temperatures, poor aeration, or chemical imbalances. Observations of the depth of rooting, and of root growth, and measurements of predawn water potential are biologically important indicators of whether problems may exist with water absorption by roots.

WATER STORAGE AND TRANSPORT IN SAPWOOD

The water-conducting xylem elements in sapwood have thick lignified secondary cell walls and are dead, i.e., they no longer contain protoplasts. Two types of conducting cells are distinguishable in the xylem: Vessels are found in angiosperms and evolutionarily more primitive tracheids are found in conifers and other gymnosperms. The tracheids are much longer than vessels and are usually of smaller diameter. The ends of tracheids are tapered and connected through special pits that may temporarily seal off one cell from another. The shorter and broader vessel elements have blunt perforated ends that connect end-to-end to form a continuous tube through the wood that reaches lengths of 30 m in some trees. Angiosperm trees have two distinctive types of wood based on the relative variation in diameter of vessels produced during the growing season. Species that produce large-diameter vessels in the spring and smaller ones later in the summer are called ring-porous. Species that produce vessels of similar diameter throughout the growing season are called diffuse-porous. Ring-porous trees may have vessels up to 30 m long; diffuse-porous trees have vessels less than 3 m in length (Skene and Balodis, 1968). Seasonal changes in the size and number of xylem cells produced result in tree rings (Chapter 11).

If vessels represented perfectly cylindrical tubes, the rate of fluid movement could be calculated using Poiseuille's law. Water movement at a given water potential gradient would be proportional to the fourth power of vessel diameter (Nobel, 1983; Zimmermann and Brown, 1971). Water flow through a capillary of 20 μm diameter with a given water potential gradient would be less than 1% of the water flow through capillaries of 75 μm diameter under the same conditions. To supply the same amount of water, trees with vessels of the smaller diameter would require 100 times more conducting tissue.

Large-diameter elements are obviously favored for efficiency in water transport. In fact, water transport in ring-porous trees with large-diameter vessels often occurs only in the outermost one or two annual rings. Only small gradients in water potential are necessary to move water in these species as long as the conducting columns remain filled. Where winter freezing and summer droughts are common, diffuse-porous species with very small vessel diameters and conifers with tracheids tend to dominate (Carlquist, 1975; Zimmermann, 1978; Hinckley *et al.*, 1981).

Even within a tree, the vessel length and diameter decrease with height and differences in the conducting properties of the wood are noted (Skene and Balodis, 1968). In *Malus* and *Prunus*, the permeability of the vascular system in the roots is 5 times greater than in the large branches and 10 times greater than in twigs (Baxter and West, 1977). Constrictions at the base of branches are also important in reducing the efficiency of water flow (Zimmerman and Brown, 1971). These differences favor the main stem in situations where there is a severe drought in which leaves, twigs, and branches will be lost progressively (Zimmermann, 1978). Among shoots, the uppermost are favored because there is sufficient growth to provide more conducting elements, at least until maximum height is approached.

In Chapter 2 we noted that the cross-sectional area of sapwood at the base of the live crown supported a fixed amount of leaf area for a given species (Table 2.4). Below the live crown, toward the base of the tree, sapwood includes progressively older cells with lower permeability. The decrease in permeability must be accommodated by an increase in sapwood area. This idea led Whitehead *et al.* (1984) to question whether different species growing in the same environment might not compensate for large differences in sapwood area by having proportional differences in wood permeability. They found spruce (*Picea sitchensis*) had half the sapwood area and twice the wood permeability of pine (*Pinus contorta*) with equivalent leaf area in plantations in Scotland.

We might conclude from this discussion that ring-porous trees, and hardwoods in general, commit less resources to their conducting system. This is possible because hardwoods generally have larger diameter elements and fewer restrictions to flow between cells than in the wood of conifers. It should be remembered, however, that conifers can continue to use their conducting tissue for many years. Moreover, if the outer portions of the tree were injured by fire, disease, or insect attack, hardwoods are more prone to develop water stress (Hinckley *et al.*, 1981).

Sapwood Water Storage

Water can be withdrawn from the conducting sapwood of trees to help meet short-term requirements for transpiration. Cavitation occurs when gas fills the void left following the removal of water (Milburn, 1979). When water columns are interrupted, no shrinkage in diameter occurs because the lignified cell walls remain rigid until structural water is lost (Siau, 1971); this only happens in dead trees.

In conifers, particularly those of large stature, sapwood storage may approach 300 t/ha of water, equivalent to the amount transpired by a forest over a period of 5–10 days (Waring and Running, 1978). In some cases this water supply may be quickly recharged following rainstorms (Chalk and Bigg, 1956; Waring *et al.*, 1979), whereas in other cases it is only refilled after the growing season (Clark and Gibbs, 1957), if at all (Hinckley *et al.*, 1981). Withdrawal of water from a mature Douglas fir forest may be nearly continuous during a summer drought (Fig. 4.3) with periodic partial recharge following rains. Only during the winter, however, when the air is nearly saturated, does the sapwood completely rehydrate. Water caught on the surface of branches, twigs, and leaves enters through cuts and the leaf stomata. This reverses the normal water potential gradient and allows cells to rehydrate (Milburn, 1979; Cremer and Svensson, 1979). Ray parenchyma cells adjacent to each cavitated element apparently aid the process (Wodzicki and Brown, 1970).

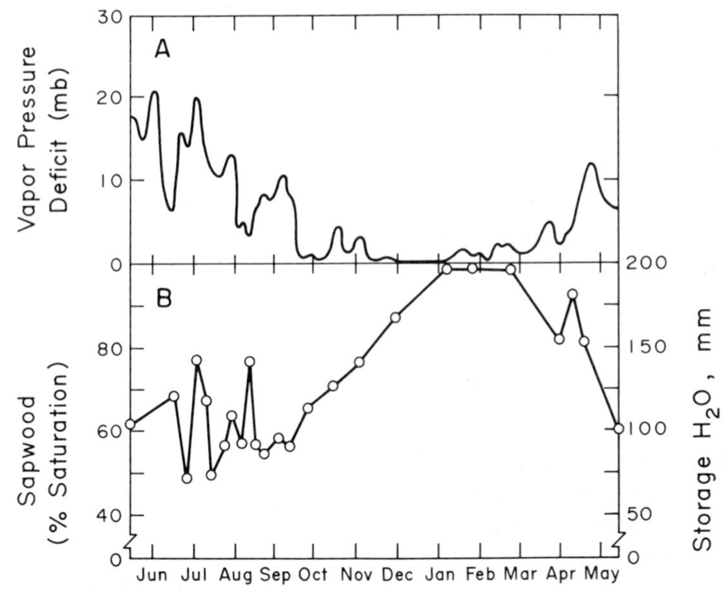

Fig. 4.3. Withdrawal of water from the sapwood of 500-yr-old Douglas fir growing in the Pacific Northwest (B) began when the water vapor gradient in the atmosphere averaged less than 5 mb for the day (A). During summer rains in June and late August, partial recharge of the sapwood began but was not completed until January, following extended periods of precipitation. Maximum daily rates of transpiration did not exceed 5 mm/day. (From Waring and Running, 1978.)

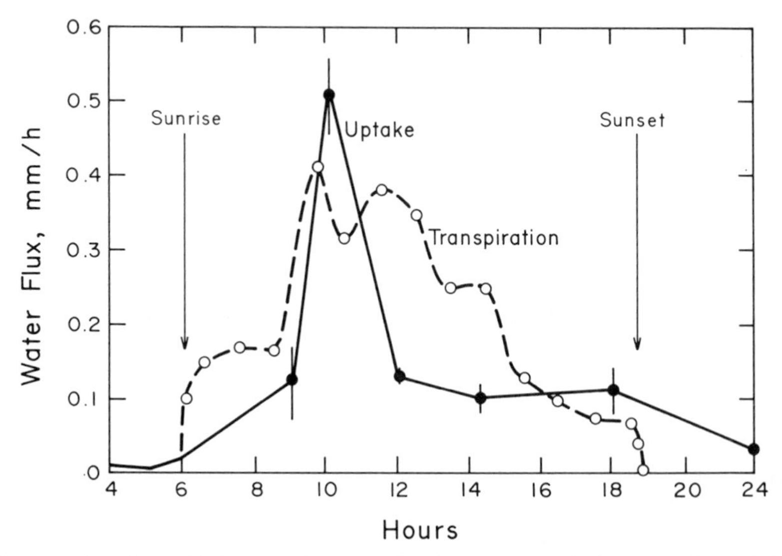

Fig. 4.4. In a 36-yr-old Scots pine forest, uptake of water over the daylight hours lagged behind transpiration by a third. Over a 24-h period, however, uptake balanced transpiration within 7%. Data derived from a study where uptake was estimated by radioisotope tracers injected into trees and transpiration were calculated from knowledge of canopy leaf area, stomatal conductance, and meteorological conditions. (After Waring *et al.*, 1980.)

Even where sapwood water recharges each night, water removed from sapwood during the day decreases the size of midday water deficits in the foliage and may provide as much as one-third of the daily requirements for transpiration as illustrated by independent measures of transpiration and water uptake in a pine forest in Scotland (Fig. 4.4).

Effects of Cavitation on Water Conduction

Unless water extracted during the day is recharged at night, the volume of functional conducting sapwood is progressively reduced. Because the larger conducting elements are usually drained first, sapwood permeability is reduced exponentially with falling water content (Puritch, 1971). At about 80% of saturation, the sapwood conducts less than a third as efficiently as at saturation and a proportional increase in the water potential gradient is required to maintain a constant flow (Waring and Running, 1978). Such high water potential gradients, however, adversely affect leaf turgor and cause stomata to at least partially close.

Low temperatures may also constrain water movement through the sapwood because fluid viscosity near freezing is approximately twice that at 25°C (Slatyer, 1967). Increases in the water potential gradient may partly offset the effects of temperature (Kaufmann, 1977), but eventually cavitation in the sapwood will occur. If trees are frozen and unable to refill cavitated cells during the winter, new sapwood must be produced before buds can break (Zimmermann and Brown, 1971).

Stem temperatures just a few degrees below freezing will stop all water transport (Zimmermann and Brown, 1971). If temperatures fall below −40°C, permanent damage is done to the ray parenchyma cells of many temperate hardwoods, preventing refilling of any cavitated vessels (Wodzicki and Brown, 1970). This effectively limits the present distribution of many species to lower latitudes and elevations (Burke *et al.*, 1976). On the other hand, boreal species such as *Picea mariana* and *Populus tremuloides* have more permeable membranes that permit rapid transport of water out of living cells, preventing formation of intracellular ice crystals (Fig. 4.5). This is an extreme kind of desiccation, but in laboratory experiments species with these capabilities may withstand temperatures down to

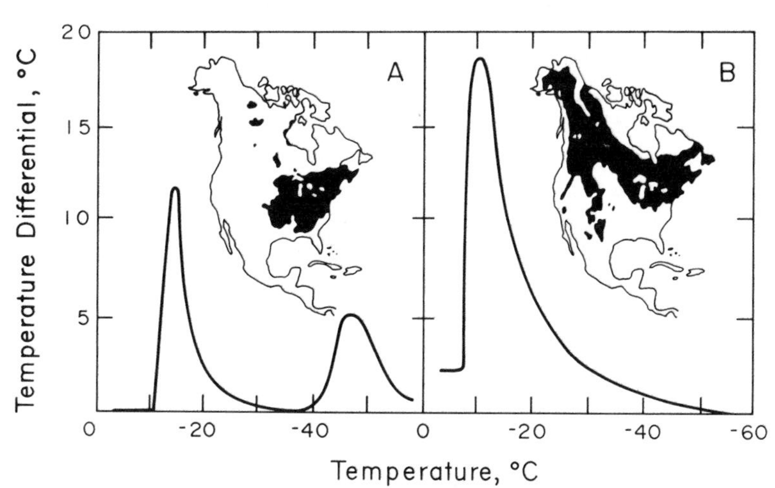

Fig. 4.5. Most temperate species such as red oak (*Quercus rubra*) have restricted latitudinal and elevational ranges because they are not adapted to winter temperatures below −40°C. During the process of freezing, such species are unable to remove water from their cells and are therefore killed. The graphs show that when extracellular water first freezes around −13°C, the heat of fusion provides for a temperature increase. Species that experience intercellular freezing, such as red oak, have a second heat of fusion peak (A). In contrast, species adapted to extreme cold such as aspen (*Populus tremuloides*) are able to transfer nearly all free water from their living cells into extracellular spaces and have only one indication of water freezing (B). (After Burke *et al.*, 1976. Reproduced, with permission, from the *Annual Review of Plant Physiology*, Volume 27, © 1976 by Annual Reviews Inc.)

−200°C during the winter without permanent damage (Burke *et al.*, 1976; Lyons, 1973).

As a result of blockage of the conducting pathway with gas bubbles, stable flow is maintained only with a steeper gradient in water potential. Beyond a certain limit, however, the flow rate is reduced as stomata close. Problems from cavitation of conducting elements are likely to be acute when soils remain cold, water vapor deficits are high, nights are of short duration, or drought occurs.

WATER RELATIONS OF FOLIAGE

As water is lost through the stomata, the water content of leaves and phloem cells decreases from the top of the tree downward throughout the day (Lassoie, 1979). The decrease in volume often represents 10–20%, but in terms of actual water contributed to the daily transpiration, this is equivalent to 1–2 h worth from the foliage and 1–3 h from the phloem along the branches and stem (Whitehead and Jarvis, 1981; Hinckley *et al.*, 1981). This water is of considerable importance to the plant's water economy if recharge during the night is possible, because well hydrated leaves allow stomata to remain open, facilitating CO_2 uptake (Chapter 2) and nutrient transport (Chapter 7). Small changes in water content may be partly compensated by corresponding increases in solute concentration (Eq. 1). When water reaches the leaves, it flows through cells from the vascular bundle to the epidermis where it may evaporate through the cuticle or out of the stomatal cavity. This process is called transpiration, and its rate is controlled by the gradient in water vapor and the stomatal conductance.

Constraints on Transpiration

The concentration gradient of water vapor between the leaf and the atmosphere at the leaf surface defines the maximum possible rate of transpiration. The concentration of water vapor at the surface of a leaf may differ from that of air, particularly when there is little wind and the leaves are broad. Under such conditions the temperature of the leaf exceeds that of air because all of the absorbed radiation cannot be quickly dissipated (Fig. 4.6).

Knowing the temperature of the air, its relative humidity, and the temperature of the leaf permits us to calculate the water vapor gradient using relationships depicted in Fig. 4.7. For example, suppose we assume that the relative humidity of air is 60%, which is equivalent to a vapor concentration of 14 mb when the air

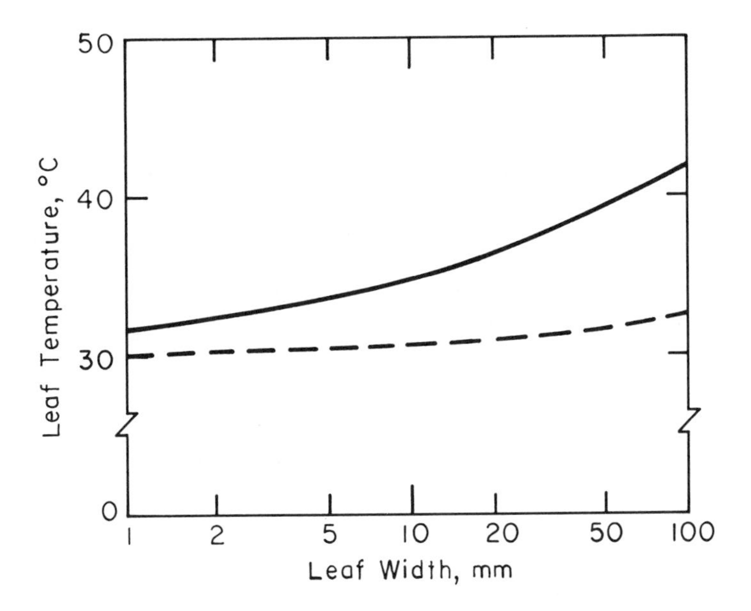

Fig. 4.6. The width of leaves greatly influences their ability to dissipate heat. For example, nontranspiring leaves significantly increase their temperature above that of the air if they are more than 5 mm wide and they receive high radiation under conditions with little air movement (——). If stomata remain open, on the other hand, leaf and air temperatures are fairly close (- - -). (After Campbell, 1977.)

temperature is 20°C and the saturated vapor pressure is 23.4 mb. If the leaf temperature were 25°C, water vapor at saturation within the leaf would be 31.7 mb, and the water vapor gradient, representing the difference, would be 17.7 mb. If we had assumed no difference in leaf and air temperatures, we would have underestimated the gradient by 90%. In Chapter 5 we discuss ways of calculating evaporative demand without knowledge of leaf temperature, and we assess the integrated effect of leaf dimensions, wind speed, and radiation on the evaporative gradient.

The shape of leaves and the thickness of the leaf cuticle affect vapor exchange, but once the leaves are formed, it is the changes in turgor of stomatal guard cells that largely control the actual rate of transpiration. When stomata are wide open, they permit transpiration at 20–40% of the potential rate of evaporation of free water from an exposed surface; when closed, losses are often less than 1% of the potential. For the stomata to be open, the adjacent guard cells must have a higher turgor (ψ_p) than subsidiary cells. Because the water potential gradient is exceedingly small, active transport of potassium ions into the guard cells changes the osmotic potential (ψ_s) and indirectly the turgor (Outlaw, 1983). The hormone abscisic acid (ABA) also plays a role. In well-watered plants, inactive forms of ABA are present in the mesophyll (Mansfield *et al.*, 1978; Davies *et al.*, 1981). As turgor is reduced, absiscic acid is released from the mesophyll and diffuses to the guard cells where it interferes with the transport of potassium and may lower guard cell turgor for a number of days. Mesophyll cells continue to produce ABA until their turgor is restored (Mansfield *et al.*, 1978). Guard cell turgor clearly plays a key role in the control of transpiration and more indirectly, photosynthesis. Thus we shall now focus on how specific environmental variables, through their influence on guard cell turgor, affect stomatal conductance.

When temperatures are high and the humidity is low, stomata in most species of plants close, regardless of the water status of the leaf (Losch and Tenhunen,

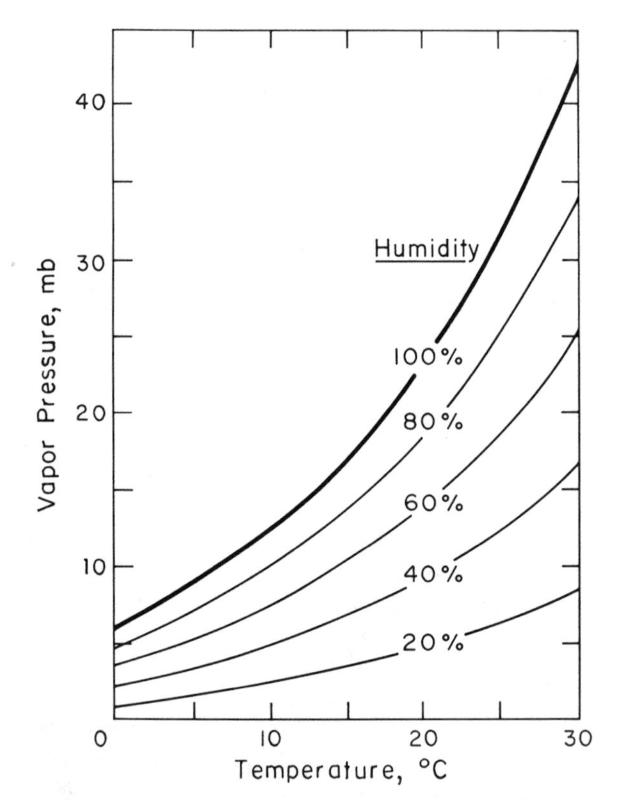

Fig. 4.7. The amount of water vapor that may be held in the air increases exponentially with temperature. The saturated vapor pressure in millibars of Hg (heavy line) is equal to: $6.1078^{(17.269\,T/237\,+\,T)}$, where T is temperature in degrees Celsius. The water vapor deficit of the air is the difference between the saturated value and that actually held at a given temperature. Here data are presented as relative humidity. (After Lowry, 1969.)

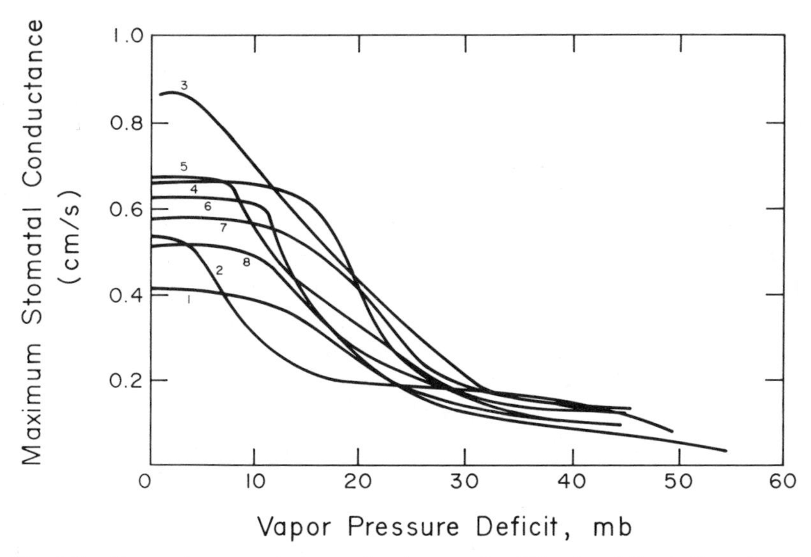

Fig. 4.8. The maximum conductance of various species differs in relation to the water vapor deficit of the air but most species close their stomata at high deficits. Conifers: (1) Douglas fir (*Pseudotsuga menziesii*); (2) western hemlock (*Tsuga heterophylla*). Deciduous trees: (3) dogwood (*Cornus nutallii*); (4) big-leaf maple (*Acer macrophyllum*. Evergreen broadleaf trees: (5) chinquapin (*Castanopsis chrysophylla*). Deciduous shrub: (6) vine maple (*Acer circinatum*). Evergreen broadleaf shrub: (7) rhododendron (*Rhododendron macrophyllum*); (8) salal (*Gaultheria shallon*). (After Waring and Franklin, 1979.)

1981). In fact, the turgor of the leaf as a whole may actually increase (Schulze *et al.*, 1974). For a number of conifers, broadleaf deciduous and evergreen hardwoods, complete stomatal closure was observed at air vapor pressure deficits greater than 35 mb (Fig. 4.8). There are some species such as *Ceanothus, Alnus,* and *Arctostaphylos* adapted to colonizing disturbed environments that can maintain open stomata even under high evaporative demand if soil water is available (Conard and Radosevich, 1981; Marshall and Waring, 1984).

The sensitivity of stomata to high evaporative demand has recently been explained through the discovery of extremely permeable, cuticle-free areas located on the inner walls of guard cells (Edwards and Meidner, 1978; Appleby and Davies, 1983). The stomata of leaves grown in exposed environments generally are more sensitive to water vapor gradients than those grown in shaded environments. In an extreme case, *Tsuga heterophylla* leaves grown in the shade and later exposed to full irradiation were killed in less than 1 h, because stomata only responded to increased light by opening (Keller and Tregunna, 1976).

Low temperatures in the soil may reduce the membrane permeability of roots, as previously mentioned. The resulting reduction in water potential affects leaf turgor and leads to stomatal closure. Subjecting leaves to temperatures near or slightly below freezing may also affect guard cell turgor and cause stomatal closure (Drew *et al.*, 1972). Increasing temperature may favor opening of stomata if humidity deficits are not extreme (Schulze *et al.*, 1973). This response would be expected from our earlier discussion of how photosynthetic capacity is influenced by temperature and the depletion of internal CO_2 (Chapter 2).

Most plants require a minimum level of light to induce stomatal opening. This level often matches the compensation point for uptake of carbon dioxide (Mansfield *et al.*, 1981; Chapter 2). Stomata usually are wide open between light levels equivalent to 5–20% of full sunlight. Stomata of shade-tolerant species open at lower light levels than those of shade-intolerant species (Pereira and Kozlowski, 1977; Woods and Turner, 1971). Sometimes these differences are exhibited on the same tree, together with differential responses to humidity (Keller and Tregunna, 1976). The presence of chloroplasts in the guard cells explains the sensitivity to both light and CO_2 levels; differences in humidity response are associated with changes in guard cell and leaf geometry (Mansfield *et al.*, 1981; Appleby and Davies, 1983; Tucker and Emmingham, 1977).

As roots extract water from soil they may eventually remove all that is available. When this happens, plants no longer recover full turgor at night as indicated by decreasing predawn water potentials (Fig. 4.2). Associated with these changes is a reduction in the maximum stomatal conductance observed during the day (Fig. 4.9) until eventually stomata remain closed all day.

As soil water becomes less available, the stomata may become more sensitive to a particular water vapor gradient (Fig. 4.10). Many plants, on the other hand,

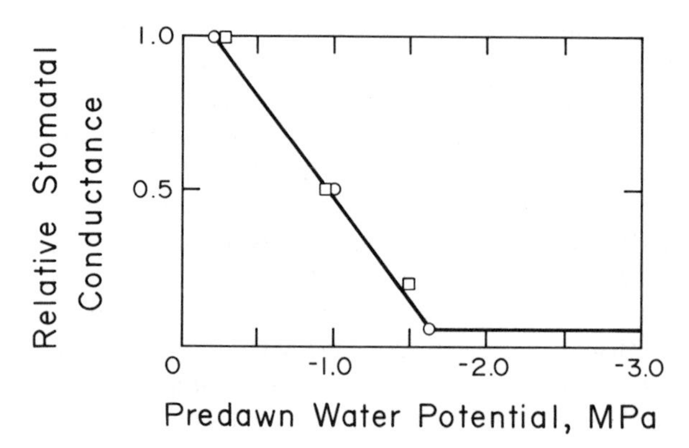

Fig. 4.9. As soils dry, the predawn water potentials of both Douglas fir (□) and tulip poplar (*Liriodendron*) (○) decrease. Maximum relative stomatal conductance during the day also decreases until complete closure is attained (−1.6 MPa here). (From Waring *et al.*, 1981b.)

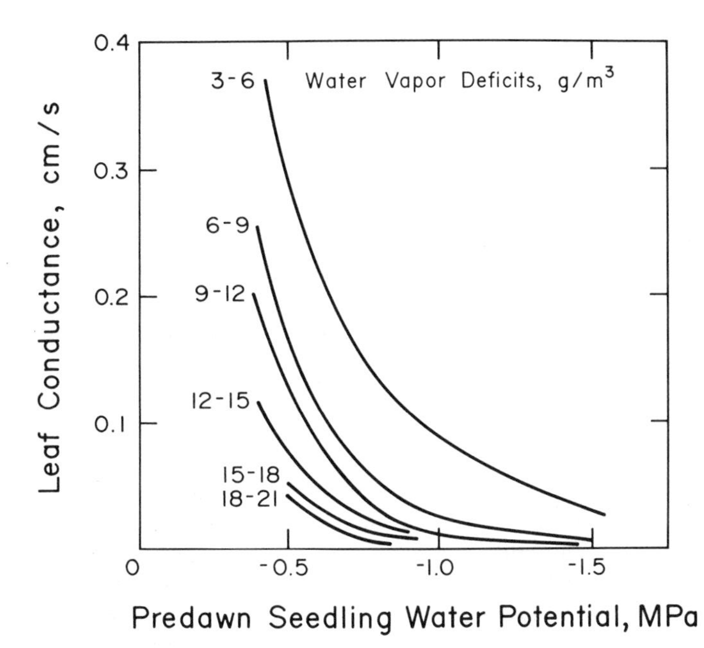

Fig. 4.10. As predawn water potentials decrease, young seedlings of Douglas fir close stomata more at a given water vapor deficit, here expressed in grams of water per cubic centimeter of air. (After Lassoie, 1982.)

become acclimated to dry conditions by increasing osmotic concentrations in cells to maintain turgor, by reducing cell wall rigidity to permit growth, and by increasing membrane permeability to facilitate the diffusion of water (Bradford and Hsiao, 1982). After extended exposure to drought, *Quercus alba,* for example, was able to open stomata to half of maximum at relative water contents and water potentials that caused complete closure earlier in the season (Parker *et al.*, 1982).

In summary, stomatal behavior may reflect previous as well as present conditions. This is particularly true in the case of drought where relationships between relative water content, water potential, and turgor are likely to change. Flooding of the roots or extensive cavitation of the vascular system may induce behavior similar to drought. For predicting stomatal response, water potential measurements need to be converted to turgor potential from knowledge of osmotic potential and relative water content relationships. The importance of cavitation in the vascular system can be assessed by comparison of diurnal or seasonal changes in transpiration versus water potential gradients or by direct assessment of sapwood relative water content (Dixon *et al.*, 1984).

SIMULATION MODELS OF WATER FLOW THROUGH TREES

Having identified the major environmental variables affecting the movement of water through trees and the subsequent changes in flow associated with adjustments in root uptake, sapwood permeability, and stomatal behavior, we are now in a position to combine these into predictive models. Water relations models have application at the ecosystem level in predicting changes in soil water content, mineral uptake, and transpiration. Also, if adequate supporting information were available, predicted changes in turgor in various tissues would assist in the development of growth models. To date, models have been applied to contrast how different kinds of vegetation affect soil water depletion and streamflow, e.g., the impact of establishing deep versus shallow-rooted species or deciduous versus evergreen cover on streamflow from watersheds (Chapter 5).

Tree models, as we have learned, should include an internal storage component that, when even partly depleted, has important effects on water potential gradients and stomatal conductance. Models developed by Jarvis *et al.* (1981) and Running (1984) incorporate this and other important relationships discussed in this chapter. No model that we know of incorporates the effects of flooding and other special soil-related conditions that might be important in bogs, coastal plains, and deltas.

Depending on the specific interest, the models (structure shown in Fig. 4.11) may incorporate many individual storage units in plant tissue and the root zone, or they may be generalized to represent one unit for the plant and one for the rooting zone. They require knowledge of solar radiation, temperature, relative humidity, wind speed, and precipitation averaged over periods of not more than 1 h. For each storage component, information is required on changes in hydraulic conductivity (permeability) with water content and water potential gradients, as well as the total storage capacity. With initial conditions defined, the models keep track of changes in soil water contents at various rooting depths, seepage or runoff water that is unavailable to the trees, and variation in the relative water content of tissues throughout the conducting pathway. The models predict stomatal conductance and the resulting transpiration.

Because rather involved calculations are required to predict soil and leaf temperatures and the amount of precipitation reaching the soil, it is often desirable to make direct measurements of these variables or associated ones, such as predawn plant water potentials. When applied to narrow-leaved species, such models may not require measurements of leaf temperature because leaf and air temperatures usually are similar (Fig. 4.6 and Chapter 5). Solar radiation data are still needed to predict maximum stomatal conductance at various levels in the canopy (Chapter 2).

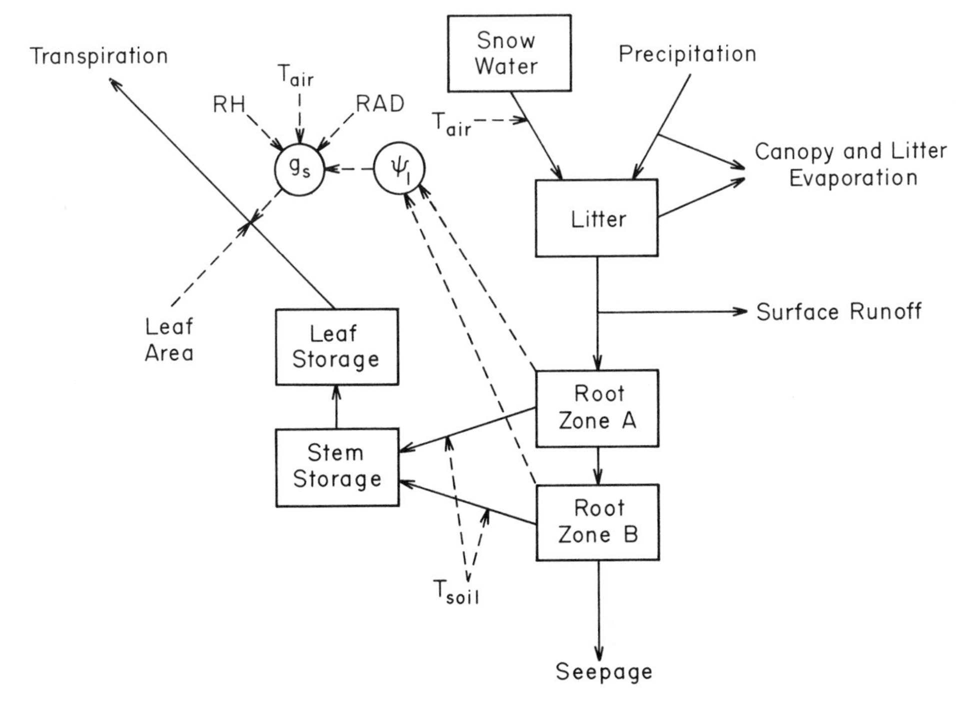

Fig. 4.11. General structure of a tree water balance model that accounts for precipitation entering into the litter or soil root zones [upper (A) or lower (B)] and its eventual loss from the system through transpiration, evaporation, runoff, or seepage. Within the tree, water may be stored temporarily in the sapwood or leaves. Transpiration depends on the potential evaporation, a function of air temperature (T_{air}), relative humidity (RH), and net radiation (RAD). Actual transpiration is less than potential, being constrained by the amount of leaf area and stomata conductance (g_s). Stomatal conductance in turn is constrained by cold soil temperatures (T_{soil}) that affect leaf water potential (ψ_l). In addition, T_a, RH, and RAD may affect stomatal conductance. Such a model may be expanded to a stand of trees and have a variable time resolution. (After Running, 1984.)

In response to potential evaporation, the models allow transpiration to remove water from a known amount of foliage at a rate determined by the initial stomatal conductance constrained by the predawn water potential or by leaf turgor and light intensity. Periodically throughout the day, water contents, water potentials, and stomatal conductance are reevaluated and adjustments made in the rate of transpiration. As the day proceeds, water is withdrawn progressively from storage and adjustments in hydraulic conductivity follow. At hourly resolution, the models define lags in the rate of water flow through various parts of the pathway (Fig. 4.12). When compared with measured values, predictions usually agree to within ±0.2 MPa for water potential and to within 10–15% of measured sto-

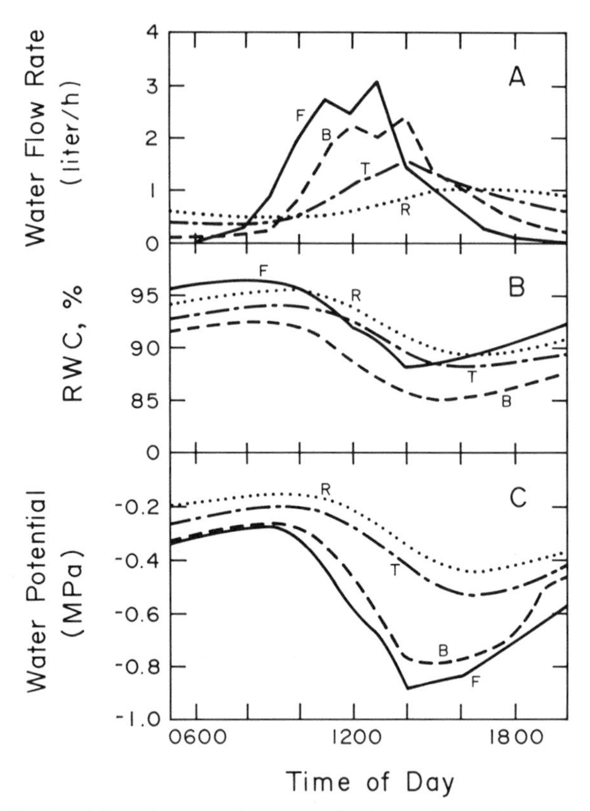

Fig. 4.12. Simulated diurnal curves of (A) water flow rate; (B) relative water content; and (C) tissue water potentials at four locations in a lodgepole pine on a clear August day in Scotland. The major tree components are (F) total foliage; (B) all live branches; (T) top of stem just below live branches; and (R) base of stem. The simulation indicates water potentials are most extreme in the foliage and least variable in the roots. More water is lost from the foliage during the day than from other tissues but rehydration of foliage is rapid at night. (After Jarvis, 1981b.)

matal conductance. Predicted changes in relative water content throughout the conducting pathway are difficult to verify but appear reasonable (Edwards and Jarvis, 1982, 1983; Dixon *et al.*, 1984).

For most ecological purposes, these models are unnecessarily refined and may be simplified. For some questions concerning the adaptation of various species to particular circumstances, such as growth at low soil temperatures, the resolution may be desirable. In many cases, if estimating stomatal conductance were our only interest, empirical correlations with irradiance and air saturation deficits might suffice (Whitehead *et al.*, 1981). The stronger our understanding of water relations becomes, however, the better our chance of predicting the growth and distribution of forest vegetation and its hydrologic effects.

WATER RELATIONS AND THE DISTRIBUTION AND GROWTH OF TREES

Although water relations have long been assumed to be important in explaining the distribution and growth of forest vegetation, growth rates also depend on the availability of carbohydrates and nutrients and their distribution within the plant (Bradford and Hsiao, 1982; Chapter 2). In any environment where trees grow, an eventual halt in height growth is observed. In the redwood region of northwestern California, *Sequoia* situated on the alluvial flats, where the soil profile remains well supplied with water throughout the summer, exceed 100 m heights. Progressing upslope to the limits of summer fog, trees become more stunted, reaching heights of less than 30 m (Waring and Major, 1964). At high elevations in the tropics, where cloud cover is prevalent and potential evapotranspiration is low, trees may grow to only a few meters in height (Grubb, 1977).

The distribution of tree species within a drought-prone region reflects both differences in the water supply and tree rooting depth. As an ecological index of relative water availability, we recommend measuring predawn water potentials on a widely distributed tree species with well-established roots (Griffin, 1973). When 2-m tall Douglas fir were compared over a range of environments in southwestern Oregon, predawn water potentials varied from about −0.6 to nearly −3.0 MPa following a period of summer drought (Fig. 4.13). The maximum height of Douglas fir varied accordingly from about 70 m to less than 30 m. Distributions of other species and their growth rates were predictable from knowledge of the Douglas fir response (Waring, 1970).

Because the differences observed in maximum height along such gradients are paralleled by reductions in growth rates at earlier ages, we shall ask the question whether the rate of wood formation determines its conducting properties and effectively limits tree height? Drought or other factors that might decrease turgor will also reduce the number of conducting elements produced and their diameters, while increasing the thickness of cell walls (Whitehead and Jarvis, 1981; Sheriff and Whitehead, 1984). In an 80-m tall *Sequoiadendron giganteum*, Rundel and Stecker (1977) reported a linear decrease in the diameter of branchwood tracheids in association with decreasing predawn water potentials from the bottom to the top of the crown. They further determined that maximum height was limited in that species when predawn water potentials could not be maintained above −2.0 MPa. This suggests that eventually some factor limits the construction of a sufficient number and size of wood-conducting elements so that turgor pressures in expanding buds are not sufficient for further height growth. The reduction in turgor is likely to affect hormone production in the bud, which determines the rest of the cell construction sequence (Zimmermann and Brown, 1971).

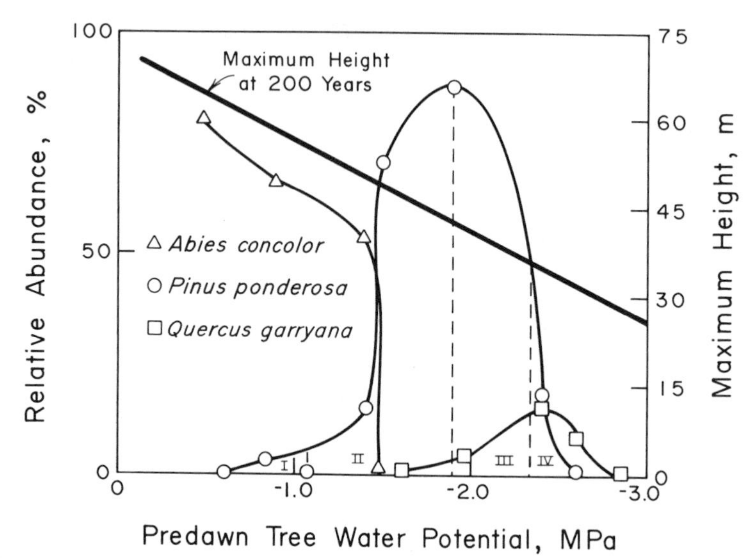

Fig. 4.13. Predawn water potentials measurements on 2-m tall Douglas fir defined a water stress gradient developed during a summer drought in a mountainous region of southwestern Oregon. The water potential gradient so defined related closely to (1) the maximum height reached by Douglas fir; (2) the relative frequency of ponderosa pine and other tree species; and (3) the rate of height growth on a range of environments, varying from class I (height of 30 m at a 100 yr) to class IV (height of 20 m at 100 yr). (After Waring, 1970.)

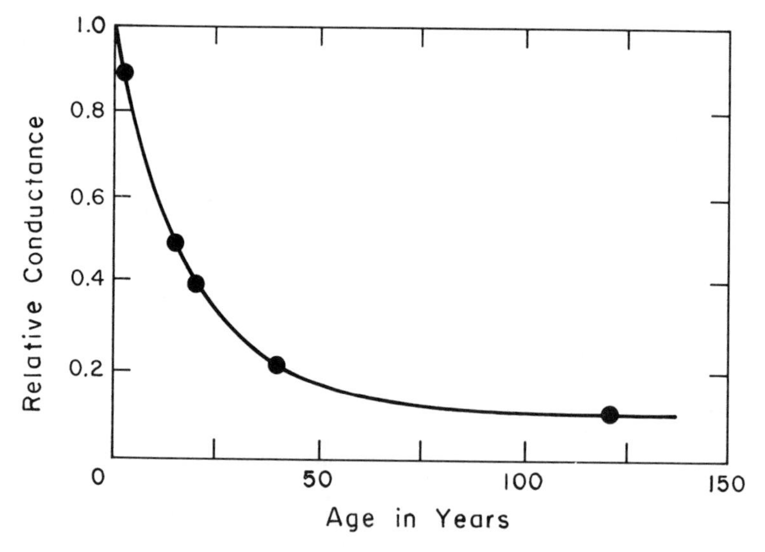

Fig. 4.14. As trees grow larger, the ability to conduct water through sapwood decreases and may eventually limit height growth. The relative conductance of Scots pine trees 5 yr old and 0.2 m high was 90% greater than that observed in 125-yr-old, 18-m tall trees on a sandy soil in central Sweden. (After Mattson-Djos, 1981.)

Although many observations have been made concerning the smaller size of foliage and conducting elements in tall trees, few measurements have been published concerning the change in conducting properties as trees age on a particular site. In a Scots pine forest of central Sweden, Mattson-Djos (1981) compared the integrated conduction efficiency of trees ranging in age from 5 to 125 yr and from less than 0.2 to 18 m in height. The oldest trees exhibited a water conducting efficiency of only about 10% of the saplings (Fig. 4.14). This level of conducting efficiency was estimated to have been reached by the time the trees are 80 yr old, corresponding with their cessation in height growth. Because the potential height of Scots pine is much higher than 18 m, we can speculate that improving the availability of nutrients or perhaps warming the soils by creating wider spaces among trees might improve the growth potential and initiate renewed height growth on older trees.

SUMMARY

In this chapter we identified major environmental constraints to water flow through trees: (1) near exhaustion of available soil water; (2) cold temperatures; (3) anerobic conditions; (4) impervious soils; (5) high potential evaporation; and (6) inadequate nutrition. All of these factors indirectly may limit growth rates and thereby reduce wood permeability. We introduce the idea that the ultimate height of trees in any environment may be dependent on maintaining a minimum growth rate sufficient to permit adequate water conduction through the sapwood to the tree top. The sapwood is an important water reservoir, but once drawn upon it may be difficult to refill. Reduction in water content makes the sapwood less efficient in conducting water and nutrients from the roots to the leaves.

Different types of trees show various adaptations to environmental stresses that affect tree–water relations. As indicators of general water stress, we recommend monitoring the degree of cavitation in the sapwood of a few widely distributed tree species, calculating changes in water conduction at comparable water potential gradients, and determining whether seasonal changes in predawn water potential occur. By coupling key environmental factors to their effects upon root water uptake, flow through the sapwood, and transpiration from foliage, general models of tree water relations now exist that can assist in developing further integrated models to predict tree growth, mineral cycling, and hydrologic effects.

5

Hydrology of Forest Ecosystems

INTRODUCTION

Water represents the greatest flow of any material substance through an ecosystem. A cubic meter of water weighs 1000 kg; thus 1 mm of precipitation weighs 10,000 kg/ha. Forest ecosystems lose water by evaporation from wet surfaces, by transpiration through leaves, and by drainage. When water is freely available, evaporation or transpiration may reach 50–60 t/ha/day. The way in which water is stored or travels through an ecosystem is as important to the availability of moisture as the amount of precipitation itself. If soils are porous and deep, or if snow is shielded from direct radiation and melts slowly, the supply of water may be sufficient to sustain rates of photosynthesis, decomposition, and other processes for months of dry weather. On the other hand, shallow or compacted soils may store little water, and runoff may be diverted rapidly to streams and rivers.

Forests generally grow in climates where precipitation delivers more water than the vegetation can use immediately or than soils can store. Runoff, however, seldom appears on the surface (Hewlett and Hibbert, 1967). How much water is used by the vegetation and how much becomes streamflow are common questions asked by forest hydrologists. Normally these questions are answered empirically from experience gained locally. However, the processes that affect water movement are general, so a more fundamental approach is warranted.

Why do a wide variety of vegetation types growing in different climates transpire at the same maximum rate? Why does snow sometimes melt more rapidly on overcast than on sunny days? Why does forest removal generally, but not always increase streamflow? The answers lie in understanding the processes that control water movement. In this chapter we (1) identify key structural and environmental variables that affect the hydrologic processes in forest eco-

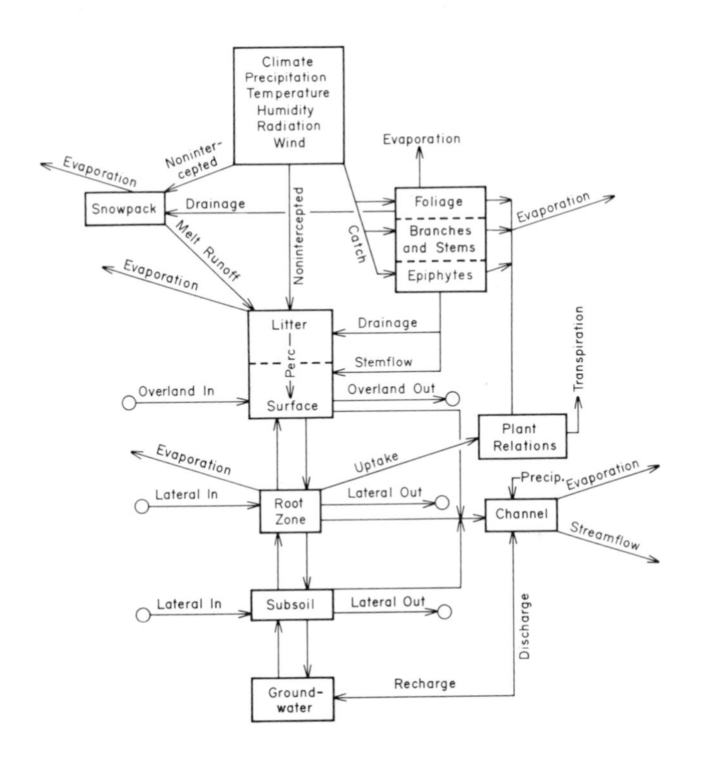

Fig. 5.1. Structure of a hydrologic model indicates five important climatic variables, which drive the system. Water moves through storage cells and finally leaves the system by drainage into stream channels or through evaporation and transpiration. Water may also move laterally in and out of an ecosystem. (After Waring *et al.*, 1981b.)

systems; (2) demonstrate the predictive power of specific process models; and (3) expand the scope to consider a hydrologically linked network of adjacent ecosystems. As justification for the approach, we illustrate that an integrated process model, run on a daily resolution, accurately predicts annual streamflow for a range of vegetation types growing in three climatically distinct regions.

Figure 5.1 provides an overview of all possible routes that water may take through a forest ecosystem. When one forest is adjacent to another, water movement through the soil links these systems. The climatic variables indicated in Fig. 5.1 drive the entire system. Of particular importance are humidity and radiation, variables too often ignored but essential to estimate transpiration, evaporation, and the energy content of snow, litter, and soil. It should be noted that the water storage capacity in various parts of the system is defined by biological criteria such as rooting depth, leaf area, or sapwood volume. The amount of litter on the forest floor is also dependent on biological factors.

CLIMATIC VARIABLES

There are five climatic variables important to hydrology: (1) precipitation; (2) temperature; (3) humidity; (4) solar radiation; and (5) wind speed. Interactions between these five climatic variables determine the microclimate of a particular forest. In some mountainous regions the climate varies significantly even over small watersheds. Snow may accumulate on one slope and not on another; and cold air may drain from ridges to lower slopes. Often meteorological stations are located on flat areas or ridges so there is a problem in extrapolating climatic data even locally.

The occurrence, total duration, and total amount of storm precipitation seems to be largely stochastic though predictable by probabilistic methods (Eagleson, 1970; Grace and Eagleson, 1966; Duckstein *et al.*, 1973). The distribution of precipitation over an area about the storm center is also predictable (Eagleson, 1970), depending on the particular kind of storm. Separation into type of precipitation (snow or rain) may be done on the basis of temperature, with a dividing line of 1.1–1.7°C (Eagleson, 1970).

Ambient temperature typically decreases with increasing elevation at a rate between 5 and 8°C/km. This gradient is referred to as the adiabatic lapse rate of temperature and can be safely used only if the mountainous localities being compared have approximately similar environments and surface profiles. The presence of nighttime cold air drainage or the formation of inversion layers may result in warmer temperatures at higher elevations than at lower. Large bodies of water or glacial ice may also influence local temperature patterns. Open areas adjacent to plains and foothills are known to cool more rapidly than the higher elevations in more mountainous topography (Kuzmin, 1961).

The water vapor deficit of the air, important in helping to determine the potential for evaporation or condensation, is dependent on temperature and pressure. The water vapor capacity of air decreases with falling temperature and increasing elevation. The air's water-holding capacity at 1000 m is only 63% of that at sea level; at 2000 m the relative capacity is reduced to 40% and at 3000 m to 25% (Kuzmin, 1961). How changes in temperature affect the air's ability to hold water is shown in Chapter 4, Fig. 4.7.

Solar energy consists of both short- (0.3–4 μm) and long-wave (4–80 μm) components. The short-wave component containing visible light is particularly

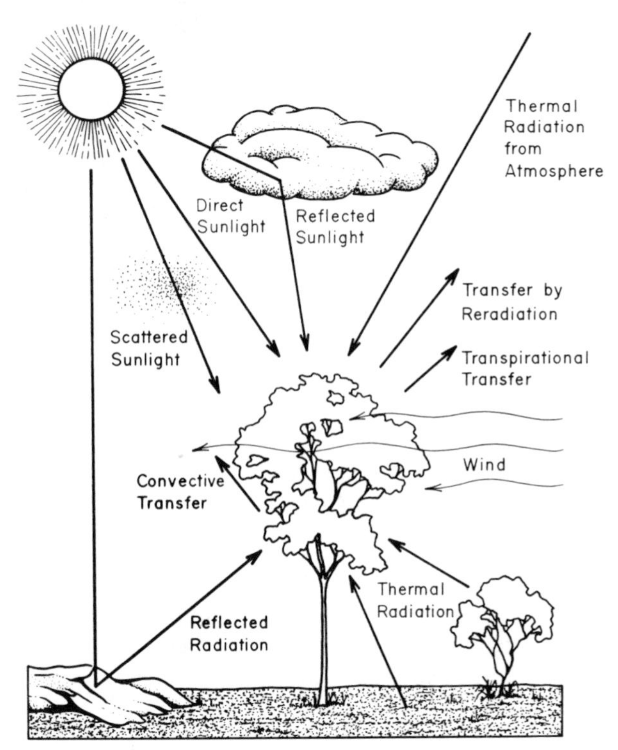

Fig. 5.2. Exchange of energy between vegetation and the environment. Solar radiation reaches plant canopies as direct, scattered, and reflected sunlight, all of which contain some short-wave components important for photosynthesis. In addition, plant surfaces absorb thermal radiation emitted at long-wave, infrared frequencies by the atmosphere, the ground, and other plants. The bulk of the heat load on plants is reradiated; evaporative cooling by transpiration and heat transfer by convection remove the rest. Some heat is stored temporarily in the soil and plant tissue and is later reradiated. (After Gates, 1980.)

important for many biological functions, including photosynthesis and stomatal opening. The long-wave component contributes heat and becomes very important in some water balance studies such as snowmelt. The total radiation incident on any surface is the sum of (1) direct short-wave radiation from the sun; (2) diffuse short-wave radiation from the sky; (3) reflected short-wave from nearby surfaces; (4) long-wave radiation from atmospheric emission; and (5) long-wave emitted from nearby surfaces (Fig. 5.2).

The short-wave radiation on a plane surface is dependent on the slope, aspect, and latitude of the surface, the declination of the sun, time of day, and transmissivity of the atmosphere. Given this information, the direct short-wave radiation on any slope may be estimated for any time interval (Garnier and Ohmura, 1970; Buffo *et al.*, 1972). Short-wave radiation can vary significantly with topography. In general, long-wave radiation from atmospheric emission is uniformly distributed over a watershed and is affected only by extremely steep terrain where reflection from adjacent slopes may be important.

Wind turbulence causes mixing and increases energy exchange rates. Wind enhances the process of fog drip and evaporation while decreasing the possibility of atmospheric inversions and ground frost. It also affects snow distribution. Wind is strongly influenced by local topography and therefore must generally be measured directly, preferably above the canopy or at least in an adequately exposed situation. Detailed discussions on wind are found in Geiger (1965) and Grace (1981).

Although many other environmental variables might profitably be monitored in developing a water balance for an ecosystem, these five climatic factors are usually sufficient for basic hydrologic studies.

HYDROLOGY OF A FOREST ECOSYSTEM

Before discussing how different portions of a watershed are linked hydrologically, we assess the operation of a single homogeneous ecosystem. The water content of such a system is the sum of water stored (1) on the foliage, branches, and stems; (2) in the snowpack; (3) in the litter; (4) on the soil surface; (5) in the vegetation; (6) in the soil root zone; and (7) in the subsoil.

Interception

Plant surfaces are the first potential obstacles encountered by precipitation in forests. As a storm begins, rain strikes the foliage, branches, and stems, or falls directly through canopy openings to the forest floor. The latter route is minor in fully closed forest stands. During the initial stages of a storm, much of the precipitation is stored on the canopy or upon the stems. As a storm continues and these surfaces reach their capacities, excess water drains to the forest floor.

Interception, the difference between rainfall recorded in the open and that collected in funnels beneath the canopy includes some evaporation. Interception has been shown to account for 10–35% of annual precipitation (Kittredge, 1948; Zinke, 1967), with interception of snow often greater than interception of rain (Hoover and Leaf, 1967; Miller, 1967). Generally ice and dew are minor components of interception on plant surfaces, although they may be of local importance (Kittredge, 1948; Fritschen and Doraiswamy, 1973). For example, rime ice contributes to the hydrological balance in some subalpine coniferous forests (Berndt and Fowler, 1969) and hardwoods (Gary, 1972). Where evergreen forests are exposed to wind-driven clouds, gross precipitation may be increased by more than 40% by condensation on plant surfaces; removal of such forests has resulted in lower streamflow (Harr, 1982; Lovett *et al.*, 1982).

The proportion of rainfall intercepted has been determined for a variety of forests by placing measuring devices in the open and at random points beneath the canopy. Other collectors placed around the stems of trees estimate stemflow (Likens and Eaton, 1970). A summary of equations for computing throughfall and stemflow from measurements of gross rainfall is presented in Table 5.1. These calculate average values when the canopy is fully developed. Variation in interception can be expected, depending on the intensity and duration of a storm.

Storage capacity of the entire canopy can be estimated from the intercept of the equations presented in Table 5.1. For a plantation of Scots pine, Rutter (1963)

TABLE 5.1 Summary of Equations for Computing Throughfall and Stemflow for Coniferous and Hardwood Forests from Measurements of Rainfall (P)[a]

Vegetation	Equations, P (in cm)		Interception (cm)
	Throughfall	Stemflow	
Red pine	$0.87P-0.04$	$0.02P$	0.15
Loblolly pine	$0.80P-0.01$	$0.08P-0.02$	0.15
Shortleaf pine	$0.88P-0.05$	$0.03P$	0.14
Ponderosa pine	$0.89P-0.05$	$0.04P-0.01$	0.13
E. White pine	$0.85P-0.04$	$0.06P-0.01$	0.14
Pine (average)	$0.86P-0.04$	$0.05P-0.01$	0.14
Spruce–fir–hemlock	$0.77P-0.05$	$0.02P$	0.26
Hardwoods, in leaf	$0.90P-0.03$	$0.41P-0.005$	0.10
Hardwoods, deciduous	$0.91P-0.015$	$0.062P-0.005$	0.05

[a] After Helvey (1971) and Helvey and Patric (1965).

found the leafy shoots retained an amount of water approximately equal to the dry weight of foliage. Storage of precipitation on foliage, branches, and stems was estimated to be 0.8, 0.3, and 0.25 mm, respectively. Swank (1972) estimated 1.8-mm storage capacity for a Douglas fir stand with about 36% retained by the stem. Zinke (1967) calculated that average rainfall storage capacities are about 2 mm for conifers and 1 mm for hardwoods. Snow storage on conifers averaged about 3.8 mm.

Large deviations from these values occur in different seasons. The distribution of storage also changes with stand development because the surface area of branches (living and dead) and stems increases with stand age (Chapter 3). The best way to characterize these changes is on the basis of total surface area available to evaporate water. In deciduous forests, the projected area of leaves per area of ground surface [leaf area index (LAI)] often ranges from 3 to 6, equivalent to a total surface area of twice these values. Similarly surface area indices for branches and stems (total surface) range from between 1.2 and 2.2 and 0.3 to 0.6, respectively (Whittaker and Woodwell, 1967). Values reported for a young eastern white pine plantation (*Pinus strobus*) were 17.8 (about 7.0 projected), 2.3, and 0.4 for the surface area of foliage, branches, and stems, respectively (Swank and Schreuder, 1973).

Epiphytes on branches and stems can greatly increase the capacity of those structures to absorb surface water and explain, in part, why some large trees may exhibit almost no stemflow even during intense storms (Rothacher, 1963). Another factor that controls stemflow is the smoothness of the bark surface. Species with smooth surfaces such as beech (*Fagus*) may transport as much as 12% of the precipitation as stemflow, whereas pine normally transfers less than 2% by that route (Kittredge, 1948). Stemflow may have special significance in distributing potassium to the area around certain smooth-barked trees, because that nutrient is easily leached from foliage (Gersper and Holowaychuk, 1971; Chapter 7). It also may concentrate water near the roots of some arid forest species such as *Acacia aneura* (Doley, 1981).

The surface area of foliage, branches, and stems may be calculated from regressions with stem diameter (Chapter 2) and physical transport models applied to calculate surface evaporation during any storm. In application of such models to both deciduous and evergreen stands and individual trees, Rutter *et al.* (1975) and Aston (1979) obtained excellent agreement to observed interception. Water evaporated from tree stems was found to represent from 20 to 30% of the total evaporation from trees in leaf and from 30 to 40% for leafless trees (Rutter *et al.*, 1975). The critical limitation to this approach is often the lack of sufficient resolution in precipitation and other climatic data.

In Chapter 4 we discussed how wide leaves create a thicker boundary layer of still air around them than do narrow leaves. Here we expand the implications of boundary layer conductance to an entire canopy and consider heat and vapor

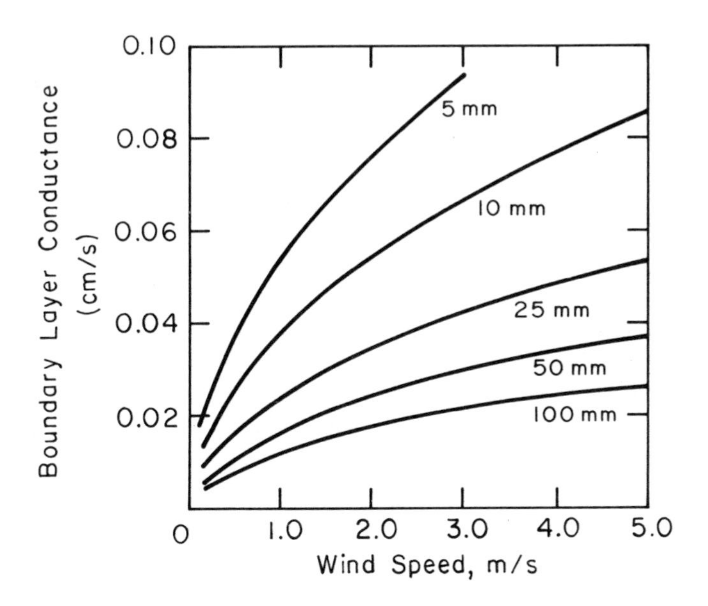

Fig. 5.3. Boundary layer conductance is a measure of how rapidly heat or water vapor may be transferred from a surface. Narrow leaves with widths of 5 mm (top curve) compared to those with widths of 100 mm (bottom curve) are at least five times as efficient in exchanging heat or water vapor from their surface. Increasing air movement improves exchange properties to some degree but the basic differences reflecting leaf dimensions still hold. (After Grace, 1981.)

exchange together. Wind is a major variable that affects how rapidly water vapor moves across the boundary layer. As wind speeds increase to above 5 m/s, the boundary layer conductance (rate of vapor transfer in centimeters per second) increases curvilinearly as shown for different dimensions of leaves in Fig. 5.3.

Rates of evaporation are directly proportional to the boundary layer conductance under identical environmental conditions, meaning water evaporates from the surface of leaves 5 mm in width nearly 10 times faster than the rate of evaporation from leaves 50 mm in width and comparable total area. Actual differences are reduced somewhat from these extremes because of mutual interference between leaves, folding of leaf margins, and differences in wind speed associated with height, canopy density, and other factors that influence exchange (Grace, 1981; Landsberg and Powell, 1973).

In addition to boundary layer conductance and its dependence on wind, the saturation deficit of the air and net radiation absorbed by the canopy are critical in calculating evaporation rates. The radiation term is of predominant importance in forests of broad-leaved species, although they tend to absorb 5–15% less short-wave radiation than coniferous forests (Jarvis *et al.*, 1975; Stanhill, 1970).

Hardwood forests, because of their canopy structure, do not scatter radiation as efficiently as coniferous canopies (Stanhill, 1981). The boundary layer conductance is also important because the slow transfer of heat away from wide leaves causes temperature to rise and creates an increased saturation water vapor deficit between the leaf and atmosphere (Gates, 1968). These interactions will be discussed in more detail when we consider canopy transpiration.

The air in forests seldom remains saturated even during a rain storm and about 30% of interception losses occur then (Jarvis and Stewart, 1979; Gash and Stewart, 1977). Continuous rewetting of coniferous canopies can often support evaporation rates of 0.3 mm/h on completely overcast days (Jarvis and Stewart, 1979). For this reason, surface evaporation can reach 6 mm/day. Under similar climatic conditions, therefore, it is not surprising to find that the wet canopy of a Scots pine forest may evaporate somewhat more than twice the amount of water lost by a grass field with similar canopy area (Jarvis and Stewart, 1979).

Snowpack

The accumulation of a snowpack and the rate of melt are two critical variables measured by forest hydrologists interested in predicting the amount and timing of streamflow. In this section we seek to understand the processes that affect the snowpack, which include: (1) heat conduction; (2) density changes resulting from compaction; (3) percolation of rain or melt water; (4) wind; and (5) variation in temperature and water vapor (U. S. Army Corps of Engineers, 1956; Eagleson, 1970). Fresh snow has a density between 0.06 and 0.34 g/cm^3, but the density increases toward a maximum with age. Temperature, water-holding content, and grain size also become similar throughout an old snowpack. Water will not flow from a snowpack unless the snow is saturated and isothermal, at 0°C throughout. Only at the surface does the temperature change, resulting in daytime thaw and melt, followed by crust formation during nighttime freezing. As the snow cover ages and a crust accumulates, the reflectivity decreases from about 0.9 to about 0.4. Snowmold and plant materials that fall on the snow increase the absorbing properties of the snowpack (Waring *et al.*, 1981b).

A number of heat-transfer processes are involved in the dynamics of snowmelt. From an energy balance standpoint, the change in heat storage is the sum of the (1) gain of the latent heat of vaporization (597 cal/g at 0°C or 2.4×10^3 J/g) by condensation of water, or loss during evaporation or the sublimation of snow; (2) convective transfer of sensible heat from air; (3) gain of heat in precipitation or loss in runoff water; (4) net radiation transfer; (5) conduction of heat from the underlying ground; and (6) gain of latent heat by freezing of liquid water within the pack (or loss by melt) (Anderson, 1968). These interrelations are indicated in diagramatic form in Fig. 5.4.

With the exception of freezing of liquid water within the pack and conduction of heat from the ground, all components of the energy balance can be assumed to take place in a thin surface layer of snow (Anderson, 1968). The surface temperature of the snowpack is an important variable, sensitive to and affecting the net absorption of radiant energy and the transfer of both latent and sensible heat.

The transfer of latent heat to the surface of a snowpack is controlled by the water vapor deficit and turbulence of the atmosphere. If the water content of the air is greater than can be held at the surface temperature, moisture condenses on the surface. This results in a release of latent heat to the surface. If the gradient is reversed, moisture and heat are lost. If the snow surface is at 0°C, it is generally assumed that the phase change is from vapor to liquid or vice versa and the latent

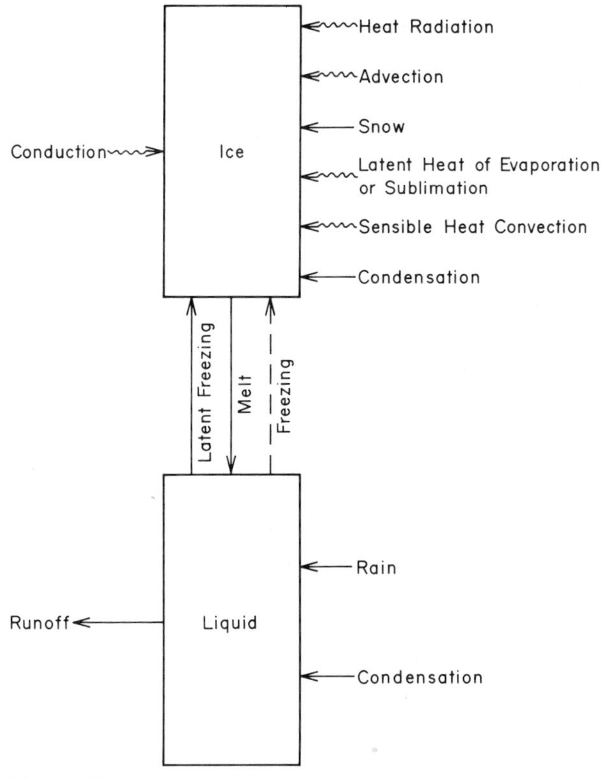

Fig. 5.4. Diagram illustrates components of an energy and water balance for snow showing major processes that affect transfer of water (——) and heat (~~). (After Waring *et al.*, 1981b.)

heat of vaporization applies. If below 0°C, the phase change is from vapor to solid and the latent heat of sublimation (677 cal/g or 2.835 $\times$ 10^3 J/g) is involved. Latent heat transfer can be important in areas where warm, moist, turbulent air occurs over snowpacks, since 1 cm of condensate can produce about 7.5 cm of melt and 8.5 cm of runoff (Anderson, 1968; Eagleson, 1970).

The transfer of sensible heat to the surface is also a turbulent exchange process, but in this case the direction of transfer is controlled by the temperature gradient. If the air temperature is warmer than the snow surface, there is a direct heat transfer from the air to the snow. The quantities of melt by convection from warm turbulent air can be similar to those from condensation melt.

Rain is warmer than the snowpack and hence there is a transfer, by advection, of heat from rain to the snowpack. Because the specific heat of water is 1 cal/g (4.2 J/g), the amount of heat added is easily calculated from the amount (in millimeters) and temperature (in degrees Celsius) of the rain. The temperature of rain can be assumed to be air temperature. Because the latent heat of fusion is only 80 cal/g (335 J/g), the snowpack's temperature is not much affected by the process of freezing water. For example, 4.0 cm of rain at 20°C must be added to a ripe snowpack (at 0°C and maximum water-holding capacity) to produce 1.0 cm of melt. When the snowpack is below freezing, however, rain adds its small heat content and gives up the latent heat of fusion as it freezes within the pack. This may quickly bring a cold snowpack to a uniform (isothermal) temperature of 0°C, ripe for melting. When melt in the snowpack or soil occurs, water percolates down where it may refreeze and release the latent heat of fusion. This quickly warms the pack or soil until it becomes isothermal. Additional melt then is held as liquid water until the water-holding capacity is reached.

Conduction of heat from the ground generally is the result of heat stored in the soil during the summer months. This may produce melt from the bottom of the snowpack. If the soil is frozen before a snowpack has accumulated, frost can penetrate very deeply since the thermal conductivity of ice is almost four times that of water. The freezing process drives all but a small amount of soil water downward creating ice lenses that are a major cause of frost heaving. These changes in water content and depth of frost have major significance to the porosity of soils, spring runoff, and nutrient availability (Fahey and Lang, 1975; Halldin *et al.*, 1980; Knight *et al.*, 1985).

Net radiation heat transfer includes both short-wave and long-wave radiation balances. In forested areas this is affected by the amount of canopy and in more open woodlands by the coloration of tree bark (light colors reflect, dark absorb). The Beer-Lambert law may be applied to calculate incoming radiant energy through a canopy of known leaf area (Chapter 2), and with knowledge of the reflectivity of the snow surface, the balance of energy absorbed by the snowpack can be estimated.

Part of the long-wave radiation from the atmosphere is absorbed by the forest cover; however, the canopy also emits long-wave radiation. Hence, the portion of the snowpack covered by the canopy is assumed to receive radiation from the canopy, while that in the open receives it from the atmosphere. The snowpack is usually assumed to absorb all incident long-wave radiation. It in turn reemits long-wave radiation in proportion to its surface temperature.

Although the foregoing is rather complicated physics, simulation models of snowpack accumulation, soil freezing, and melt have been developed and found to apply well in different regions (Anderson, 1968; Eggleston *et al.*, 1971; Leaf and Brink, 1973; Rogers, 1973; Halldin *et al.*, 1980). With characterization of the forest and soil structure, the accumulation, melt, and infiltration of water can be predicted from only five climatic variables. The absence of even one, however, requires many additional empirical measurements because basic physics cannot be applied.

Predicting under what stand conditions maximum snowpack will occur requires knowledge of the snow's temperature, the radiation environment, and wind conditions. Cold dry snow can be blown from a canopy or drift in from an open area and accumulate under a dense stand of trees. Wet heavy snow may be quickly shed from a canopy before much evaporation occurs. Condensation may be much greater in openings exposed to the wind than under the forest. Overcast weather conditions minimize differences in microclimate, but condensation on such cloudy days can melt more snow than during a clear day. Condensation, as much or more than rain, may play a key role in snowmelt in cloudy, maritime regions. A change from an evergreen to a winter deciduous canopy permits more snow to fall to the ground and may provide an environment similar to an open field. We emphasize that differences observed in the rate of snowmelt in various regions are predictable if the key variables required to solve energy-exchange equations are monitored and the relative importance of various processes identified. This is an alternative to present empirical approaches.

Litter

Water balance processes in the litter layer, i.e., infiltration, percolation, and lateral flow, are important to litter decomposition (Chapter 8) and to runoff. Hence it is necessary that hydrologic models include this small but important segment of the ecosystem. Infiltration is the process by which water enters the soil; percolation is the process by which water moves downward under gravitational forces through the litter, soil, or other porous media. Lateral flow is the movement of water under gravitational forces generally parallel to the slope.

The forest floor may be divided into several layers, based on stages of decomposition (Chapter 8). For purposes of this discussion, we divide the forest floor into two main layers: (1) an upper, litter horizon in which the origin of most material is easily identifiable; and (2) a lower, humus horizon in which the origin of material is not easily recognized. The hydrology of the two layers is different.

The water-holding capacity of the surface horizon depends on the surface area of the material, analogous to storage on foliage, stems, and branches. The upper horizons of ponderosa pine litter are reported to hold about 175% of their dry weight at field capacity (Clary and Ffolliott, 1969). In the lower layers, water is held by capillary force and the capacity is reported to be 210% of the litter mass. For the entire layer, the litter in coniferous forests may have an average water content of 215% at field capacity and about 330% at saturation (Helvey and Patric, 1965). When the surface litter dries, the normal minimum moisture content for both hardwood and conifer litter is about 30% (Schroeder and Buck, 1970), but under extreme conditions it may approach 15% (Reynolds and Knight, 1973).

The litter layer under humid temperate forests generally intercepts less than 5% of the total precipitation (Helvey, 1964). Under more arid conditions, however, litter may intercept proportionally much more precipitation (Aldon, 1968; Reynolds and Knight, 1973). During a storm, water falling on the litter will gradually increase the litter moisture level to field capacity. Some water can

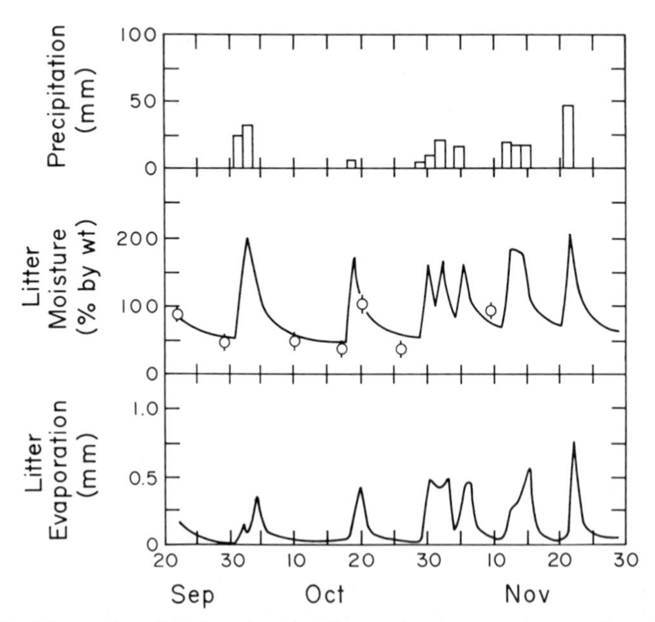

Fig. 5.5. Measured precipitation, simulated litter moisture content (compared to observed ϕ), and estimated evaporation for an 80-day period under a hardwood forest in the southeastern United States. (After Moore and Swank, 1975.)

reach the soil and infiltrate before all the litter is wet. Only very rarely does overland flow occur in a forest except in the vicinity of road surfaces and drainage ditches. Under these conditions, some water may be stored on the soil or road surface. With knowledge of precipitation and storage characteristics of litter, Moore and Swank (1975) developed a hydrologic model that predicted evaporation and litter moisture content (Fig. 5.5).

Evapotranspiration

Vascular plants, particularly trees, exert a unique influence on the hydrologic properties of terrestrial ecosystems by extracting water from the soil below the shallow depth that is usually affected by evaporation. This in turn influences the amount of water remaining for seepage and streamflow. The supply of water in the root zone and, to a lesser extent, in vascular-conducting tissue, may constrain water uptake and ultimately transpiration. Irradiance and the atmospheric saturation deficit are the major environmental variables that influence transpiration and evaporation. Frozen and cold soils also restrict transpiration (Fahey, 1979; Chapter 4). In this section, principles that applied to transpiration from individual trees are extended to whole stands and then combined with a consideration of evaporation from the canopy and litter to estimate the total flux of water vapor from forests. The latter is termed evapotranspiration.

Very low irradiance levels are sufficient to stimulate stomatal opening in the morning, but during the day increasing atmospheric water vapor deficits can lead to a progressive decrease in stomatal conductance, as illustrated for a Scots pine canopy in Fig. 5.6. Other species of conifers and hardwoods, both temperate and tropical, also have been reported to be sensitive to an increasing atmospheric saturation deficit (Whitehead *et al.*, 1981; Waring and Franklin, 1979; Edwards and Meidner, 1978; Hinckley *et al.*, 1975; see also Chapters 2 and 4). The fact that stomatal conductance is progressively reduced as the evaporative demand increases is the major reason why transpiration rates by forests do not exceed about 0.7 mm/h or 6 mm/day (Whitehead and Jarvis, 1981). For example, in a dense canopy of Sitka spruce (*Picea sitchensis*), transpiration rates on a sunny day remained below 0.4 mm/h although the atmospheric saturation deficit increased from 8 to 16 mb (Fig. 5.7). If stomata had remained open at the level observed shortly after dawn, transpiration rates would have continued to increase during the day to values exceeding 0.8 mm/h. Species that exhibit less sensitivity to atmospheric saturation deficits support relatively low LAIs and tend to close stomata when shaded (Pereira and Kozlowski, 1977; Marshall and Waring, 1984; Chapter 4). In either case, maximum rates of transpiration are limited to the same degree.

Extraction of water from the sapwood may contribute measurably to the water transpired by forests. For forests with a large volume of sapwood, 30–50% of the total water transpired during the day may be extracted from stem or branch wood. An additional 1–2% may come from foliage or other living tissue (Whitehead and Jarvis, 1981; Hinckley *et al.*, 1981). In some coniferous forests, sapwood represents a reservoir equivalent to more than 20 mm of precipitation (Whitehead and Jarvis, 1981). Hardwoods may also extract some water from sapwood but the storage capacity is much smaller, representing less than 5 mm of water (Federer, 1979).

Unless water extracted during the day is recharged at night, the volume of functional conducting sapwood is progressively reduced. Because the larger conducting elements are usually drained first, sapwood conductivity is reduced exponentially with falling water content (Chapter 4).

Transpiration from forests may be quite similar over a wide range in species composition, size, and stocking density, particularly if the canopy LAI exceeds 3.0 (Jarvis, 1981a; Doley, 1981). A small number of large, widely spaced trees may transpire more than many small, closely spaced individuals. As an extreme

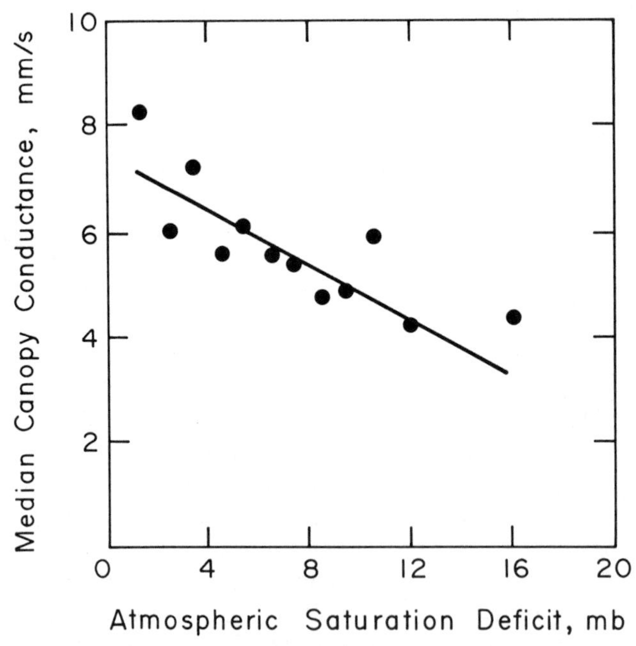

Fig. 5.6. Stomatal conductance of an entire Scots pine canopy with an LAI of 3.1 decreased progressively as the atmospheric saturation deficit increased to 16 mb (1.6 kPa). Other environmental variables remained favorable for gas exchange through stomata. (After Jarvis, 1981a.)

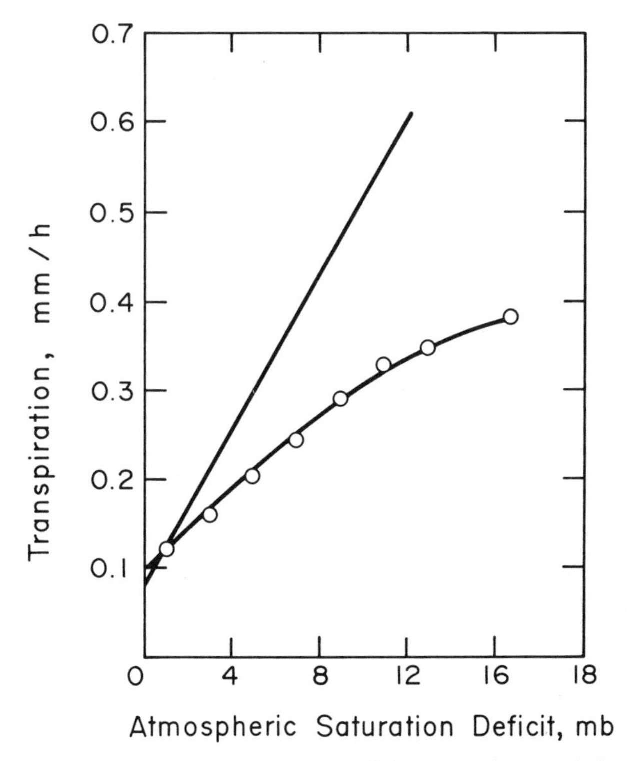

Fig. 5.7. Because the atmospheric saturation deficit tends to increase during clear days as temperature rises, partial stomatal closure is induced, reducing transpiration rates below those predicted if stomatal conductance remained near maximum (upper line) observed shortly after sunrise in a dense *Picea sitchensis* forest growing in Scotland. The actual transpiration, compiled from a large number of hourly observations under different saturation deficits, was much reduced (lower curve) remaining below 0.4 mm/h even with increasing atmospheric saturation deficits (1.0 mb = 0.1 kPa). (After Jarvis and Stewart, 1979.)

example, to increase grass cover the canopy was reduced to less than 10% of the original leaf area in some eucalyptus forests of western Australia, yet transpiration was nearly half the rate of a complete canopy. This was sufficient to prevent shallow saline groundwater from rising and damaging forage for livestock (Greenwood and Berestord, 1979). In general, exposed trees maintain more open stomata and have higher air–leaf water vapor deficits than trees growing in closed stands (Whitehead and Jarvis, 1981). This is particularly true for broadleaved species that have higher leaf temperatures at the same radiation (Campbell, 1977; Jarvis *et al.,* 1976; Chapter 4).

These interactions are too complicated to be visualized graphically, but Monteith (1965) developed an equation that combines the factors affecting the transfer of momentum, heat, and water vapor from dry, transpiring canopies. The equation is similar to an earlier one developed by Penman (1948), which described evaporation of water from soil, water, and plant surfaces. The difference is that the combined formula, known as the Penman–Monteith equation, includes a term for stomatal conductance. The equation has the form:

$$E_t = \frac{k_5 R_n + k_2 k_3 D g_a \text{LAI}}{k_1 [k_5 + k_4 (1 + g_a/g_s)]}$$

The transpiration rate, E_t, may be calculated for an entire canopy or partitioned into various layers. In the numerator of the equation are terms that affect the evaporative demand: net radiation (R_n) absorbed by the canopy; the integrated effect of atmospheric saturation deficit (D); and the boundary layer conductance (g_a) for the entire canopy. In the denominator of the equation are those terms that restrict vapor transfer from the leaf, specifically, the ratio of boundary layer conductance (g_a) to stomatal conductance (g_s).

There are a number of physical constants in the equation that change slightly with temperature: (1) the latent heat of vaporization, k_1; (2) the specific heat of air, k_2; (3) the density of air, k_3; and (4) the psychrometric constant, k_4 (see Table A3 in Monteith, 1973). Leaf temperature is not necessary because it is included in a relationship between temperature and the water vapor content of air held at saturation, k_5. This last physical parameter varies considerably with temperature (as shown in Chapter 4, Fig. 4.7) and has values of 0.61, 0.83, 1.45, and 2.44 mb/°C at 5, 10, 20, and 30°C, respectively (Monteith, 1973, Table A4).

Environmental variables required to utilize the equation are the same as mentioned previously: air temperature, humidity, solar radiation, and wind speed above the canopy. These, together with certain characteristics of vegetation—LAI, average width of leaves, reflectivity of the canopy to solar radiation, and stomatal conductance (g_s)—permit calculation of boundary layer conductance (g_a), net absorbed radiation (R_n), the atmospheric saturation deficit (D) and, finally, transpiration (E_t).

Without the term for stomatal conductance, the equation reduces to that required to estimate evaporation of water from wet leaf surfaces. Here we shall illustrate the general application of the Penman–Monteith equation and compare the relative importance of R_n, wind speed, and D upon transpiration by broad- and narrow-leaved canopies.

From inspecting the equation, we see that if the boundary layer conductance (g_a) in the numerator were large, it would dominate the potential for evaporation under most conditions. This is generally the case with narrow-leaved canopies. If leaves averaging 5 mm in width in a canopy with an LAI of 5.0 were exposed to low wind speeds (1 m/s), high R_n (1.0 cal/cm^2/min or 698 W/m^2), and modest saturation deficit (5 mb or 0.5 kPa), the deficit term (Dg_aLAI) would account for

70% of the evaporative demand. In contrast, if leaves averaged 100 mm in width, the boundary layer conductance would be reduced by about 80% and R_n would control more than 90% of the potential evaporation. Further details concerning the relationship between leaf width and wind speed are presented in Fig. 5.3.

Because leaves of many coniferous species average only a few millimeters in width, the boundary layer conductance term usually dominates. In the denominator of the equation, the ratio of g_a to g_s is often in the range from 5 to 20 or even higher. Because of this, boundary layer conductance, net absorbed radiation, and the effects of wind speed can often be ignored in calculating transpiration from coniferous forests. The Penman–Monteith equation then reduces to:

$$E_t = \frac{k_2\, k_3}{k_1\, k_4}\, D\, g_s\, \mathrm{LAI}$$

Thus, transpiration from coniferous forests only requires knowledge of air temperature and humidity to calculate the atmospheric saturation deficit (D) and a precise determination of the four physical parameters (k_1 through k_4), which are available from standard tables or calculated as simple functions of temperature (Monteith, 1973). The stomatal conductance (g_s) can also be predicted for a particular species from knowledge of D, but can be expected to be lower depending on the soil water potential in the rooting zone or decreases in predawn plant water potential (Chapter 4, see Fig. 4.9).

Because the Penman–Monteith equation has all the components required to solve for both evaporation and transpiration from forest canopies, we will use it to compare the two processes. Evaporation, we must remember, occurs from the branches and stem as well as the foliage.

In regions where soil drought is pronounced, the extraction of water from various soil horizons can be monitored periodically and compared with predicted uptake and evaporative losses. An example for a Swedish pine forest is presented in Fig. 5.8 where climatic, vegetative (surface area and stomatal conductance), and soil data were available to evaluate a complete hydrologic budget. In May a major discrepancy between predicted and measured water content results from ignoring that more than 1 mm/day may be derived from storage reserves in sapwood, particularly in the spring (Waring and Running, 1978; Waring *et al.*, 1979; Running, 1980).

Simulation models of evaporation and transpiration have also been combined to estimate the influence of various kinds of vegetative cover upon streamflow (Kaufmann, 1984). In Fig. 5.9 a coniferous and hardwood forest with maximum LAI of 7.0 and 5.0, respectively, show higher rates of annual evapotranspiration compared to a recent clearcut with maximum LAI of 0.75. Note that during the summer, when canopies of the two forest types were at maximum, predicted

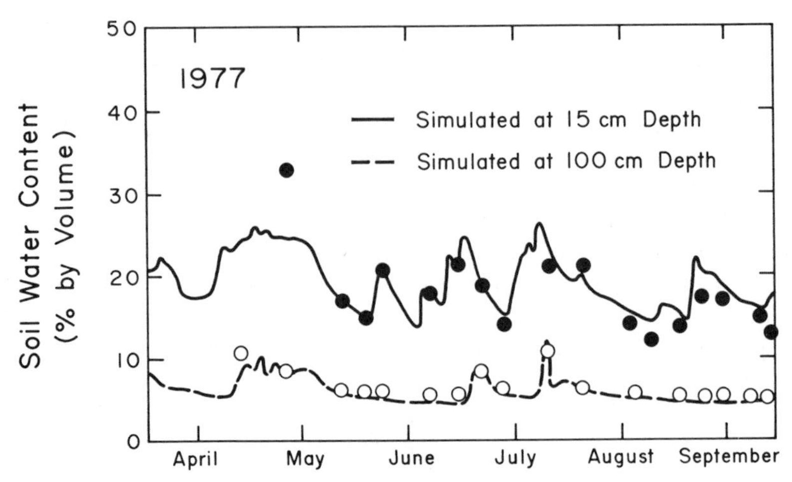

Fig. 5.8. Recharge or withdrawal of soil moisture from various depths were predicted by combining models for estimating water flux through the canopy, litter, and soil. Soil water contents were independently assessed (○, ●) with a neutron probe at four access tubes installed in the sandy soil under a 120-yr-old Scots pine forest in central Sweden. (After Jansson and Halldin, 1979.)

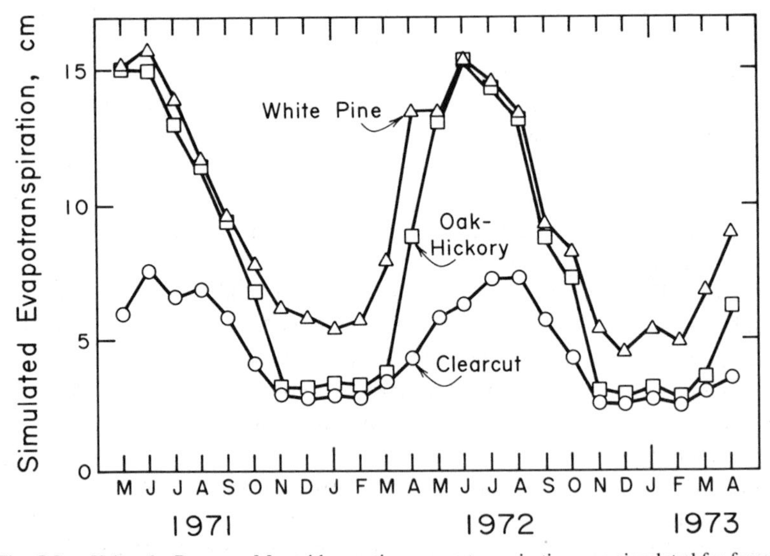

Fig. 5.9. Using the Penman–Monteith equation, evapotranspiration was simulated for forests of white pine (*Pinus strobus*), oak–hickory (*Quercus–Carya*), and vegetation reestablished on a recent clearcut. The predictions take into account seasonal changes in canopy LAI and climate. (After Swift *et al.*, 1975.)

evapotranspiration rates were similar, averaging 0.5 mm/day. The recently cleared area with a cover 15% that of the mature hardwoods utilized water at about 50% the maximum rate. During the winter months, the coniferous leaf area was reduced by 50%, the mature hardwoods by 90%, and the recent clearcut by 33%. The evergreen forest continued to intercept and evaporate more water than the other vegetation types. On an annual basis, the coniferous forest was estimated to utilize 110 cm out of a total of 192 cm in rainfall with about 30% partitioned into evaporation and 70% into transpiration (Swift *et al.*, 1975). The model predicted that the hardwood forest would use about 90 cm of water during the year and would have a similar partitioning ratio because evaporation increased from bark surfaces during the winter. The recently cleared area used only about 50 cm of water during the year.

Water Movement and Storage in Soils

Once water reaches the soil surface, the rate and direction of flow are determined by soil and topographic characteristics. Infiltration is important to the growth of plants and microorganisms, while runoff affects adjacent downslope terrestrial and aquatic ecosystems. Surface runoff may sweep litter and pollutants directly into streams but this rarely occurs on forested watersheds unless they have a large area in roads (Rothacher, 1971). Of more significance are large, natural channels that permeate most forest soils and permit water to flow rapidly into streams without first wetting all of the soil. The kind of vegetation and its management affect the amount and rate of water flowing through an ecosystem.

In this section we review major processes that control the movement of water after it reaches the ground. The amount of water reaching the soil is often a function of the kind of litter; certain types are known to be hydrophobic when dry and to absorb water very slowly (Meeuwig, 1971; DeBano and Rice, 1973). Infiltration of water into the soil, particularly forest soils, is not a uniform process, as woody debris and stems may channel or divert the entry of water. Actual infiltration may take two major routes: (1) through the capillary system of small pores that vary with soil texture and compaction; and (2) through a channel system of macropores created by clay shrinkage, cracking, roots, soil organisms, and sometimes the presence of depressions on the surface (Sharma and Luxmoore, 1979; Beven and Germann, 1982; Bouma and Wösten, 1984). The channel system drains and fills by gravity; the capillary system drains by gravity and as a result of water uptake by roots (Dixon, 1971; Beven and Germann, 1982). Although the size of macropore channels may increase with drying, their importance to drainage is during the wet season. Infiltration capacity, if variable seasonally, should be measured when drainage is significant (Dixon, 1971; Luxmoore, 1983). Models of infiltration that include both kinds of drainage are mainly empirical (reviewed by Childs, 1969; Whisler and Bouwer, 1970; Luxmoore, 1983) compared to theoretical models for flow through homogeneous porous media (Bear *et al.*, 1968; Philip, 1969).

Water movement through capillaries in the soil is controlled by the strong surface and capillary forces that hold water against the force of gravity. Soil water potential is an expression of these forces, in comparison to the energy status of free water. Soil water potential varies primarily with water content (Chapter 4, Fig. 4.1) but is also influenced by temperature and chemical composition of the soil water (Corey and Klute, 1985).

Hydraulic conductivity is a measure of the ability of a soil to conduct water in response to gradients in water potential, analogous to the conductivity of sapwood (Chapter 4). Capillary conductivity in soils, as in wood, decreases exponentially as water content decreases because the larger-diameter pores are drained first. In a typical silt loam, a decrease in volumetric water content from 0.4 to 0.3 might result in a reduction of capillary conductivity of two orders of magnitude (Elrick, 1968).

Compaction of fine textured soils, as often happens during logging and road construction, may reduce pore volume and capillary conductivity, even at saturation, by one or more orders of magnitude. Impervious layers force water to be diverted from normal channels and concentrate flow along roads or through surface horizons, rather than allowing for normal percolation throughout the entire soil matrix. This can lead to substantial increases in peak stormflow (Harr *et al.*, 1979) and accelerated erosion (Chapter 10). In general, conductivity decreases with soil depth but the transition may not be gradual. Quite often at the maximum depth of rooting, a relatively impermeable layer of soil is found. Colluvial or alluvial layering may also result in abrupt changes in pore size. Infiltration through a layer characterized by small pore size may require that those above be nearly saturated (Miller, 1969; Whisler *et al.*, 1972). Conversely, if a stratum of low conductivity overlies a coarse stratum of high conductivity, the upper strata must be nearly saturated to provide sufficient percolation to the lower one. In any case, percolation of water to deeper layers is controlled by the conductivity of the limiting layer and it is important to identify that layer.

Generally, forest soils have higher combined conductivities in a direction parallel to the slope than at right angles to it (Bear *et al.*, 1968). This situation is clearly favorable to lateral downslope movement of water in the upper, high conductivity layer of the soil (Hewlett and Hibbert, 1967; Dunne and Black, 1970; Hornbeck, 1973; Harr, 1977; Beven and Wood, 1983). Overland flow is extremely rare on forested watersheds if soils are not frozen (Freeze 1972b; Dunne, 1983).

The modeling of subsurface flow has received considerable attention because of its importance in maintaining streamflow and its sensitivity to alterations in surface drainage patterns associated with road construction (Freeze, 1972a,b;

Stephenson and Freeze, 1974; Dunne, 1983; Beven, 1982). Subsurface lateral flow makes a significant contribution to storm runoff only when convex hill-slopes feed into steeply incised channels, and the surface soil horizons have high conductivities. At low soil conductivity, lateral subsurface flow is not an important component of storm runoff. On convex slopes with low conductivities and on all concave slopes, storm runoff is dominated by precipitation infiltrating directly into the zone of saturated or near saturated soil situated around the channel (Freeze, 1972b; Troendle, 1979; Beven and Wood, 1983). The saturated zone increases during a storm and decreases during drought periods, making its area a difficult but important variable to estimate.

For our purposes, it is essential to divide the soil into at least two layers: a root zone and a subsoil. Biological processes influence water movement in the root zone but purely physical processes act in the subsoil. Evaporation losses are confined to the surface layer of the root zone and are generally very small under a canopy or litter layer (Tomanek, 1969). Water can move upward in response to potential gradients within the soil from a water table some meters below the surface. This is called capillary rise, which can be very important in saturated, coarse-textured soils where water may be supplied from a depth of 1 m at rates approaching 1 mm/day (Slatyer, 1967). If soils drain much below saturation, however, capillary rise slows or nearly stops.

The change in water content of the root zone is thus the sum of (1) infiltration; (2) rise of water from the subsoil; (3) subsurface lateral flow in; minus (4) evaporation; (5) uptake by roots; (6) percolation into subsoil; and (7) subsurface lateral flow out.

The change in water content of the subsoil is the sum of (1) percolation from the root zone; (2) subsurface lateral flow in; (3) rise from regional groundwater; minus (4) rise to the root zone; (5) subsurface lateral flow out; and (6) percolation (or seepage) to the regional groundwater.

Often some of these important variables have not been considered in studies that attempt to predict changes in soil water content. Certainly where predictions of water use are higher than the remaining stores of water, a consideration of additional sources of water to the rooting zone is warranted.

This completes our discussion of the hydrology of an individual ecosystem. We found that precipitation entering a unit passes first through a canopy consisting of foliage, branches, and stems. Some is intercepted and evaporated and the rest directly or indirectly reaches the ground surface. Precipitation may be stored in a snowpack that eventually melts and delivers water to the surface. At the surface it may be held in the litter and evaporate, or percolate through the litter to the soil. Excess water on the surface may run off through the litter as overland flow, but this is rare in forested watersheds. Once in the soil, water can percolate downward and move laterally in response to potential gradients. Percolation can reach the regional groundwater. Within the root zone, water in excess of −1.5 MPa potential is available for uptake by most trees. This water may not immediately leave the system by transpiration and can be stored in the sapwood, or to a much smaller extent in other tissue reservoirs. Transpiration withdraws water from the sapwood and the soil, reducing the conductivity of both, which may create a water deficit in the foliage. The water deficit, together with the saturation deficit of the atmosphere (and critically low temperatures), reduces stomatal conductance. Other physical variables such as irradiance and wind also contribute to determining the potential for evapotranspiration from an ecosystem.

HYDROLOGY OF A WATERSHED

Overland and subsurface lateral flow provides the basis for coupling terrestrial units to one another as well as to aquatic ecosystems. Surface topography, clearly defined on good topographic maps, can be employed to determine the path of overland flow and is an indication of the probable paths of subsurface flow. The actual paths of subsurface flow may be modified by local micro-relief.

Simulation models have been developed to calculate water budgets for various hydrologic units (U.S. Forest Service, 1972; Rogers, 1973; Troendle, 1979; Beven and Wood, 1983). Knowledge of the flow paths is incorporated into maps and the vegetative cover and soil storage capacities are defined for various areas. When excess water is available in a particular topographic unit, it is routed downhill to the next terrestrial unit, and eventually on to the stream channel.

In this section we incorporate the principles used to describe the water budget of individual terrestrial units to evaluate interactions between units and to predict streamflow from entire watersheds. In doing this we employ a simulation model developed by Rogers (1973) and modified (Waring *et al.*, 1981b) to include submodels for all of the water balance processes depicted in Fig. 5.1 except for groundwater. The model evaluates climatic, vegetation, and soil conditions daily and accumulates information (weekly, seasonally, or yearly) on changes in water storage and movement through and out of the watershed by way of streamflow or evapotranspiration. Various components of this model and related models have successfully estimated seasonal changes in soil, litter, and plant water content as well as streamflow (e.g., Swift *et al.*, 1975; Waring *et al.*, 1981b; Sollins *et al.*, 1981; Kaufmann, 1984; Knight *et al.*, 1985). Models that use hourly environmental data are required to predict water movement accurately during individual storms.

In this analysis, we report annual results that allow generalized comparisons in three distinct climatic regions: (1) the Coweeta Hydrologic Laboratory in the Southeast; (2) the H.J. Andrews Experimental Forest in the Pacific Northwest; and (3) the Beaver Creek Experimental Watershed in the Southwest. These represent sites with significantly different climates, vegetation, soils, and geology. In each area, a comparison was made of the effects of changing the original vegetation by thinning or clearing, or by complete replacement with another form of vegetative cover.

The Coweeta Hydrologic Laboratory located in North Carolina receives approximately 200 cm of precipitation, evenly distributed seasonally, with only 2% in the form of snow. The dominant vegetation is deciduous oak and hickory (*Quercus* and *Carya*) with a growing season LAI of 5.0. The H. J. Andrews Experimental Forest in Oregon has about 230 cm of precipitation, about 75% of which falls between October and March, some as a temporary snow cover. Fog drip is not as important as elsewhere in Oregon (Harr, 1982). The vegetation is dominated by 450-yr-old Douglas fir 50–60 m tall with a complement of other conifers and a few evergreen hardwoods. The LAI in undisturbed stands ranges from 10 to 12. The Beaver Creek Watershed is located in central Arizona. Precipitation averages about 64 cm annually with about 65% falling during the winter. Winter snow packs are quite variable but may be continuous during some years. The forest is composed of a mixture of scattered ponderosa pine, oak, and juniper with a LAI of about 1.5.

A number of comparisons of simulated hydrologic responses are presented in Table 5.2. For example, a 1-yr-old clearcut at Coweeta, representing a community of shrubs and hardwood saplings (LAI of 0.5–0.75), was predicted to have a streamflow of 199 cm, which is 36.0 cm, or 35% more than obtained from the original cover of hardwoods. Inspecting components of the water budget, we note that the increase is primarily due to a 65% reduction in canopy evaporation and a 33% reduction in transpiration. A comparison with actual, paired watersheds showed that streamflow agreed within 2% of that predicted. A similar exercise showed a 57% increase in streamflow following clearcutting of the Douglas fir forest in Oregon. This prediction was confirmed when the old-growth forest on the watershed was harvested (R. L. Fredriksen, Pacific Northwest Forest and Range Experiment Station, Corvallis, Oregon, personal communication).

At Coweeta, changing the cover from hardwoods to conifers (*Pinus strobus*), with a maximum LAI of 7.0, reduced streamflow by 18% over the original forest cover, mainly due to increasing interception losses (+27%) and dormant season transpiration (+24%). Again, the prediction of streamflow agreed almost exactly with that observed for a watershed with pine cover (Swank and Douglas, 1974).

In the more arid Southwest, thinning the already sparse stands of pine by 75% of the initial basal area reduced evaporation and transpiration by 20–25%. Converting to shrub and grass cover with growing season LAI of 2.0 and a dormant season LAI of 0.5 was predicted to increase streamflow by 27%, or 10% more than the thinning treatment. Again these predictions were confirmed by the observed values at gauged watersheds in which these treatments were applied (H. E. Brown, 1970; J. J. Rogers, Rocky Mountain Experiment Station, Tucson, Arizona, personal communication).

The conversions to different vegetative cover had a notable effect upon with-

TABLE 5.2 Components of Annual Water Budgets Simulated for a Range of Watersheds and Vegetation[a]

			Precipitation		Evaporation (E)			E
Watershed	Vegetation	Year	Rain	Snow	Canopy	Litter	Snow	Total
Beaver Creek	Pine	1965	48	50	10.6	1.2	2.5	14.3
Beaver Creek	Pine	1973	45	74	13.7	1.5	3.9	19.1
Beaver Creek	Shrubs	1973	45	74	9.7	1.6	3.6	14.9
Coweeta	Hardwoods	1972	192	3	24.4	1.4	0.0	25.8
Coweeta	Hardwoods	1973	235	5	24.2	1.3	0.0	25.3
Coweeta	Clearcut	1972	192	3	8.6	2.4	0.0	11.0
Coweeta	Pine	1972	192	3	30.9	1.4	0.0	32.3
Coweeta	Pine	1973	235	5	30.4	1.3	0.0	31.7
Andrews	Douglas fir	1973	119	48	32.9	0.6	−1.5	32.0
Andrews	Douglas fir	1974	206	98	22.3	0.7	−2.2	20.8
Andrews	Clearcut	1973	119	48	2.3	3.3	−2.4	3.2
Andrews	Upper 37%	1973	119	48	21.0	1.7	−1.7	21.0
Andrews	Lower 10%	1973	119	48	29.9	0.8	−1.7	29.0

[a] All values in centimeters. After Waring *et al.* (1981b).

drawal of water from the lower rooting depths, suggesting that subsurface flow might be considerably increased following such treatments. Thus, removing vegetation from the upper or lower slopes of a watershed might make a considerable difference in the availability of water to downslope terrestrial and aquatic units. Because the structure of the simulation models was designed to test these possibilities, predictions were made for partially cut watersheds.

On the Oregon watershed, clearcutting the upper slopes, which represent the drier environments, would encompass 37% of the total area and should, on a straight proportional basis, increase streamflow from 82.7 to 100.2 cm (+17.5 cm). The simulation, however, predicted streamflow of only 98.0 cm because an additional 2.2 cm was presumed to be transpired by the midslope forests during the summer drought. On the other hand, removal of vegetation growing adjacent to the stream, where soils maintain higher moisture contents, gave a predicted increase in streamflow that was proportional to the area involved. Of course, transpiration is not normally restricted in that zone during the summer.

When a similar simulation was conducted on the Coweeta watersheds, no difference was found regardless of where vegetation was removed. This prediction has been independently confirmed (W. T. Swank, Southeast Experiment Station, Franklin, North Carolina, personal communication) and can be explained by the more even distribution of precipitation throughout the growing season that maintains soil water supply near field capacity across the watershed.

In Fig. 5.10, the predicted streamflow for extremely wet and dry years was

Infiltrate	Transpiration (T) from soil horizons A	B	C	D	T total	Total ET	Streamflow Estimated	Observed
69.7	33.7	8.8	8.3	6.4	57.2	71.5	26.0	33.1
70.3	28.6	7.8	8.3	4.9	49.5	68.6	56.3	55.4
67.5	18.8	5.8	6.0	0.4	31.0	45.9	71.7	70.1
168.0	53.7	4.4	4.3	0.0	62.4	88.2	103.0	104.3
205.0	51.0	6.6	0.4	0.0	58.0	83.3	147.0	136.0
182.0	41.9	0.0	0.0	0.0	41.9	52.9	139.0	142.5
163.0	73.4	4.2	0.0	0.0	77.6	109.9	84.4	84.1
200.0	65.3	6.3	1.8	0.0	74.2	105.9	122.0	117.7
134.0	30.0	7.5	5.0	0.0	42.5	74.5	82.7	80.1
269.0	19.0	7.5	7.1	0.0	33.6	54.4	267.0	258.5
163.0	18.8	0.1	0.0	0.0	18.9	22.1	130.0	133.4
145.0	25.9	5.0	2.7	0.0	33.5	54.5	98.0	—
137.0	28.9	6.6	4.5	0.0	40.1	69.1	87.9	—

compared with values obtained from gauged watersheds at the three experimental forests. Predictions of annual changes in streamflow resulting from manipulations of vegetation were accurate. Although the general approach is sound, further efforts are necessary to predict peak stormflow, interactions with nutrients and toxic compounds, and some aspects of soil erosion (Chapters 6 and 10).

SUMMARY

We emphasize that common underlying processes control water movement in forest ecosystems: evaporation, transpiration, snow pack energy balance, infiltration, percolation, lateral flow, and capillary rise. From a detailed analysis of these processes, we justify the measurement of five climatic variables, seasonal changes in canopy leaf area, and specific characteristics of the soil that relate to the amount of water that may be stored, extracted, or conducted through various soil layers.

Computer simulation models based on the common processes and driven by the specified climatic variables are recommended for helping to predict the relative importance of various factors, and to assess changes in hydrology at-

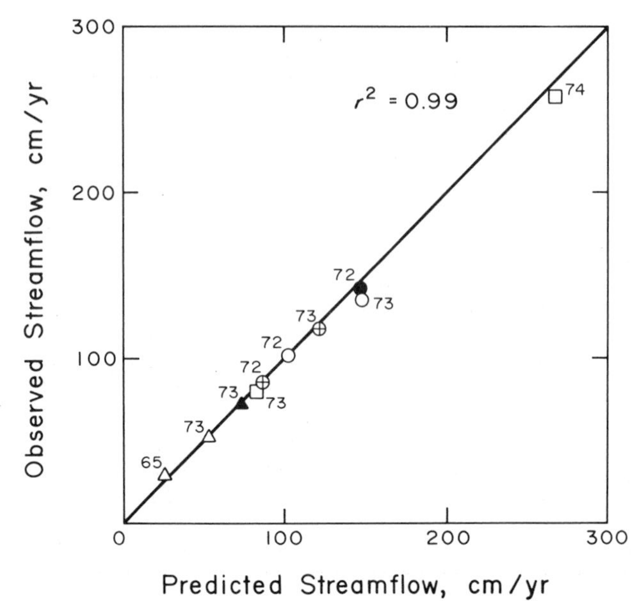

Fig. 5.10. Predicted annual streamflow from application of a watershed hydrologic model (Rogers, 1973) are in close agreement with measured values for a range of sites, climatic conditions, and vegetative cover. The slope of the regression line is 1.0 and the $r^2 = 0.99$. Gauged watersheds for which independent verification of streamflow could be made included those at Coweeta in the Southeast with cover of oak–hickory (○), white pine (⊕), and a recent clearcut (●); the Beaver Creek area in the Southwest with ponderosa pine–oak–juniper (△) and conversion to shrub and grass (▲); and the Andrews Experimental Forest in the Pacific Northwest dominated by 450-yr-old Douglas fir (□). Examples are provided in some cases for extremely wet and dry years. (After Waring *et al.*, 1981b.)

tributed to various forest practices that modify the vegetative cover, soil permeability, and routes that water moves through watersheds.

Many of the changes reported in the rate of snowmelt or streamflow upon forest cutting are—following a process analysis—logical results of basic interactions. Knowledge of basic principles simplifies many problems. It is not difficult, for example, to calculate the maximum evaporation or transpiration from any forest; evapotranspiration cannot exceed 6 mm/day. Similarly, the presence of wind-driven clouds should be noted, for in such situations we can expect gross precipitation and atmospheric chemical inputs to increase substantially if forests are present or to decrease if forests are absent.

The hydrologic principles described in this chapter aid in evaluating the leaching and cycling of nutrients (Chapter 6), movement of organic and inorganic material into streams and rivers (Chapter 10), and the potential effects of clearing large areas of forests in various regions (Chapter 11).

6

Nutrient Cycling through Forests

INTRODUCTION

What Is Biogeochemistry?

Every year the Amazon River carries an average of 3.6 million tons of nitrogen (N) and 0.3 million tons of phosphorus (P) from tropical rainforests to the sea (Salati *et al.*, 1982; Chase and Sayles, 1980). As a major avenue of nutrient movement from land to the oceans, riverflow is one process that links forest ecosystems to the large-scale circulation of chemical elements at the surface of the Earth. Studies of chemical processes occurring in the atmosphere, oceans, and land ecosystems and of transfers of elements between these compartments comprise the science of biogeochemistry. Some biogeochemical processes are physical events, such as the release of volatile elements, including sulfur (S), from volcanoes to the atmosphere. Others are the direct result of biological activity. For example, N is lost from wetland forests to the atmosphere as a result of bacterial activity in waterlogged soils. Of course, various elements that exist as particles or gases in the atmosphere are free to circulate and may be deposited in forest ecosystems in other locations. Thus, many elements move in biogeochemical cycles from land to sea and back again.

Forest Biogeochemistry

A model for understanding the major biogeochemical processes in forest ecosystems is shown in Fig. 6.1. Forests receive nutrient elements from the atmosphere. These may be deposited as particles or as dissolved components in rainfall, or they may be absorbed directly as gases. Rock weathering also yields available nutrients, such as phosphorus, for forest growth. Biological processes such as nitrogen fixation (N fixation) may result in nutrient input, and the excrement from migratory birds is an input when it contains nutrients that are derived from outside the specific forest ecosystem of study. Nutrients may leave by geological processes, such as streamwater runoff or seepage to groundwater; biological processes, such as denitrification; and meteorological processes, such as windborne dust. In this chapter we examine each of the major processes of input and loss of nutrients in forest ecosystems. Because many of the pathways of nutrient movement also involve movements of water, it is convenient to follow the transfers through forest ecosystems by use of the hydrologic watershed as the basic unit of study (Bormann and Likens, 1967; Chapter 5).

Inter- and Intrasystem Cycles

The gains and losses of nutrients in a forest result in the exchange of nutrients with other ecosystems on the surface of the Earth, and we refer to such nutrient movements as comprising the intersystem cycle. As we shall see, the cycle is complete only when the entire globe is considered as a single unit. In many cases it is perhaps better to refer to these processes as the *intersystem transfers* that cross the boundaries of the forest ecosystem and link forests to other systems

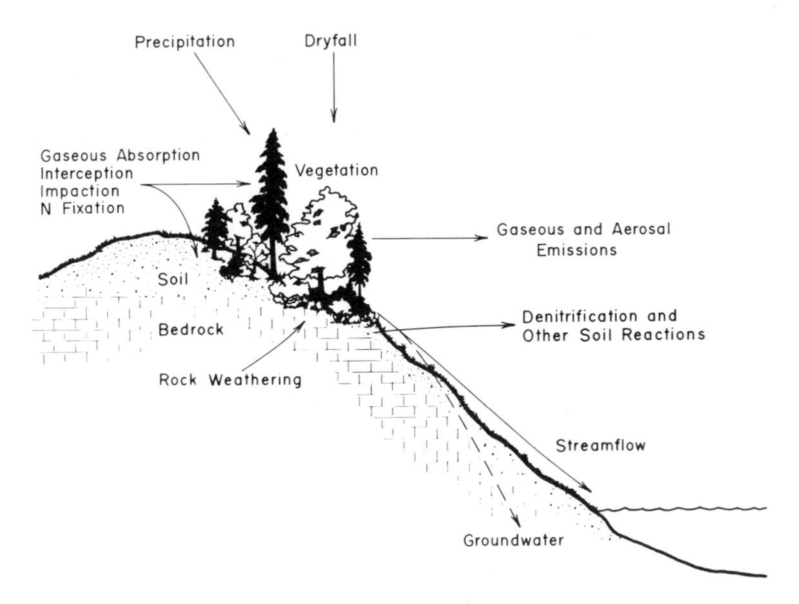

Fig. 6.1. A model of the pathways of nutrient movement through undisturbed forest ecosystems.

(Likens and Bormann, 1974). On a time scale of geological events, land ecosystems are degrading systems, because rock weathering and erosion are active on all land above sea level. However, on a shorter time scale, forest ecosystems may retain nutrient elements. Phosphorus released by rock weathering may be used by the vegetation for many years before being lost from the site in streamwater. In this chapter we examine nutrient storage in vegetation as a factor that affects the nutrient budget of a forest. Changes in storage during forest development or succession (Chapter 3) affect losses from the ecosystem (Gorham *et al.*, 1979). The movement of nutrients within a forest is called intrasystem cycling and includes plant nutrient uptake and nutrient losses in the death and decomposition of plant parts (Duvigneaud and Denaeyer-De Smet, 1970). The intrasystem cycle is treated in detail in Chapters 7 and 8.

ECOSYSTEM INPUTS

The Atmosphere

We know, of course, that through photosynthesis the atmosphere is the source of the carbon (C) content of forest ecosystems (Chapter 2). Photosynthesis is thus a biogeochemical process by which C is stored on land. The atmosphere is also a source of plant nutrients. Rainfall and snowfall contain dissolved and particulate constituents, including important quantities of nitrogen for forest growth. Leaf and branch surfaces act as efficient filters for atmospheric particles and gases that contain nutrients. Nearly all of the pool of nitrogen (N), sulfur (S), and chlorine (Cl) in forest ecosystems is derived from the atmosphere.

The atmosphere contains trace gases, particularly oxidized forms of N and S, that are sources of plant nutrients for land ecosystems. These gases are derived from a variety of natural and man-made sources. In wetland ecosystems and local zones of waterlogged soils, bacterial activity releases reduced gases of N and S to the atmosphere (Table 6.1). These gases are quickly changed to oxidized form (NO_2 and SO_2) by atmospheric reactions that are often catalyzed by sunlight. In addition, there are direct additions of oxidized N and S gases to the atmosphere. Lightning produces gaseous oxides of N (Hill *et al.*, 1984; Levine *et al.*, 1984) and SO_2 is released from volcanic eruptions (Stoiber and Jepsen, 1973). Normally, these trace compounds are rather minor constituents of the atmosphere because they are highly reactive (Rodhe *et al.*, 1981). Both oxidized and reduced forms of many of these trace gases readily dissolve in rainwater and are transported to the ground as dissolved ions available for plant uptake. For example, the reaction sequence:

$$2SO_2 + O_2 \rightarrow 2SO_3$$

$$2SO_3 + 2H_2O \rightarrow 2H_2SO_4$$

$$2H_2SO_4 \rightarrow 4H^+ + 2SO_4^{-2}$$

yields S in available form for plant uptake and circulation in the ecosystem. In recent years, man's industrial activities have also added many gases to the atmosphere, including large amounts of both NO_2 and SO_2.

TABLE 6.1 Some Examples of Bacterial Biochemical Reactions that Yield Reduced Forms of Nitrogen and Sulfur Gases from Anaerobic Soils[a]

Biochemical reaction	Example species
Nitrogen	
Denitrification:	
$4NO_3^- + 2H_2O \rightarrow 2N_2 + 5CO_2 + 4OH^-$	*Pseudomonas*
$2NO_3^- + H_2O \rightarrow N_2O + 2CO_2 + 2OH^-$	*Pseudomonas*
Sulfur	
$SO_4^{2-} + 6H_2O \rightarrow H_2S + 5OH^-$	*Desulfovibrio*

[a] In every case an organic energy source such as carbohydrate is available as a reactant.

Rainfall constituents are also derived from the ocean. Some 3–4% of the ocean surface is covered with bubbles at any given moment, and each year 10 billion tons of sea salt is injected into the atmosphere as these bubbles burst and tiny water droplets are held aloft by air currents (MacIntyre, 1974). These droplets evaporate and leave behind small particles, aerosols, which contain the important chemical constituents of seawater, especially sodium (Na), magnesium (Mg), chloride, and sulfate. Aerosols are produced over a broad range of sizes, but those that are smaller than 1 μm may remain in the atmosphere for days and may be transported over great distances.

Dust derived from windstorms in arid and semiarid regions also produces atmospheric aerosols that are particularly rich in calcium (Ca), potassium (K), and sulfate (Macias *et al.*, 1981; Moyers *et al.*, 1977; Munger, 1982; Gupta *et al.*, 1981). In addition, volcanic eruptions and fires can inject large quantities of particulate matter and gases into the atmosphere, some of which can be transported over great distances (Fruchter *et al.*, 1980).

In the atmosphere, cloud droplets begin to form by condensation on particles that are usually greater than 0.1 μm in diameter. These, of course, might be seasalt aerosols or soil dust particles. As raindrops enlarge and fall to the ground, they collide with other particles and absorb atmospheric gases. Thus, rainfall is efficient at scavenging both particles and gases from the atmosphere.

Those constituents of precipitation which serve as nuclei for the initial condensation of raindrops constitute the rainout component of precipitation, whereas those constituents cleansed from the atmosphere during the raindrops' descent comprise the washout component. The concentration of an ion in rainfall varies depending on the duration and rate of precipitation and whether the ion in question is largely of rainout or washout origin. Concentrations are highest in the early minutes of rainstorms and decline with time. The decline is greatest for ions derived from washout, as the atmosphere is progressively cleansed over the course of a storm (Schlesinger *et al.*, 1982). Small raindrops generally have higher ionic concentrations than larger raindrops, an important consideration in ecosystems where a substantial portion of the precipitation may enter as small fog droplets. Snow is usually less effective at scavenging than rainfall.

Atmospheric scientists measure the relative efficiency of scavenging in rainwater by calculating a washout ratio:

$$\text{Washout} = \frac{\text{Ionic concentration in rain (mg/liter)}}{\text{Ionic concentration in air (mg/m}^3\text{)}}$$

Some representative washout ratios are shown in Table 6.2. Large ratios are generally the result of ions that exist as relatively large aerosols or as highly water-soluble gases in the atmosphere.

Except in unusual circumstances, it is likely that nearly all the sodium in

TABLE 6.2 Some Representative Washout Ratios for Rainfall in Central England during 1971[a]

Element	Concentration in rain (mg/liter)	Concentration in air (mg/m^3)	Washout ratio
Na	2.3	0.0068	338
Cl	4.1	0.0148	277
Ca	1.1	0.0044	250
Fe	0.2	0.0019	105
Mn	0.008	0.0001	80
Pb	0.039	0.0007	56

[a] Note the large values for Na and Cl, which are derived from sea-salt aerosols, and smaller values for some trace metals, which are often derived from anthropogenic pollution. Calculated from data in Peirson *et al.* (1973), reprinted by permission from *Nature* (*London*) **241,** 252–256. Copyright 1973 Macmillan Journals Limited.

rainfall is derived from the ocean. One can use the ratio of Na to other ions to deduce their origin in precipitation. For example, the calcium to sodium (Ca/Na) ratio in seawater is 0.04. For rainfall with Ca/Na ratios close to this value, one would suggest that most of the Ca was also of marine origin. Such ratios are found in forests close to the ocean (e.g., Art *et al.*, 1974; Schlesinger *et al.*, 1982). In the eastern continental United States, however, the typical Ca/Na ratio is 1.3 (Likens *et al.*, 1977). Here the Ca content of rainfall has been increased relative to Na, presumably because the airflow that brings precipitation to this region has crossed large continents from which calcium is derived from soil dust and other sources. These observations are significant in interpreting the differences in nutrient input to forests in maritime compared with continental locations and the seasonal differences within forest ecosystems. For example, Lewis (1981) measured extremely high rainfall concentrations in the first storms that followed the seasonal dry period in Venezuela. Presumably, these storms washed the atmosphere of particulate ash derived from local forest fires during the dry season.

The deposition of nutrients by precipitation is often called wetfall; dryfall is the result of gravitational sedimentation of particles during periods without rain. With the design of collectors that were electronically sensitive to rain, i.e., collectors that close during rainfall, forest ecologists were able to separate the nutrient input received from these sources. Dryfall is an important process in many ecosystems, particularly those with long dry seasons. Even in the humid forests of the southeastern United States, Swank and Henderson (1976) reported that 19–64% of the total annual atmospheric deposition of ions such as Ca, Na,

K, and Mg and up to 89% of the deposition of P were derived through dryfall. Ions derived from continental sources are usually more important in dryfall than those derived from the ocean. Again, the process is particle-size dependent, and its relative importance for various nutrients can be estimated by calculating an index of dry deposition velocity:

$$\text{Deposition velocity} = \frac{\text{Rate of dryfall (mg/cm}^2\text{/sec)}}{\text{Concentration in air (mg/cm}^3\text{)}}$$

Accurate measurements of dryfall are difficult to obtain. Collections are frequently contaminated by dust resuspended from the local area, particularly when soils are dry. Similarly, there is some evidence that particles are released from the surface of vegetation (Lawson and Winchester, 1979). When these local processes contribute to the nutrient content measured in dryfall and precipitation, one must not assume that the measured deposition from the atmosphere represents the input of new ions to the ecosystem from beyond its boundaries.

In many early studies of atmospheric inputs to forest ecosystems, open funnels were used to collect both wet and dry fallout. Together, these were called bulk precipitation (Likens *et al.*, 1977). In natural communities, however, vegetation and other obstacles may capture fog (cloud) droplets and aerosols from the horizontal airstream. The nutrient content of the cloud droplets captured by horizontal interception on vegetation is also an atmospheric input to forest ecosystems. Fog water collected by vegetation is an important moisture input in many coastal and montane forests, especially during periods without general precipitation (Azevedo and Morgan, 1974; Lovett *et al.*, 1982). Similarly, the capture of dry aerosols on the surface of vegetation is called impaction, and also results in nutrient input (Art *et al.*, 1974). Reactive gases such as SO_2 and NH_3 are absorbed directly from the atmosphere by plant leaves and soil. In each case, the atmospheric input of nutrients caught in open funnels is an underestimate of the annual deposition of ions that are available for forest growth.

Horizontal interception, impaction, and gaseous absorption are difficult to measure with any degree of accuracy at the ecosystem level, but they may be very significant processes. Eriksson (1955, 1960) found that river runoff in Sweden contained 3.2 times more chloride than could be explained by precipitation. He linked the excess Cl to capture of sea-salt aerosols by forest vegetation surfaces. Nearly one-third of the atmospheric deposition of S in the northern hardwoods forest at Hubbard Brook, New Hampshire appears to be due to impaction and gaseous absorption (Eaton *et al.*, 1978). Similarly, the deposition of lead and zinc in forests in central Tennessee is dominated by impaction (Lindberg *et al.*, 1982).

Various workers have attempted to measure interception and impaction using plastic foliage (Schlesinger and Hasey, 1980; Tjepkema *et al.*, 1981), aerosol impactors (White and Turner, 1970), and ionic ratios (Art *et al.*, 1974; Gosz *et al.*, 1983), but none of these methods has been entirely satisfactory. Laboratory studies of gaseous absorption are difficult to extend to the ecosystem level in the field. Processes of impaction and absorption are likely to be dependent on canopy leaf area, and thus may increase in importance through succession (Chapter 3). Certainly, more work is needed to understand all of these processes of nutrient capture by the forest canopy.

Nitrogen Fixation

Over 78% of the atmosphere is composed of nitrogen. Various gases, including oxygen, argon, carbon dioxide, and trace constituents comprise the remaining composition. In its dominant form, N_2, atmospheric nitrogen is inert as far as biological processes are concerned. Thus, plants are bathed in a "sea of nitrogen" that they can not use (Delwiche, 1970).

Several types of bacteria and blue-green algae possess the enzyme nitrogenase, which converts atmospheric N_2 to NH_4^+, a form readily available to biota. Some of these exist as free-living forms (asymbiotic) in soils, whereas others, such as *Rhizobium* and *Frankia,* form symbiotic associations with the roots of higher plants. Symbiotic bacteria reside in root nodules that can be recognized in the field. Nitrogenase activity can be measured using the acetylene-reduction technique, which is based on the observation that this enzyme also converts acetylene to ethylene in experimental conditions. Plants or nodules are placed in small chambers or small chambers placed over field plots, and the conversion of injected acetylene to ethylene over a known time period is measured using gas chromatography. The conversion of acetylene in moles is not exactly equivalent to the potential rate of fixation of N_2 because the enzyme has different affinities for these substrates. Appropriate conversion ratios can be determined using other techniques. Unfortunately, these ratios are known to vary in different situations, and are seldom specifically determined in many studies. Although most of the N-fixation data are from acetylene-reduction studies, the rates must be interpreted with caution. As an alternative, investigators have applied $^{15}N_2$, the heavy stable isotope of N, in chambers and measured the increase in organic-^{15}N in test plants or soils through time. This method is much more direct, but also more expensive to use on a large-scale basis.

Nitrogen that enters terrestrial ecosystems, including forests, by fixation is an intersystem input from the atmosphere. This input can be very important in certain situations, particularly in N-poor successional sites. For example, in the postfire development of Douglas fir (*Pseudotsuga menziesii*) forests of Oregon, Youngberg and Wollum (1976) found that N fixation by the nodulated successional shrub, *Ceanothus velutinus,* contributed up to 100 kg/ha/yr of N on some sites (cf. Binkley *et al.*, 1982). For comparison, rainfall added only 1–2 kg/ha/yr in this region. Many alder (e.g., *Alnus rubra*) ecosystems show N-fixation inputs of a similar magnitude (Binkley, 1981). This N is eventually added to the soil organic matter and then becomes available to other species within the system. Proper forest management sometimes includes the encouragement of such early-successional, N-fixing shrubs on harvested sites to restore the soil nitrogen for forest growth (Binkley, 1983).

The N-fixation process is an energy-consuming reaction for plants with symbiotic bacteria. Silvester *et al.* (1979) found that N fixation by lupine, an understory shrub, was directly related to available light in the lower canopy of pine

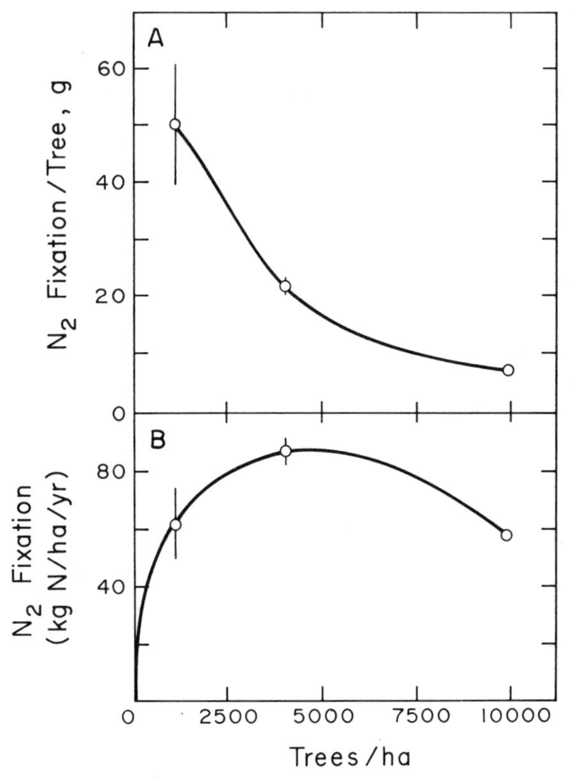

Fig. 6.2. Nitrogen fixation in red alder (*Alnus rubra*) plantations of varying density. In (A), N fixation per tree declines with increasing tree density, as shading limits the photosynthetic production available to symbiotic bacteria. In (B), total N input is seen to peak at intermediate plantation density, as the effect of greater numbers of trees partially compensates for the reduced fixation per tree. (From "Stand density effects in young red alder plantations: Productivity, photosynthate, partitioning, and nitrogen fixation" by B. T. Bormann and J. C. Gordon, *Ecology*, 1984, **65**, 394–402. Copyright © 1984 by the Ecological Society of America. Reprinted by permission.)

forests. The highest rates of fixation (in kilograms per hectare per year) by alder are found at intermediate stocking densities, such that the efficiency of net primary production is maximized (Fig. 6.2). Thus, N fixation is limited by rates of photosynthesis and C allocation, as outlined in Chapter 3. When alder (*Alnus incana*) seedlings were grown with limited N, moderate additions of N stimulated photosynthesis which allowed greater rates of nodule formation and N fixation (Ingestad, 1980). However, fixation is usually inhibited at high levels of available soil nitrogen.

In the absence of nodulated trees and shrubs, N fixation is of limited importance in forest ecosystems. Nitrogen fixation by asymbiotic bacteria typically ranges from 0 to 3 kg/ha/yr in most temperate forest soils (Roskoski, 1980; Tjepkema, 1979; Nohrstedt, 1982) but there have been very few studies of this process. Nitrogen fixation occurs in blue-green algae that exist on leaf surfaces (Bentley and Carpenter, 1984) and as symbionts in the canopy lichens of tropical rain forests (Forman, 1975) and Douglas fir forests of the Pacific Northwest. In boreal forests, a small amount of N fixation by blue-green algae occurs within the layer of *Sphagnum* peat moss (Basilier, 1979; Rosen and Lindberg, 1980). Nitrogen fixation also occurs in the gut of termites and may be an important input in tropical forest ecosystems where termites are predominant as decomposer organisms (Breznak *et al.*, 1973).

In most ecosystems, our understanding of the importance of N fixation is incomplete, as a result of the tendency for investigators to concentrate on only a portion of the system, e.g., fixation in logs, while fixation may be occurring in several compartments. There are few studies of N fixation among the leguminous species of tropical forests. Collectively, these might provide an important source of N in these ecosystems, and account for the abundant N circulation in tropical forests. The nitrogenase enzyme requires molybdenum, cobalt, and iron as co-factors (Jurgensen, 1973) and thus N fixation may be affected by limited quantities of these elements in some ecosystems. Other workers have suggested that N-fixing species have a high demand for P. There has been little investigation of any of these limitations at the ecosystem level.

Rock Weathering

The crust of the Earth consists of a variety of rock types. Nearly all of these contain primary minerals that were formed under conditions of greater temperature and pressure than found at the surface of the Earth. Upon uplift and exposure, all rocks undergo weathering, a general term that encompasses a variety of geological processes by which parent rocks are broken down. Mechanical weathering is fragmentation of materials with no chemical change. Chemical weathering occurs when parent rock materials react with water and mineral constituents are released as dissolved ions. Chemical weathering also includes the formation of new, secondary minerals that are more stable in the physical conditions at the surface of the Earth. Weathering is closely involved with the formation of soils, since the bulk of the physical structure of soils is composed of fragmented and weathered rock materials. Of interest here, weathering is an important source of nutrients for forest ecosystems. Soluble ions are often released in weathering reactions and are available for uptake and cycling within the forest. From 80 to 100% of the input of Ca, Mg, K, and P is derived from weathering in forest ecosystems. Because these nutrients are derived from outside the biotic ecosystem, they are part of the intersystem cycle of a forest. Of course these ions as well as those received from the atmosphere are eventually lost to rivers and to the ocean where they may again be deposited and transformed into rock. Thus, nutrients derived from weathering complete the biogeochemical cycle only in very long periods of time.

Mechanisms of mechanical weathering include rock splitting by the freezing of water and by the growth of roots in rock crevices. Mechanical weathering is important in extreme and highly seasonal climates and in areas with much exposed rock. Fragmented rock often forms the lower horizon of soil profiles (Chapter 8). The sand and silt fractions of soils are largely derived from the mechanical weathering of primary minerals, especially quartz. Thus, mechanical weathering is important in exposing parent rock to other weathering processes. Finely divided rock and soil can also be removed by erosion, the transport of particulate solids from the ecosystem. Massive mechanical weathering occurs in infrequent events such as landslides (Chapter 10), but these events do not result in nutrient input to the forest ecosystem.

When we consider the weathering of parent materials as a source of nutrients for forest growth, the important processes consist of chemical reactions, which continuously release dissolved ions to the soil solution and to streamwaters. Chemical weathering reactions occur most rapidly in warm temperatures and with large amounts of precipitation. Thus, on a worldwide basis, climate is the major determinant of weathering rates. Chemical weathering is more rapid in tropical forests than in temperate or boreal forests and occurs more rapidly in forests relative to grasslands and deserts.

Rates of chemical weathering and nutrient release are strongly dependent on rock type. Metamorphic rocks (e.g., gneiss, schist, quartzite) and many igneous rocks (e.g., granite) were formed deep in the Earth. These consist of primary silicate minerals that are crystalline in structure. Quartz is the simplest silicate mineral, consisting only of silicon and oxygen in a tetrahedral crystal. It is very resistant to chemical weathering. Other primary minerals are silicates in which various cationic elements (e.g., Al, Ca, Na, K, and Mg) and trace metals (Fe, Mn, etc.) are substituted in the crystal lattice. These minerals include feldspars, micas, olivines, and hornblende. Substitutions make these primary minerals less stable and more prone to chemical attack. In rocks of mixed composition, such as granite, chemical weathering may be concentrated on the relatively labile minerals, while others such as quartz are left unchanged. In the process of chemical weathering, primary minerals are altered to more stable forms, ions are released, and secondary minerals are formed as weathering products.

Sedimentary rocks underlie 75% of the land area and include shales, sandstones, and limestones. These rocks are formed relatively close to the surface of the Earth, usually as sediment accumulations under water. Shales and sandstones may consist largely of secondary minerals or resistant primary minerals eroded during earlier weathering epochs, and they are often rather weakly cemented rocks. In many instances, sedimentary rocks are easily eroded, but the component minerals, stable end products of earlier weathering, may not be easily susceptible to chemical attack. In these cases, one must recognize that rapid erosion does not necessarily imply abundant nutrient availability in the soil profile.

The most important chemical weathering process in forest ecosystems is that of carbonation. The reaction is driven by the formation of carbonic acid, H_2CO_3, in the soil solution:

$$H_2O + CO_2 \rightleftarrows H^+ + HCO_3^- \rightleftarrows H_2CO_3$$

Because plant roots and decomposing soil organic matter release CO_2 to the soil air, the concentrations of H_2CO_3 are often much greater than that in equilibrium with atmospheric CO_2 at 0.035% concentration. Carbonation reactions are dominant in the weathering of limestone,

$$CaCO_3 + H^+ + HCO_3^- \rightarrow Ca^{2+} + 2HCO_3^-$$

yielding soluble Ca. Silicate minerals also weather by carbonation, for example,

$$\text{K-Al-silicate} + H^+ HCO_3^- \rightarrow K^+ + H_4SiO_4 + \text{Al-silicate} + HCO_3^-$$

This simplified reaction represents the weathering of a potassium feldspar. During this process, a primary mineral is converted to a secondary mineral by removal of K and soluble silica. Except in unusual circumstances, the dominant anion in runoff waters is HCO_3^-, reflecting the importance of carbonation weathering conditions.

Other weathering reactions include simple dissolution of minerals, oxidations, and hydrolysis. Organic acids released from plant roots can weather biotite mica, a primary mineral that contains K (Boyle and Voigt, 1973; Boyle *et al.*, 1974; Jackson and Voigt, 1971). Lichens are important in rock weathering through the release of phenolic acids (Tansey, 1977; Ascaso *et al.*, 1982). Fungi release oxalic acid, which weathers soil minerals and affects the solubility of other soil minerals (Cromack *et al.*, 1979; Fisher, 1972).

Weathering of silicate rocks yields nutrient cations, e.g., K, Ca, Mg, and Fe in varying proportions depending on the initial rock composition and the environmental conditions for weathering reactions. Weathering of carbonates is more rapid and yields soils and streamwaters that are dominated by Ca and Mg. In these neutral to alkaline soils, the availability of other soil nutrients, particularly P, may be limited. Weathering of an unusual metamorphic rock called serpentine yields soils with unusually high concentrations of Mg, Fe, and trace metals relative to Ca. On some serpentine areas in Oregon and California, forest growth is stunted or impossible (Whittaker *et al.*, 1954; Proctor and Woodell, 1975).

Many types of secondary minerals can form in soils through weathering processes. Temperate forest soils are dominated by layered silicate or clay minerals. These exist as small (< 0.002 mm) particles that provide a great deal of the structural and chemical properties of soils (Chapter 8). In general, two types of layers characterize the crystalline structure of these minerals—Si layers and layers dominated by Al, Fe, and Mg. These layers are held together by shared oxygen atoms. Clay minerals and the size of their crystal units are recognized by the number, order, and ratio of these layers. Moderately weathered soils are often dominated by secondary minerals such as montmorillinite, which have a 2 : 1 ratio of Si- to Al-dominated layers. More strongly weathered soils, such as in the southeastern United States, are dominated by kaolinite clays with a 1 : 1 ratio of layers. Secondary minerals such as illite can incorporate other elements, such as K, into their crystalline structure.

In tropical rain forests, high temperatures and rainfall cause a relatively high proportion of silicon to be removed from soils. The highly weathered soils of tropical forests are dominated by crystalline oxide minerals and hydrous oxides of Fe (e.g., goethite, hematite) and Al (e.g., gibbsite, boehmite). Removal of Ca, K, and other basic elements is nearly complete as a result of long periods of intense leaching through the soil profile.

Recognition of secondary minerals is important when one estimates the nutrient inputs from the weathering of primary minerals in an ecosystem. When a soluble ion (e.g., K) is subsequently incorporated into a secondary mineral, it is not available for circulation or biotic uptake within the system. A classic study of New Zealand soils illustrates an analogous phenomenon for phosphate (Fig. 6.3). As soil develops, phosphate is weathered from the primary mineral apatite $\{[Ca_{10}(F, OH, Cl)_2(PO_4)_6]\}$ and is initially available for plant uptake. Through time, however, more and more of this phosphate becomes complexed with iron and is unavailable. Total phosphate in the ecosystem declines continuously because of erosion and leaching losses, so that nearly all of the remaining phosphate is eventually contained in complexed form. In these soils, weathering initially releases P, but secondary mineralization removes it from circulation. Thus, studies of weathering must recognize the proportions and chemical compositions of the weathering materials and the more resistant materials that may be accumulating as the soil develops.

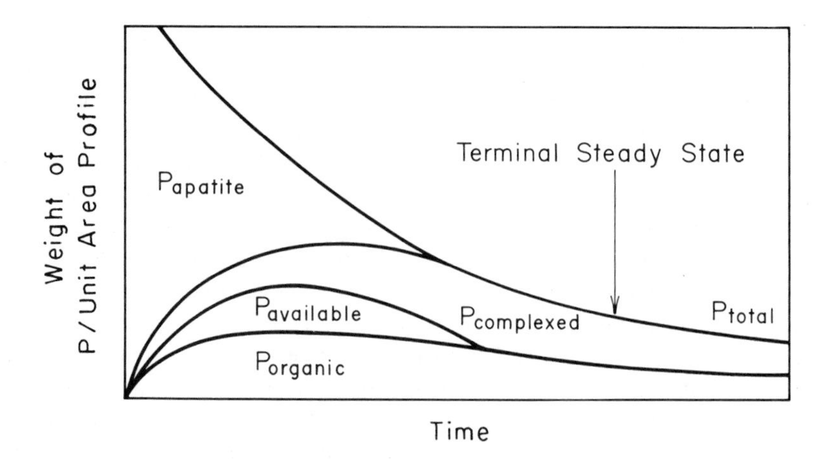

Fig. 6.3. Changes in the forms of soil P during the profile development of soils on basaltic rock in New Zealand. (Modified from Walker and Syers, 1976.)

Weathering is difficult to study in forest ecosystems. Few data are available on the annual inputs from weathering on an area basis, i.e., in kilograms per hectare per year. Most workers have used one of two approaches to measure weathering: watersheds or lysimeters.

The watershed approach was developed through long-term studies beginning in 1963 by Gene Likens, Herbert Bormann, and Noye Johnson working at the Hubbard Brook Experimental Forest in New Hampshire (Bormann and Likens, 1967; Likens *et al.*, 1977). Here a large number of comparable watersheds are underlain by an impermeable bedrock with no flow to groundwater. The entire area is covered with a late successional northern hardwoods forest. These workers reasoned that if atmospheric inputs of nutrients were subtracted from streamwater losses, the difference should reflect the annual nutrient release from rock weathering. They were able to calculate the rate of rock weathering and release of various plant nutrients, using an equation analogous to Fick's law of diffusion:

$$\text{Flux} = \frac{\text{Gradient}}{\text{Resistance}} \text{ or conductance} = \frac{\text{flux}}{\text{gradient}}$$

Thus,

$$\text{Weathering} = \frac{(\text{Ca lost in streamwater}) - (\text{Ca received in bulk precipitation})}{(\text{Ca in parent material}) - (\text{Ca in residual material in soil})}$$

The solution of this equation shows rather different amounts of bedrock weathering when the calculations are performed using different rock-forming elements

TABLE 6.3 Calculation of the Rate of Primary Mineral Weathering, Using the Streamwater Losses and Mineral Concentrations of Cationic Elements[a]

Element	Annual net loss (kg/ha/yr)	Concentration in rock (kg/kg of rock)	Concentration in soil (kg/kg of soil)	Calculated rock weathering (kg/ha/yr)
Ca	8.0	0.014	0.004	800
Na	4.6	0.016	0.010	770
K	0.1	0.029	0.024	20
Mg	1.8	0.011	0.001	180

[a] Data from Johnson *et al.* (1968). (Reprinted with permission from *Geochim. Cosmochim. Acta,* **32,** 531–545. Rate of chemical weathering of silicate minerals in New Hampshire, Copyright (1968), Pergamon Press Ltd.)

(Table 6.3). (You may wish to test your understanding by verifying the calculations in the table.) Calcium and sodium losses imply higher rates of weathering than potassium and magnesium losses. Johnson *et al.* (1968) suggest that the latter elements are accumulating in secondary minerals in the soil. In addition, some nutrients may accumulate in long-term biotic storage (e.g., wood growth) in the ecosystem. All estimates of weathering using this approach are likely to be too high because the method assumes that a kilogram of fresh rock yields a kilogram of residual material. Thus, watershed approaches to weathering studies are complicated by incongruent releases of nutrients from parent material, secondary mineral formation, and biotic uptake.

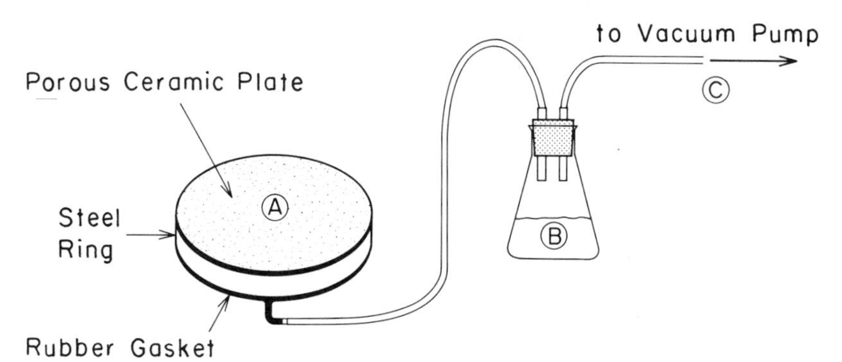

Fig. 6.4. A typical soil lysimeter consists of a porous ceramic plate (A) that is placed in the soil profile with minimal disturbance. The soil solution is pulled through the plate and collects (B) using a vacuum pump (C), normally maintained aboveground. The collection bottles (B) are placed in an access-trench and replaced at daily to monthly intervals.

Watershed studies permit the calculation of the total nutrient input from weathering processes that occur in both the upper soil horizons and the deeper horizons of fragmented parent rock. However, weathering in the lower soil may have little consequence in terms of nutrient input for forest growth. Various workers have used lysimeters to study the weathering of soil minerals in the plant rooting zone (e.g., Sears and Langmuir, 1982). A wide variety of lysimeters are available (Fig. 6.4), all of which are designed to extract the soil solution at various depths with minimal disruption of the water flow and chemical exchange reactions that occur in the soil profile. Table 6.4 shows some lysimeter analyses of the soils in a silver fir (*Abies amabilis*) forest in the Central Cascades of Washington. By the time rainwater enters the soil it has become acidic as a result of organic acids leached from vegetation and decaying organic matter in the forest floor. This solution mobilizes Fe, Al, and other cations from weathering of the soil at 15 cm. These are largely precipitated in a lower soil layer (30 cm), but carbonation weathering commences again at 60 cm, resulting in losses of cations and silica in runoff waters. Thus, processes of chemical weathering change through the soil profile. Such details are missed in many watershed studies that examine only streamwaters.

It is traditional to think of weathering of the underlying bedrock as the source of nutrient input and soil development in forest ecosystems. However, over large areas, soil profiles are developed from materials that have been transported to the site. For example, many forests in the northeastern United States occur on glacial

TABLE 6.4 Chemical Composition of Precipitation, Soil Solutions, and Groundwater in a 175-yr-old *Abies amabilis* Stand in Northern Washington[a]

		Total cations	Soluble ions (mg/liter)			Total (mg/liter)	
Solution	pH	(mEq/liter)	Fe	Si	Al	N	P
Precipitation							
Above canopy	5.8	0.03	<0.01	0.09	0.03	0.60	0.01
Below canopy	5.0	0.10	0.02	0.09	0.06	0.40	0.05
Forest floor	4.7	0.14	0.04	3.50	0.79	0.54	0.04
Soil							
15 cm A2	4.6	0.12	0.04	3.55	0.50	0.41	0.02
30 cm B2hir	5.0	0.08	0.01	3.87	0.27	0.20	0.02
60 cm B3	5.6	0.25	0.02	2.90	0.58	0.37	0.03
Groundwater	6.2	0.26	0.01	4.29	0.02	0.14	0.01

[a] Data from Ugolini *et al.* (1977), *Soil Sci.* **124,** 291–302. Copyright (1977) The Williams & Wilkins Co., Baltimore.

deposits. In other areas, volcanic ash, windborne soil or loess, or streamwater alluvium have resulted in deep and fertile soils. In all these cases, weathering may largely be of minerals in the deposited horizons and not of parent bedrock.

Hydrologic Inputs

Most studies of forest ecosystems consider streamwaters only as a means of nutrient output. However, studies of wetland and swamp forests require us to recognize that streamwaters can also bring nutrients to a forest site. Substantial nutrient input may occur in the annual deposition of soil alluvium in floodplain forests (Mitsch *et al.*, 1979; Mitsch and Rust, 1984). In cypress (*Taxodium distichum*) swamp-forests of the southeastern United States, net primary production shows a strong correlation with P inputs from upland ecosystems—ranging from 500 g/m^2/yr in closed bogs without streamwater input to 2000 g/m^2/yr in forests receiving polluted waters (Fig. 6.5). Streamwater inputs are a good example of an intersystem transfer of nutrients from one ecosystem to another.

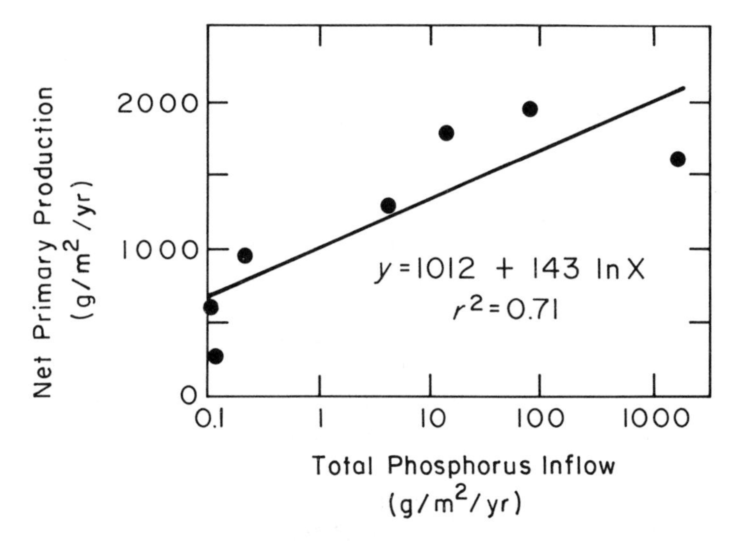

Fig. 6.5. Aboveground net primary production in cypress (*Taxodium distichum*) swamp forests as a function of the annual phosphorus received in streamwaters from upland areas. (From "A comparison of the structure, primary productivity, and transpiration of cypress ecosystems in Florida" by S. Brown, *Ecol. Monogr.*, 1981, **51,** 403–427. Copyright © 1981 by the Ecological Society of America. Reprinted by permission.)

BIOTIC ACCUMULATION AND STORAGE

Nutrient Accumulation during Forest Development

The growth of trees has direct and profound influence on the intersystem cycling of nutrients through forest ecosystems. As a forest grows on a disturbed site, nutrients are accumulated in short-lived tissues such as leaves and in long-lived tissues such as wood. Similarly, when soil profile development parallels forest growth, there is storage of nutrients in the surface litter, or forest floor, and in the soil organic matter, or humus, dispersed through the lower, mineral soil horizons. The leaf area and forest floor accumulations usually reach a steady state within a few years of forest growth; relatively rapid decomposition of these materials allows their nutrient content to be recirculated through the ecosystem. However, the accumulation of wood and soil humus may continue for centuries. The nutrient stores in wood and humus are isolated from biogeochemical circulation until the organic matter is decomposed, burned, or harvested for human use.

In Chapter 3, we examined the pattern of biomass accumulation during forest growth. Patterns of nutrient accumulation are roughly similar, since nutrient concentrations in wood do not change appreciably as trees age. However, some species of early successional trees and shrubs show especially rapid accumulations of readily available nutrients after forest burning or cutting (Marks, 1974). In general, nutrients are accumulated rapidly during the growth of early successional vegetation and more slowly as forest biomass develops. Theoretically, nutrient storage in vegetation reaches an asymptotic or cyclic-asymptotic value in old-growth forests that have reached a steady state in the accumulation of biomass (Fig. 6.6). Here, the annual growth of wood is balanced by the death of older trees.

There is a similar pattern of nutrient storage as the forest floor and soil organic matter accumulate with forest growth during primary succession. In most instances, the forest floor is likely to reach steady state before the organic matter in the underlying mineral soil layers. Such is the case for organic C during the development of forest soils on volcanic mudflows in northern California (Fig. 6.7). However, the storage of N in this soil was initially more rapid in the lower soil, presumably because the successional vegetation included several N-fixing shrubs that were rooted in this zone. In this case several centuries were needed before the soil layers approached steady state, i.e., when litter inputs were nearly balanced by decomposition.

During secondary succession after cutting, a different pattern of nutrient stor-

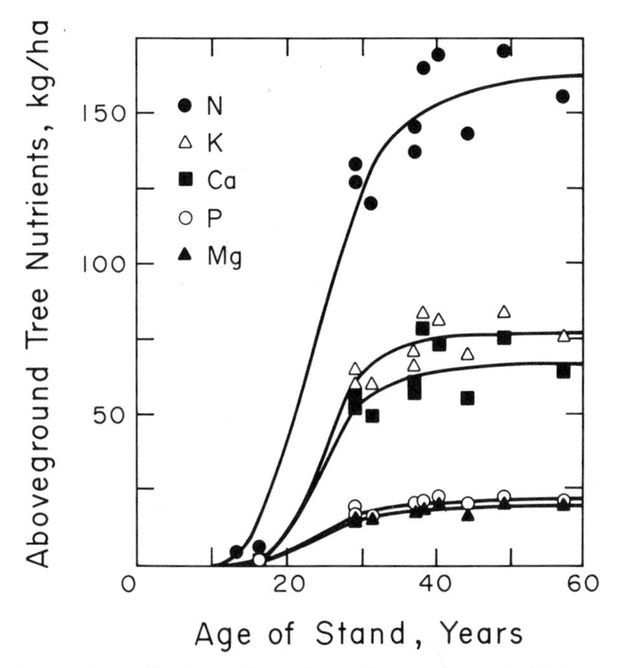

Fig. 6.6. Accumulation of nutrients during the postfire development of jack pine (*Pinus banksiana*) stands in New Brunswick, Canada. (From MacLean and Wein, 1977.)

age in soil is likely. Such disturbance usually causes little change in the organic matter or the nutrient pool in the mineral soil, but the forest floor may decline for several years after cutting because the rate of litter input from the regenerating vegetation does not equal the rate of decay of litter already present in the soil. During this time there can be large nutrient losses as a result of the decomposition of organic accumulations on the site. Covington (1981) suggested that the decline in forest floor mass continues for 15–30 yr after cutting hardwood forests in New England. However, he found that a total of only 60–80 yr was necessary for the forest floor to recover to precutting levels. Similar patterns are seen in tropical forests, but the recovery is more rapid (Aweto, 1981).

These patterns produce a strong relationship between net ecosystem production and the retention of limiting nutrients in forest ecosystems during development (Fig. 6.8; Vitousek and Reiners, 1975). As long as net ecosystem production is positive, plant nutrients will be stored in organic matter. Thus, losses of nutrients to streamwater will be low—lower than the sum of inputs. In old-growth forests, there is little net nutrient accumulation in organic matter, living and dead, and total ecosystem losses should balance inputs. These hypotheses

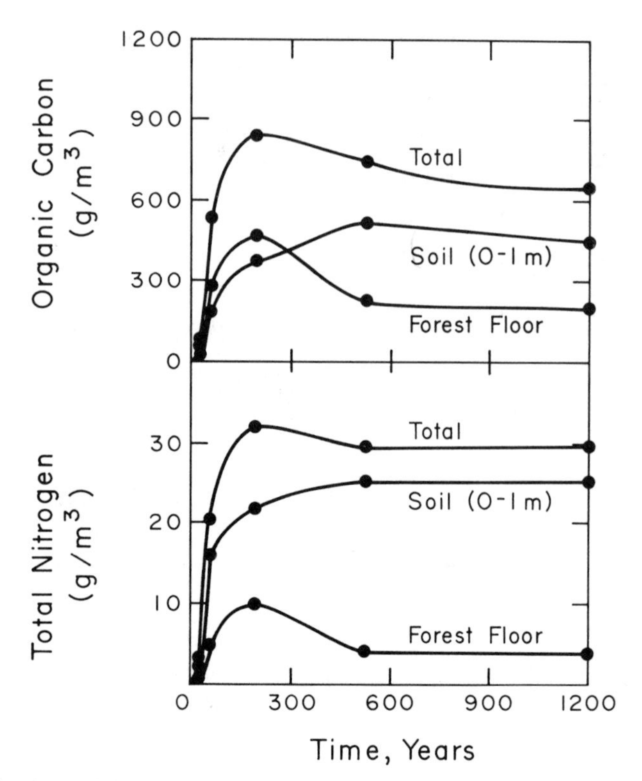

Fig. 6.7. Accumulation of organic C and N in soil developed on volcanic mudflows of varying age on Mount Shasta, California. (Modified from Dickson and Crocker, 1953.)

gain support from comparative observations of different elements in streamwater. In growing forests there is retention of N and K; streamwater losses of these elements are low, particularly during the growing season (Vitousek and Reiners, 1975; Henderson *et al.*, 1978). In contrast, there is little biotic control over the loss of Na, a nonessential element for most plants. The losses of ions that are not essential for plant growth can be predicted by understanding only the physical and chemical reactions in the soil. Thus, the growth of forest ecosystems affects the intersystem transfers of essential nutrients by determining the rate at which the inputs are stored in the ecosystem.

Nutrient Storage in Forest Ecosystems

There are a large number of studies of the nutrient content of forest vegetation. Table 6.5 provides some mean values for mature forests arranged by world biome type. In the aboveground vegetation, storage increases in the order: boreal < temperate < tropical forests. In contrast, both the mass and nutrient content of the forest floor increase from tropical forests to boreal forests, as a result of the slow decomposition in the cold conditions of higher latitudes. Similarly, the total mass of organic matter in the soil profile increases from tropical to boreal forests, though less dramatically than the change in forest floor alone (Schlesinger, 1977).

Aboveground biomass comprises 20–50% of the total (living and dead) organic matter in most forest ecosystems, increasing from boreal forests to the tropics. In Chapter 3 we examined the allocation and dynamics of above- and belowground biomass in forests. Similarly, Table 6.6 compares the nutrient storage among the components of a 50-yr-old *Liriodendron tulipifera* forest in Tennessee. Foliage comprises a small percentage of ecosystem organic matter, but leaves contain 5–20% of the nutrients in vegetation. The percentage nutrient storage in vegetation is low compared with that stored in soil organic matter, as a

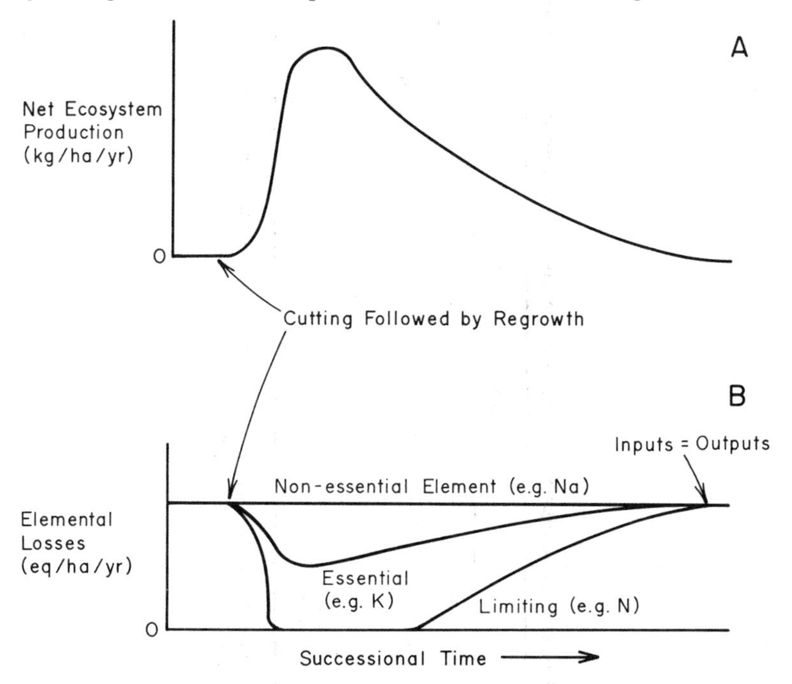

Fig. 6.8. Trends of net ecosystem production and nutrient losses during successional development of forest ecosystems. Before disturbance, intersystem nutrient inputs are equal to ecosystem losses. During succession, when net ecosystem production is positive, nutrients are stored in vegetation; thus, losses in streamwater are less than inputs—falling to zero for limiting elements (e.g., N). In contrast, nonessential elements show little change in streamwater losses during the successional sequence. (Modified from Vitousek and Reiners, 1975. Copyright © 1975 by the American Institute of Biological Sciences.)

TABLE 6.5 Storage of Organic Matter and Nutrients in the Aboveground Biomass and Forest Floors of World Forest Types[a]

Ecosystem	Aboveground vegetation						Forest floor					
	Biomass	N	P	K	Ca	Mg	Mass	N	P	K	Ca	Mg
Boreal coniferous	51,300	116	16	44	258	26	113,700	617	115	109	360	140
Temperate												
Coniferous	307,300	479	68	340	480	65	74,881	681	60	70	206	53
Deciduous	151,900	442	35	224	557	57	21,625	377	25	53	205	28
Tropical	292,000	1,404	82	1,079	1,771	290	27,300	214	9	22	179	24

[a] All data are in kilograms per hectare. Data for boreal and temperate forests are the means of 36 studies summarized by Cole and Rapp (1981); tropical forest data are means of 5 sites from Edwards and Grubb (1982).

TABLE 6.6 Mass and Nutrient Content of Compartments of a 50-Yr-Old Tulip Poplar (*Liriodendron tulipifera*) Forest in Tennessee[a]

Compartment	Organic matter[b]	N	P	Ca	K
Tree					
Foliage	3,200 (1)	54 (1)	6	35	52
Branches	27,100 (8)	79 (1)	31	144	46
Trunks	94,400 (28)	172 (2)	10	277	75
Roots	36,000 (11)	123 (2)	19	151	112
Understory	8,800 (3)	24 (0)	6	115	32
Forest floor	6,000 (2)	78 (1)	5	100	9
Soil organic	159,000 (48)	7,650 (93)	2,840	8,130	38,960
System total	334,500	8,180	2,917	8,952	39,286

[a] All data are in kilograms per hectare; organic matter and N, percentage of system total given in parentheses. Data are rounded from Cole and Rapp (1981, p. 393).

[b] Numbers in parentheses indicate percentage.

result of the rather low nutrient concentrations in wood. For N, aboveground biomass contains only 4–8% of the total quantity within most temperate forest ecosystems (Cole and Rapp, 1981). This percentage is greater in the tropics (Edwards and Grubb, 1982; Hase and Folster, 1982), but in most forests, the largest nutrient storage pool is in soil organic matter. The greatest accumulations occur in boreal forests and other areas with impeded decomposition, such as swamp and peat-forming forests. However, the growth of many forests is limited by nutrients, because most of the pool in soil organic matter is not readily available for plant uptake (Chapter 8).

Nutrient storage in roots, particularly in fine roots, is seldom measured accurately. Many fine roots are included in soil organic matter, particularly in collections of the forest floor. A large portion of the nutrient content of the forest floor and soil organic matter may actually be comprised of soil microbes, including mycorrhizal fungi. Studies of soils must distinguish between the nutrient content of the soil organic matter and the nutrients that are available as ions in the soil solution or on the cation exchange sites (Chapter 7). For the latter, there are various extraction techniques. Some of these enhance the weathering of soil minerals, which makes estimates of available nutrient contents unrealistically high.

Short-lived tissues such as leaves and fine roots are the sites of active physiological function in trees. The dynamics of these tissues and of the available soil nutrient pools determine the major characteristics of the annual nutrient circulation within a forest (Chapter 7). Nevertheless, in considering intersystem cycling, it is changes in the size of the long-lived, large storage pools (i.e., live wood, fallen logs, and humus) that affect the patterns of nutrient transfer through the ecosystem.

ECOSYSTEM LOSSES

Streams and Groundwaters

The transport of dissolved and particulate substances in streamwater is perhaps the most obvious means by which nutrients are lost from forest ecosystems. Streamwater concentrations are directly related to the weathering of soil minerals and bedrock in a forest watershed. In addition, streamwater concentrations of Na, Cl, and SO_4 can often be traced to atmospheric deposition in the ecosystem. As we have seen, streamwater also reflects the actions of biota on nutrient transfer and storage. Thus, streamwater chemistry can often be used to monitor a number of aspects of ecosystem function.

Nutrient transport in streams is often divided into two fractions: that carried in the form of dissolved ions and that carried as particulates. The products of chemical weathering are largely seen in the dissolved load. The particulate load, dominated by the products of mechanical weathering, represents erosion and sediment transport (Chapter 10). Particulate contents include materials ranging in size from colloidal clays to large boulders. These are also examples of inorganic particulates; leaves and floating logs are two kinds of organic particulates. Particulate losses include material suspended in the water and material that moves along the bottom of the stream channel.

Streamwater transport of clay minerals and undecomposed organic matter removes the nutrient elements in these materials from the watershed. Loss of larger rocks removes the nutrients contained in primary minerals, even though these may never have been made available for circulation in the ecosystem. Particulate losses increase strongly with the rate of stream discharge. At low flows organic particles predominate; however, in undisturbed forests, annual losses of organic matter are small, usually about 1% of net primary production (Schlesinger and Melack, 1981). Large amounts of inorganic transport may occur during runoff from infrequent, large storms. Particulate losses of all types increase after fires and forest cutting and are examined in detail in Chapter 10.

Except for Fe and P, which are only sparingly soluble in waters and easily adsorbed on particles, the annual losses of plant nutrients from undisturbed forest land are predominately in dissolved form (Table 6.7; cf. Martin and Meybeck, 1979). The concentrations of dissolved substances are often related to the volume of flow. The concentrations of many ions (e.g., Na) decrease with increasing discharge, whereas the concentrations of others may increase or show little change. In undisturbed forests, these concentration changes are often relatively small; thus, streamwaters have the appearance of relative constancy in chemical

TABLE 6.7 Average Annual Gross Output of Some Elements as Dissolved (D) and Particulate (P) Substances from Watershed 6 of the Hubbard Brook Experimental Forest, New Hampshire[a]

Element	Particulate + dissolved element (total kg/ha)	Particulate		Dissolved	
		kg/ha	(%)	kg/ha	(%)
Al	3.37	1.38	40.9	1.99	59.1
Ca	13.93	0.21	1.7	13.7	98.3
Cl	4.58	—	0	4.58	100
Fe	0.64	0.64	100	—	0
Mg	3.34	0.19	5.7	3.15	94.3
N	4.01	0.11	2.7	3.90	97.3
P	0.019	0.012	63.2	0.007	36.8
K	2.40	0.52	21.7	1.88	78.3
Si	23.8	6.19	26.0	17.6	74.0
Na	7.48	0.25	3.3	7.23	96.7
S	17.63	0.03	0.2	17.6	99.8
C	12.3	3.98	32.4	8.35	67.5

[a] From Likens *et al.* (1977).

composition. For individual ions, concentrations and concentration patterns differ depending on watershed characteristics, particularly soil reactions. Concentration patterns can often be understood by considering streamflow as a changing mixture of dilute waters from rainfall and concentrated waters from the soil solution (Johnson *et al.*, 1969; Lewis and Grant, 1979). General patterns often show seasonal changes, particularly for nutrients that are in short supply for forest growth. Concentrations of NO_3 and K are often lower during the growing season, despite the tendency for higher evapotranspiration to increase the concentration of most ions remaining in the soil solution at that time (Sears and Langmuir, 1982).

The effect of high discharge volume predominates over most changes in concentration; thus, total removal of nutrients is greater during high flows or years of greater discharge (Fig. 6.9). In temperate forests, losses are greatest during the spring snowmelt and winter months, when little water is being lost in evapotranspiration. Overall cation losses often follow the order: Ca > Na > Mg > K, though this will vary depending on bedrock (Feller and Kimmins, 1979). The order reflects the tendency for carbonates and Ca-silicates to weather easily and for little involvement of Ca in secondary mineral formation. Losses of cations and anions must show chemical equivalence, i.e., the sum of positively and negatively charged ions must balance in milliequivalent units. The total loss of cations is determined by the availability of anions within the ecosystem (Gorham

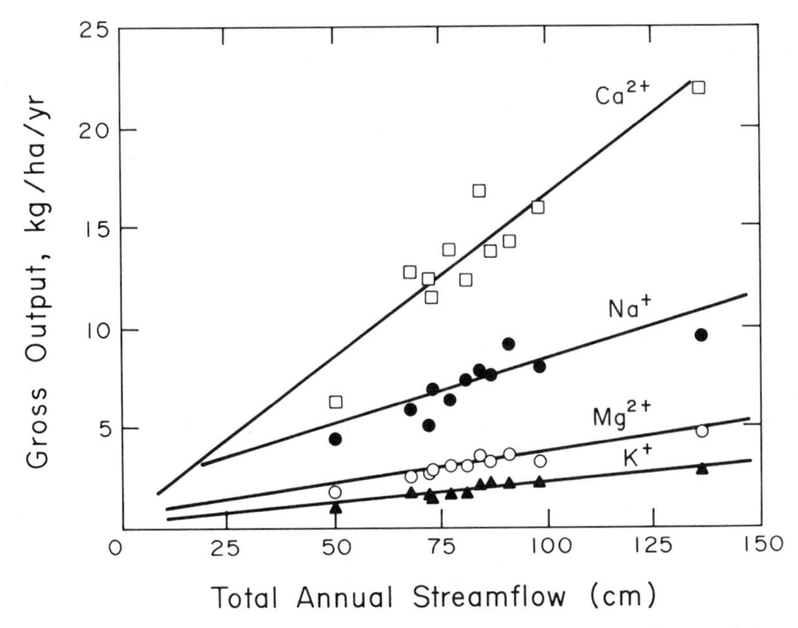

Fig. 6.9. Annual streamwater loss of major cations as a function of annual stream discharge in the Hubbard Brook forest, New Hampshire. (From Likens *et al.*, 1977.)

et al., 1979; Johnson and Cole, 1980). In most undisturbed forests, bicarbonate is the dominant anion in streamwaters as a result of carbonation weathering reactions. Small quantities of sulfate and nitrate are present from atmospheric inputs and microbial releases of these ions from decomposing organic matter (e.g., proteins). In areas that receive acid rain, however, the input of SO_4^{-2} may control the weathering reactions and streamwater cation losses. As we shall see, the importance of NO_3^- as a streamwater anion may increase greatly after forest cutting or fire.

Thus far our consideration has focused on surface runoff in streams. In many regions there are also nutrient losses to groundwater, which are difficult to measure. Usually the volume of flow is estimated by the residual in the watershed hydrologic budget:

$$\text{Precipitation} - (\text{streamflow} + \text{evapotranspiration}) = \text{groundwater flow}.$$

In these cases, evapotranspiration is calculated using the equations described in Chapter 5. The nutrient concentrations in groundwater are those measured in deep soil lysimeters or wells. Because large errors are potentially associated with all of these assumptions, most of the reliable nutrient budgets have been developed for forest sites that have minimal groundwater loss. Watersheds with impermeable bedrock simplify nutrient-budget studies, because groundwater losses appear as surface runoff in streams and are more easily monitored.

Gaseous and Particulate Losses to the Atmosphere

Methane and hydrogen sulfide gases are frequently released from wetland soils, as anyone who has visited a swamp or marsh will readily attest! These reduced gases are produced by specialized, generally anaerobic soil bacteria (Table 6.1). As we have seen, some reduced gases are an important source of the dissolved nutrient content of rainfall.

Recently ecosystem ecologists have recognized the presence of similar bacterial reactions in undisturbed, upland forests. During periods of heavy rainfall, anaerobic conditions may occur temporarily in thick layers of soil organic matter (Sexstone *et al.*, 1985) or in soil pits produced by the fall of canopy trees. It also appears that these bacterial reactions may persist in the center of soil crumbs, which remain as anaerobic microsites even in well-drained soils, due to the slow diffusion of oxygen through the soil pore spaces (Rosswall, 1981). In any of these locations, soil nitrate can be reduced to N_2 and N_2O in the reactions of denitrification by *Pseudomonas* bacteria.

It is difficult to measure the release of N_2 from soils because the concentrations produced are low compared to that of N_2 in the atmosphere. It is now known that the denitrification reaction that yields N_2 is blocked by the presence of acetylene, such that the intermediate metabolic product, N_2O, is the sole release from denitrification activity. N_2O can be measured using gas chromatography, allowing the rate of denitrification to be measured. Because acetylene has effects on other bacterial reactions, (e.g., N fixation), this acetylene-block technique is best calibrated in individual soils by using nitrogen isotopes to label the initial NO_3 and thus the denitrification products (Sexstone *et al.*, 1985). Using these methods, field measurements of denitrification are becoming feasible and more common in forest studies.

One early finding in these studies was the release of substantial quantities of N_2O from agricultural soils following fertilization (Bremner and Blackmer, 1978). It now appears that this release may be due to the action of *Nitrosomonas* bacteria, which are widespread in most soils and normally responsible for the oxidation of NH_4^+ to NO_3^- (i.e., nitrification). A metabolic intermediate of nitrification may be converted to N_2O. Relative release of N_2 and N_2O from forest soils may depend on moisture, aeration, pH, and other soil variables (Robertson and Tiedje, 1984; Goodroad and Keeney, 1984). Thus, in many forests both denitrification products, N_2 and N_2O, may be forms by which available N in the soil is transformed and lost from the ecosystem (Melillo *et al.*,

1983). At the present time, however, there are no forest ecosystem studies that have measured these losses as part of a complete N budget. The rate of denitrification is limited by the rate at which NO_3^-, the necessary substrate, is available in the soil (Fig. 6.10). In steady-state ecosystems, denitrification losses should not exceed inputs of N from the atmosphere.

Although substantial quantities of sulfur gases are released from bogs, swamps, and floodplain forests, much less is known about the bacterial reactions that might lead to gaseous release of S compounds from upland forests. In addition to H_2S, S may be lost as dimethyl sulfide and as other gaseous forms, some of which can be produced in aerobic conditions (Adams *et al.*, 1981). There is little agreement on the relative importance of all of these losses, particularly from terrestrial ecosystems.

Plants may be direct sources of gaseous and particulate nutrient loss to the atmosphere. Volatile emissions of hydrocarbons (e.g., terpenes) from forests have long been known to add to summertime haze. There is good reason to suspect that some volatile compounds might contain N and S. In laboratory conditions, both H_2S and NH_3 (ammonia) gases are known to be emitted from plant leaves (Winner *et al.*, 1981; Farquhar *et al.*, 1979). Similarly, aerosols containing K, S, and P appear to be emitted from tropical forest canopies (Crozat, 1979; Lawson and Winchester, 1979). Because of technical difficulties, there are no studies that estimate the actual quantities of such losses from entire

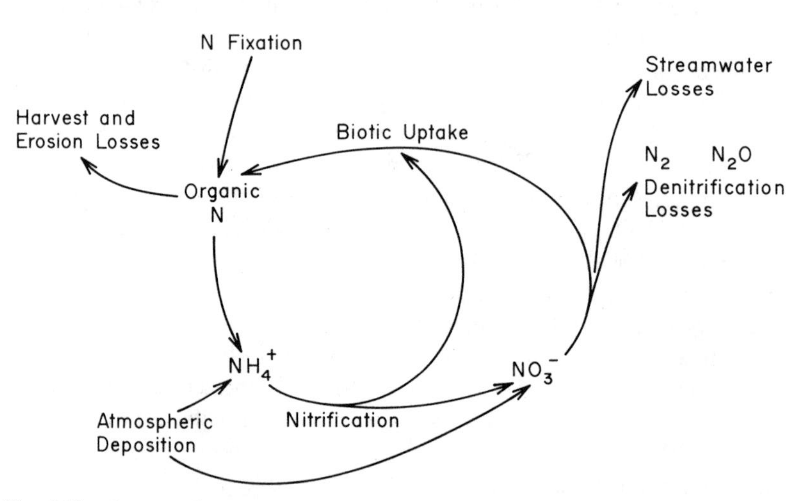

Fig. 6.10. Inputs and outputs of N which comprise the intersystem transfers in a forest ecosystem. Denitrifying bacteria must compete with biotic uptake of NO_3^- by other microbes and plants; thus the rate of nitrification sets an upper limit on denitrification losses. Streamwater losses represent the excess of available N over that taken up by biotic processes. Intrasystem processes are treated in more detail in Chapter 8.

forest ecosystems. In contrast, the gaseous release of NH_3 from the decomposition of vertebrate excreta has been incorporated into the nutrient budgets of grassland ecosystems (Woodmansee, 1978).

Nutrient Losses in Fire

During fires, nutrients are lost in the gases and particles in smoke. Moreover, following a forest fire, there is often increased runoff and erosion from bare, ash-covered soils. Before human intervention, fires were a natural part of the environment of many forests; thus, these nutrient losses occurred at infrequent but somewhat regular intervals. Normally we estimate annual nutrient budgets for undisturbed forests; however, long-term budgets for a forest site must consider the pulses of nutrients that are lost in fires. How long, for instance, does it take for natural inputs to replace the losses of nutrients from a single fire?

When leaves and twigs are burned in laboratory conditions, up to 85% of their N content can be lost, presumably as N_2 and as one or more forms of nitrogen oxide gas (DeBell and Ralston, 1970). Forest fires volatilize N in proportion to the heat generated and organic matter consumed. Typically, N losses range from 100 to 300 kg/ha, 10–40% of the amount in aboveground vegetation and surface litter (Table 6.5). Grier (1975) reported a volatile N loss of 855 kg/ha, 39% of the pool, in an intense wildfire in a montane coniferous forest in Washington. Air currents and updrafts during fire carry particles of ash that remove other nutrients from the site. These losses are usually much less significant than N losses. Expressed as a percentage of the amount present in aboveground vegetation and litter before fire, the losses often follow the order: $N \gg K > Mg > Ca > P > 0\%$. Keep in mind, however, that nutrient losses to the atmosphere in one location may result in added atmospheric deposition in adjacent locations (Clayton, 1976).

Recognition of volatile losses is important in forest management. Losses of N are particularly significant because little N is derived from rock weathering. When light, prescribed fires were used to remove undergrowth in southeastern U.S. pine forests, 11–40 kg/ha of N were released (Richter *et al.*, 1982). This was equivalent to 3–12 times the annual input from bulk precipitation in this region (Swank and Henderson, 1976). The loss of S in these fires was less than the annual replacement from the atmosphere. Significant losses of S were found when harvest residues were burned following forest cutting in Costa Rica (Ewel *et al.*, 1981). Since N loss is proportional to burning temperatures, losses are minimized when fuel loads are kept low. However, slash burning in the Pacific Northwest stimulates the germination of *Ceanothus velutinus* shrubs, which may provide important N replacement by symbiotic fixation (Youngberg and Wollum, 1976).

Depending on intensity, burning kills aboveground vegetation and transfers varying proportions of its mass and nutrient content to the soil as ash. There are a large number of changes in soil chemical and biological properties as a result of ash addition and fire (Raison, 1979), but our concern here is with the changes that affect intersystem nutrient transfers.

While burning removes nutrients from the ecosystem, nutrient availability in the ash layer is generally greater than in the undisturbed forest floor. Cations and P may be readily available in the ash, which usually yields increased soil pH. In addition, N may be released from ash by microbial activity immediately after fire. This increases the soil N that is available as NH_4^+ and NO_3^-, even though total N may be lower. Losses of N from surface soils may be overestimated by failure to recognize increases in N in the lower profile. Presumably this N is driven downward in volatile forms that condense after fire (Mroz *et al.*, 1980).

Streamwater runoff is often greater after fire because of reduced water losses in transpiration. High nutrient availability in ash coupled with greater runoff can lead to large nutrient losses from the ecosystem. These losses depend on many factors, including the season, rainfall pattern, and growth of postfire vegetation. Wright (1976) noted significant increases in the loss of K and P from burned forest watersheds in Minnesota. These losses were greatest in the first 2 yr after fire; during the third year there was actually less P lost from burned watersheds than from adjacent mature forests, presumably due to uptake by regrowing vegetation (McColl and Grigal, 1975). Percentage losses of Ca, Mg, Na, and K in runoff often exceed those of N and P, but there are exceptions to this pattern. Nutrients disappearing from the ash layer may be leached to lower soil depths and not lost from the site (Grier, 1975; McColl and Grigal, 1975) so that in some regions there is no significant increased loss of nutrients in streamwater after fire.

Over many areas, man has reduced the size of fires and the frequency at which particular sites are burned. The result is a greater intensity of fire when it occurs. These changes plus the increasing use of fire as a management tool need to be further evaluated for their effects on forest nutrient cycling.

Nutrient Losses in Forest Harvest

For the ecosystem ecologist, forest harvests range from natural "harvests," such as blowdowns, to large commercial clearcuts. Harvest of forest products represents a loss of the nutrients that they contain from forest ecosystems. Certainly the frequency and intensity of man's forest use will increase in the future, and the ecosystem approach allows us to evaluate some of the potential consequences of forest harvest on soil nutrient supplies. In most instances, the removal of woody biomass results in only a small percentage loss of the total content of nutrient elements in forest ecosystems, because the largest pools are in the soil (Table 6.6). Sawlog harvest of a mixed-oak forest in Tennessee removed from 0.1 to 7.0% of the nutrient pool of N, P, K, and Ca (Johnson *et al.*, 1982). Firewood harvest of a young rainforest in Costa Rica resulted in removal of 3% (Ca) to 31% (S) of the pool of various nutrients in the vegetation and surface soil (Ewel *et al.*, 1981). The foliage and branches that are left behind contain a large portion of the nutrient pool in vegetation, due to higher nutrient concentrations in these tissues than in wood. These materials are likely to decompose rapidly, releasing nutrients for regrowth. With conventional harvest, the number of years needed to replace nutrient removals by the annual inputs from precipitation and weathering is usually less than the harvest cycle (Table 6.8) (Johnson *et al.*, 1982c; Van Hook *et al.*, 1982).

Harvest of whole trees for pulpwood or biomass energy results in substantially greater nutrient removals, although these may still be replaced within a regrowth cycle in many instances (Silkworth and Grigal, 1982; Van Hook *et al.*, 1982). Nutrient removals in whole-tree harvests are especially large when expressed as a percentage of the available soil nutrients for forest regrowth (Freedman *et al.*, 1981; Johnson *et al.*, 1982). Whole-tree harvest removed 30% of the available Ca in a northern hardwoods forest in New Hampshire, but this was only 2% of the total in the soil (Hornbeck and Kropelin, 1982). The consequence of such nutrient removals can only be evaluated by measuring the rate at which the available pool is replenished by decomposition of soil organic matter and weathering of soil minerals. Thus, the ecosystem approach is directly applicable in the assessment of harvest effects. If the forest harvest cycle is more rapid than the rate of nutrient replenishment from the atmosphere and rock weathering, then forest fertilizations may be necessary to maintain site fertility for reforestation.

When the aboveground vegetation is removed, there are a number of changes within the ecosystem, many of which can affect nutrient transfers. Transpiration is reduced, yielding greater streamwater runoff. Soil temperatures and moisture may be higher and more favorable to microbial decay. Moreover, harvest residues leave large amounts of organic matter to decompose, including nutrient-rich tissues such as leaves. Until successional vegetation is established, available soil nutrients may be lost in runoff. The comparative losses in streamwater and product removal differ depending on the harvest technique, forest age, and other factors (Vitousek and Matson, 1984). In most instances, the streamwater losses are less than the losses in forest products (Hewlett *et al.*, 1984; Table 6.8). Nevertheless, streamwater losses are derived from the pool of available nutrients in the soil, and may have negative effects on streamwater quality.

When the Hubbard Brook Forest in New Hampshire was clearcut and regrowth prevented for several years, large increases in the loss of dissolved nutrients were observed (Fig. 6.11; Likens *et al.*, 1970). Increased losses also occurred as a result of increased erosion of particulate matter. These appeared to be due to the

TABLE 6.8 Comparative Losses of Nutrients as a Result of Timber Harvest in Different Ecosystems

Locale/process	N	P	K	Ca	Reference
British Columbia					
(Western hemlock–Douglas fir)					
Losses (kg/ha)					Feller and Kimmins (1984)
Harvested products	234	34	168	260	
Streamflow (2 yr)[a]	10	0	8	0	
Total	244	34	172	260	
Precipitation input (kg/ha/yr)	4	0	1	7	
Replacement time (yr)[b]	61		176	37	
Minnesota (aspen)					Silkworth and Grigal (1982)
Losses (kg/ha)					
Harvested products	454	43	355	1034	
Streamflow (5 yr)[a]	0	0	0	62	
Total	454	43	355	1096	
Precipitation input (kg/ha/yr)	7	3	10	5	
Replacement time (yr)[b]	65	14	36	219	
Florida (slash pine)					Pritchett and Comerford (1983); Riekirk (1983)
Losses (kg/ha)					
Harvested products	179	15	73	128	
Streamflow[a]	4		1	0	
Total	183	15	74	128	
Precipitation input (kg/ha/yr)	5	0	3	5	
Replacement time (yr)[b]	37	75	25	26	
Costa Rica					
(tropical rain forest)					Ewel *et al.* (1981); Lewis (1981)
Losses (kg/ha)					
Harvested products[c]	111	4		96	
Streamflow (11 mo.)[a]	329	11		392	
Total	440	15		488	
Precipitation input (kg/ha/yr)	7	2	4	9	
Replacement time (yr)	63	8		54	

[a] Streamflow losses in excess of losses on control watersheds during the interval indicated.
[b] Replacement of losses assuming atmospheric inputs only.
[c] Firewood harvest only.

combined effects of increased runoff and higher erodibility (Bormann and Likens, 1979a). Researchers attributed these effects to the removal of the biotic component of the ecosystem, with its internal cycling, regulation, and conservation of nutrients. Nutrient losses decreased as the forest was allowed to regrow.

After clearcutting, soil populations of nitrifying bacteria (*Nitrosomonas* and *Nitrobacter*) produced large amounts of NO_3^- from decaying litter:

$$\text{Organic N} \rightarrow NH_4^+ \rightarrow NO_3^-$$

Forest growth was prevented in this experiment and there was no plant uptake of NO_3^-. As a soluble anion, NO_3^- was leached through the soil profile and appeared in streamwater. Since ionic equivalance must be maintained in solution, the high NO_3^- concentrations were balanced by an increased removal of cations from the watershed (Likens *et al.*, 1970; Bormann and Likens, 1979a). Large

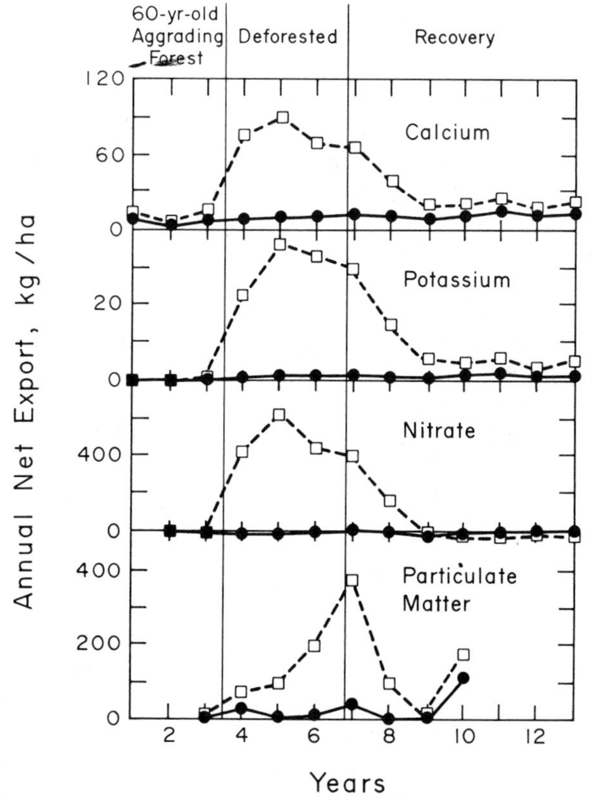

Fig. 6.11. Export patterns of dissolved substances (Ca, K, and nitrate) and particulate matter in streamwater from (□- -□) the experimentally clearcut watershed (W2) and (●—●) the forested reference watershed (W6). (From Bormann and Likens, 1979a.)

quantities of available NO_3^- also stimulate denitrification losses in streambed sediments (Swank and Caskey, 1982). When these findings were not confirmed in all cases of clearcutting, a large amount of work was devoted to understanding the process of nitrification in forest soils of different regions (Vitousek and Melillo, 1979; Vitousek *et al.*, 1982).

Nitrification occurs in most forests and provides available nitrate for plant uptake. In undisturbed conditions, nitrification rates are higher in the soils of many deciduous forests than in the soils of coniferous forests (Gosz, 1981). These differences appear to be due to differences in site fertility and to the nutrient quality of litter; coniferous litter has lower nutrient content and higher quantities of organic substances that resist or retard decomposition (e.g., phenolics and lignin) (Chapter 8). When forests are cut, nitrate losses can only be predicted by considering the natural rate of nitrate production and the rate of uptake by regrowing vegetation. Where uptake lags behind production, nitrate will appear in streamwater (Vitousek *et al.*, 1982). Without comprehensive studies, it is difficult to predict potential streamwater losses of N upon forest cutting, except to note that they are likely to be smaller on sites with low N availability before disturbance. By selective-, strip-, or patch-cutting, nutrient losses can be minimized, because streamwaters from harvested areas can pass through areas of regenerating forest where available nutrients will be taken up by plants (Jordan, 1982). Thus, from our previous considerations, various principles emerge which are useful in understanding the effects of forest harvest on nutrient transfers.

SUMMARY

Nutrient budgets for forest ecosystems are an attempt to compare and balance the inputs and outputs by the processes that we have reviewed in this chapter. Budgets can be made for a single year or an entire harvest cycle. In these efforts, we consider:

$$\text{Inputs} - \text{outputs} = \Delta \text{ storage}.$$

This mass-balance approach is often very powerful, for if some processes are not easily measured (e.g., aerosol impaction), their presence can be inferred by the accurate measurement of other terms. In a growing forest, for example, the inputs must at least equal the measured losses in streamwater and the storage in wood. Budgets are most easily constructed for elements like Cl, which have no gaseous transfers and relatively little storage in biota or secondary minerals.

Our considerations of nutrient cycling lead to several generalizations. First, weathering is the dominant source of nutrients such as Ca, Mg, K, Fe, and P, which are rock-forming elements. These nutrients may circulate through forest vegetation for many years, but they are eventually lost in streamwater. When they are deposited in ocean sediments, these elements may once again become constituents of rock. Geologic uplift of these rocks creates new land surfaces, and the cycle begins again. We speak of these processes as the sedimentary cycle, but it is completed over long periods of geologic time.

A corollary to our first generalization is that the annual inputs of N and S are largely derived from atmospheric sources in forest ecosystems. These elements circulate within the forest and may be lost as dissolved ions in streamwater, e.g., NO_3^- and SO_4^{-2}. Bacteria in anaerobic soils can reduce these ions to gases that are released to the atmosphere. These reactions may occur at some distance from the forest, for example, in estuarine mudflats and ocean sediments. Since rainfall and N fixation may return these nutrients to upland ecosystems, these elements complete a relatively rapid cycle. Often we say that N and S move in biological cycles, which are linked by the biotic activities of several global ecosystems. It is interesting to note that man's releases of NO_2 and SO_2 as air pollutants may be substantially increasing the rate of biogeochemical cycling of N and S over natural conditions. The sharp contrast between sources of elements that move in the sedimentary and biological cycles is seen in the Hubbard Brook Forest summarized in Table 6.9.

A second generalization emerges from the participation of nutrients in forest growth. While biomass is accumulating, nutrients that are in short supply are retained more strongly than those that are readily available. Thus, streamwater losses of N and P are very small percentages of total inputs (Table 6.9), whereas losses of Ca and Mg are greater (Feller and Kimmins, 1979). Streamwater losses give the forest the appearance of being "leaky," but it is important to recognize that outputs represent the excess of inputs after forest growth. Comparing forests from Oregon, Tennessee, and North Carolina, Henderson *et al.* (1978) showed strong N retention in each, despite a 10-fold difference in N input from the atmosphere. Thus, we might expect forest growth to be limited by N in many regions. Compared with N, losses of Ca in these three forests were always a greater percentage of the ecosystem pool. Calcium loss was greatest from a forest on limestone soils, where inputs from weathering were presumably much greater than biotic demands. In these comparisons, it is often useful to examine Na. As a nonessential element for forest growth, its loss often closely balances inputs (e.g., Table 6.9), despite a wide range of inputs between forests in continental and maritime regions.

Bog ecosystems are often said to be ombrotrophic, meaning to "feed on rain." This emphasizes the tendency for bogs to be isolated from streamwater inputs and rock weathering. The low productivity in bogs is probably directly related to the low nutrient inputs, which are only derived from atmospheric sources (Schlesinger, 1978). Some workers have considered boreal forests to be

TABLE 6.9 Inputs and Outputs of Elements from the Hubbard Brook Experimental Forest, New Hampshire[a]

	Inputs (%)		Output as a percent of input
	Atmosphere	Weathering	
Ca	9	91	59
Mg	15	85	78
K	11	89	24
Fe	0	100	25
P	1	99	1
S	96	4	90
N	100	0	19
Na	22	78	98
Cl	100	0	74

[a] Data from ''Some Perspectives of the Major Biogeochemical Cycle,'' Likens *et al.* (1981), Copyright (1981) by John Wiley & Sons, Ltd. Reprinted by permission of John Wiley & Sons, Ltd.

ombrotrophic in the sense that they are often underlain by permafrost. Most forest ecosystems are ombrotrophic for N since so little N is derived from rock weathering. Indeed, net primary production in temperate coniferous forests appears to show a correlation to N inputs from the atmosphere (Cole and Rapp, 1981). Downwind of urban regions, forest growth could be stimulated by excess N deposition due to pollutants (Chapter 9).

The annual nutrient requirements of forest trees are determined by the amount stored in long-lived tissues and the amount needed for the growth of seasonal tissues such as leaves and fine roots. Thus, the nutrient content of new wood growth does not represent the total annual uptake and circulation of nutrients in the forest. Atmospheric inputs of N, an intersystem transfer, generally supply only 5–32% of the total annual uptake by temperate forest ecosystems (Cole and Rapp, 1981). The remaining N must be supplied by intrasystem cycling, including efficient retention of N in trees and the release of N in decomposition of dead remains. Thus, nutrient availability for uptake is determined by both external supplies and internal recycling. We examine the intrasystem cycle of nutrients in Chapters 7 and 8.

7

Nutrient Uptake and Internal Plant Distribution: The Intrasystem Cycle

INTRODUCTION

Plant tissues contain a large number of chemical elements. Carbohydrates contain carbon, oxygen, and hydrogen and constitute the majority of the dry weight of plant tissue. However, more than a dozen other elements are essential for the growth of plants. In the previous chapter, we emphasized the sources and losses of nitrogen (N), phosphorus (P), potassium (K), calcium (Ca), magnesium (Mg), and sulfur (S) in forest ecosystems. Such processes determine the ultimate availability of these elements, which are needed in relatively high concentrations

TABLE 7.1 Nutrient Concentrations in the Leaves of Sugar Maple (*Acer saccharum*) in the Hubbard Brook Experimental Forest, New Hampshire[a]

. Nutrient	Oven dry weight (%)
Macronutrients	
N	2.19
P	0.18
K	1.01
Ca	0.60
Mg	0.12
S	0.21
Micronutrients	
Fe	0.012
Zn	0.0052
Cu	0.0009
Mn	0.17

[a] From Likens and Bormann (1970).

for plant growth. Often these elements are called macronutrients (Table 7.1), and deficiencies of N and P commonly limit the productivity of forest ecosystems. In addition, plants require iron (Fe), copper (Cu), zinc (Zn), manganese (Mn), boron (B), chloride (Cl), and molybdenum (Mo) in much smaller quantities as micronutrients. These are available in adequate quantities in most forest soils as a result of rock weathering and, in the case of Cl, from atmospheric deposition. Plants may also accumulate other elements as nonessential constituents, for example, aluminum (Al), silicon (Si), and various trace metals. In this chapter, we consider the processes by which trees accumulate essential nutrients and the distribution of these nutrients within forest biomass. The uptake of essential nutrients and their return to the soil constitute the intrasystem cycle of a forest ecosystem (Duvigneaud and Denaeyer-DeSmet, 1970).

NUTRIENT UPTAKE

Requirements

There are a large number of physiological and biochemical studies of the nutrient requirements of plants. This research comprises the science of plant mineral nutrition. Most experiments have involved herbaceous plants grown under laboratory conditions. Our knowledge of the nutrient requirements of trees in field conditions is much less advanced (Linder and Rook, 1984).

The biochemical roles of macronutrients in plants are well known (Salisbury and Ross, 1978). For example, N is a major constituent of proteins, nucleic acids, and chlorophyll; P is a component of ATP, nucleic acids, and cell membranes; and S is found in many plant proteins. Specific roles of K (in stomatal function), Ca (in cell walls), and Mg (in chlorophyll) are well established. These nutrients also stimulate the rate of enzymatic reactions. The micronutrients Fe, Cu, Zn, and Mn are widely involved in enzymes and coenzymes, specifically in the reactions of photosynthesis, whereas the essential roles of B and Cl are still poorly known. Molybdenum is essential for N metabolism in plant tissues, as well as for N fixation by symbiotic bacteria (Chapter 6). Note that we do not consider cobalt as an essential element for higher plants, though it is essential for the microorganisms involved in N fixation.

Nutrient requirements for forest growth are not simply fulfilled by the presence of the essential elements in the soil. These nutrients must occur in a form that is available for plant uptake. Moreover, nutrients must be available in an appropriate balance (Shear *et al.*, 1946). Added N may simply create or exacerbate a P deficiency. Soil chemistry includes complex reactions that are not easily replicated in laboratory experiments. We must begin our discussion of nutrient uptake by considering aspects of the soil that determine the availability of plant nutrients.

The Soil Solution

We have seen how rock weathering contributes to soil formation by the breakdown of parent materials and the formation of new, secondary minerals in the soil profile (Chapter 6). The chemical constituents of secondary minerals are not readily available to plants without further weathering, but the properties of secondary minerals greatly influence the availability of dissolved ions to plants. The silicate clay minerals possess surface negative charges that attract and hold soluble nutrient cations from the soil solution. This binding is reversible and exists in equilibrium with ionic concentrations in the soil solution. Iron and aluminum hydroxides have surface positive charge at most normal levels of soil pH. Nutrient anions, especially PO_4^{-3} and SO_4^{-2}, are held on these soil particles.

Silicate clay minerals possess several types of negative charge which contribute to soil cation exchange capacity, expressed as milliequivalents per 100 g of soil. At the edges of clay particles, hydroxide ($-OH$) radicals are often exposed to the soil solution. Depending on the pH of the solution, the H^+ ion may be more or less strongly bound to this radical. At neutral to high pH, a considerable number of the H^+ are disassociated, leaving negative charges ($-O^-$) that can attract cations (e.g., Ca^{+2}, K^+, and NH_4^+) from the soil solution. In addition, there is another source of negative charge that arises from ionic substitutions within silicate clays. For example, when Mg^{+2} substitutes for Al^{+3}, there is an unsatisfied negative charge in the internal crystal lattice. This charge is expressed as a zone of negative charge surrounding the surface of clay particles in the soil. Unlike the first source of negative charge, this second source is permanent because it originates inside the crystal structure and cannot be neutralized by covalent bonding of H^+ from the soil solution.

In most forest soils, a large amount of cation-exchange capacity is also contributed by soil organic matter. These are pH-dependent charges originating from the phenolic ($-OH$) and organic acid radicals ($-COOH$) of soil humic materials. In some sandy soils, as in central Florida, nearly all cation exchange is the result of organic matter.

Cation-exchange capacity and base saturation, the percentage of the exchange sites actually occupied by nutrient cations, both increase during the initial soil development on newly exposed parent materials. As weathering of soil minerals continues, cation-exchange capacity and base saturation decline (Bockheim, 1980). Temperate forest soils dominated by 2 : 1 clay minerals have greater cation-exchange capacity than soils dominated by 1 : 1 clay minerals (Chapter 6). Most fertile temperate soils have high cation-exchange capacity, which holds nutrient cations on clay particles and reduces losses to runoff waters. Tropical forest soils dominated by iron and aluminum oxides have low cation-exchange capacity. Exchangeable cations are often depleted in tropical forest soils as a result of long periods of intense weathering and runoff, but small amounts of cations are held on exchange sites in the soil organic components.

Anion adsorption is especially important in tropical forest soils in which it occurs as a result of association and dissociation of H^+ on iron and aluminum hydroxide radicals (Fig. 7.1). The zero-point-of-charge occurs around pH 9, so that significant anion absorption capacity is present in these soils in most field situations. Anion adsorption is sometimes present in the soils of warm temperate forests, as a result of iron and aluminum hydroxides in the soil profile (D. W. Johnson *et al.*, 1981).

Exchange of cations and anions from the surface of secondary soil minerals occurs as a function of chemical mass balance with the soil solution. Elaborate models of ion exchange have been developed by soil chemists. In general, cations are held and displace one another in the sequence:

$$Al^{+3} > H^+ > Ca^{+2} > Mg^{+2} > K^+ > NH_4^+ > Na^+$$

on the cation-exchange sites. Anion absorption follows the sequence:

$$PO_4^{-3} > SO_4^= > Cl^- > NO_3^-$$

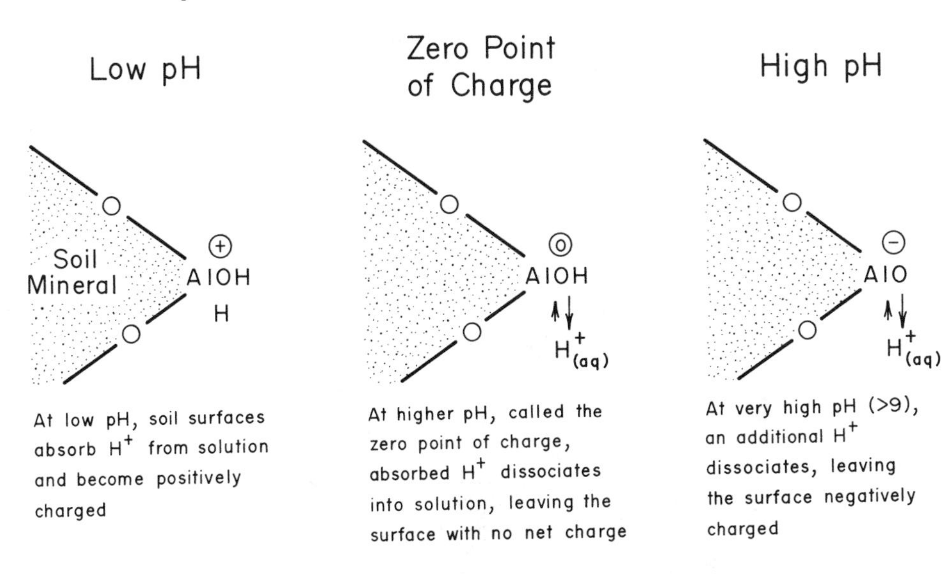

Fig. 7.1. Variation in surface charge on iron and aluminum hydroxides as a function of the pH of the soil solution. [Modified from Johnson and Cole, 1980. Reprinted with permission from (*Environ. Int.*, **3,** Johnson and Cole, Anion mobility in soils: Relevance to nutrient transport from forest ecosystems), Copyright 1980, Pergamon Press Ltd.]

Either of these sequences can be altered by the presence of large quantities of the more weakly held ions in the soil solution. Agricultural liming, for example, is an attempt to displace and neutralize H^+ ions from the exchange sites by "swamping" the soil solution with excess Ca^{+2}. Phosphorus is tightly held on anion-exchange sites and relatively immobile in most soils. However, this exchangeable P is potentially available, unlike P that is permanently complexed in secondary minerals (Chapter 6).

While these exchange reactions are the dominant chemical processes that occur in soils, they are by no means the only ones. The presence of some ions may determine the availability of others in the soil solution. For example, PO_4^{-3} is strongly bound by Fe in many soils. Graustein *et al.* (1977) suggest that the production of oxalate compounds by fungal hyphae may increase PO_4^{-3} availability by complexing the iron in the soil solution. Soluble organic compounds, including polyphenols and humic acids, may also mobilize or complex ions in the soil solution (e.g., Chapter 6, Table 6.4).

Hydrogen ions affect the solubility of most other ions in the soil solution (Fig. 7.2); thus, pH is often considered the master variable that controls soil nutrient availability. Many factors act to lower the pH of the soil solution during soil development. The release of Al^{+3} ions during weathering reactions results in the

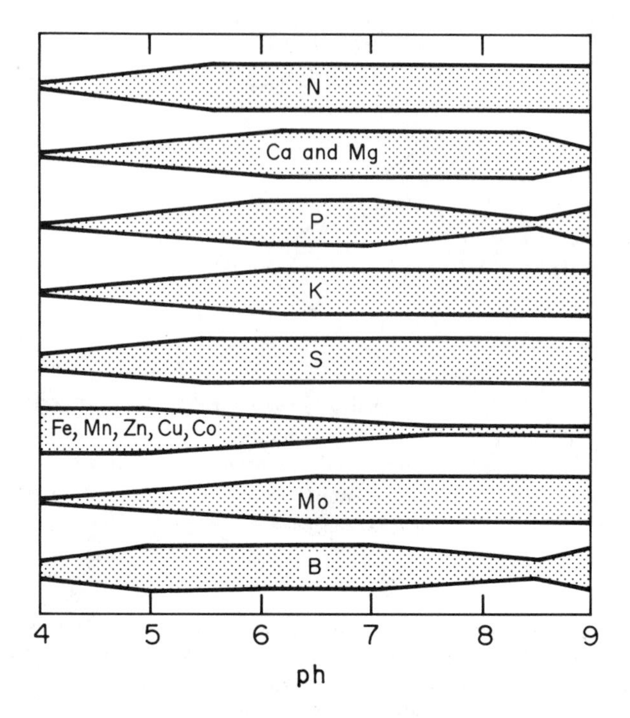

Fig. 7.2. The relative availability of plant nutrients, shown as the width of bands, as a function of soil pH. (From Buckman and Brady, 1969; reprinted with permission of Macmillan Publishing Company from "The Nature and Properties of Soils," 7th Ed. by H. O. Buckman and Nyle C. Brady. Copyright © 1969 by Macmillan Publishing Company.)

formation of H^+ in the soil solution as the Al^{+3} is precipitated as aluminum hydroxide, i.e.

$$Al^{+3} + H_2O \rightarrow Al(OH)^{+2} + H^+$$

$$Al(OH)^{+2} + H_2O \rightarrow Al(OH)_2^+ + H^+$$

$$Al(OH)_2^+ + H_2O \rightarrow Al(OH)_3 + H^+$$

These reactions account for the acid conditions in many highly weathered tropical soils. Plant roots also affect soil pH in several ways. For example, root respiration releases CO_2, forming carbonic acid in the soil solution (see Chapter 6).

Because more soil nutrients occur as positive ions than negative ions, one might expect that plant roots would develop a charge imbalance as a result of ion uptake. However, as nutrient ions such as K^+ are removed from the soil solution in excess of the uptake of negatively charged ions, the plant releases H^+ to maintain an internal balance of charge. This H^+ may, in turn, replace K^+ on a cation-exchange site, driving another K^+ into the soil solution. The high concentration of N in plant tissue causes the form in which N is taken up to dominate this process. Nye (1981) has shown how plants that use NH_4^+ as an N source tend to acidify the immediate zone around their roots. The uptake of NO_3^- has the opposite effect as a result of plant releases of HCO_3^- and organic acids to balance the negative charge (cf. Middleton and Smith, 1979; Hedley *et al.*, 1982). Microbial processes, such as nitrification, also affect soil pH, but these are considered in the next chapter. In terms of soil development and fertility, the continued production of H^+ tends to lower pH and base saturation through time, as nutrient cations are released to runoff waters or removed in harvested materials (Bockheim, 1980; Nilsson *et al.*, 1982).

Plant Uptake Processes

The diversity of soil chemical reactions means that plant roots extract nutrients from a solution that behaves very differently from the pure solution cultures of laboratory experiments. Nutrients are supplied to plant root surfaces by three mechanisms: (1) the growth of roots into the soil to intercept exchangeable nutrients; (2) the mass flow of ions with the movement of soil water as a result of transpiration; and (3) the diffusion of ions toward the root surface when uptake rate exceeds supply, resulting in a concentration gradient near the root surface (Barber, 1962). The relative mobility of nutrients in the soil solution and the rate of plant uptake determine which of these mechanisms predominates. Uptake of Ca is often the result of the interception of ions in newly exploited soil zones. Mass flow is important for Mg, SO_4^{-2} and Fe (Prenzel, 1979; Turner, 1982). Plant demand for N, P, and K often exceeds delivery by mass flow, such that diffusion is the dominant process that supplies these macronutrients (Nye, 1977). Compared to K^+, NH_4^+, and NO_3^-, the phosphate ion is particularly immobile in most soils; thus, the rate of diffusion strongly limits its supply to the plant root.

Although some nutrients enter the plant passively following the flow of water, the depletion of ions around roots indicates that nutrient uptake is also an active process. Plant physiologists have shown that uptake is controlled at root cell membranes and requires energy. Excess, nonessential, and toxic elements are often excluded from the plant or show only limited entry. For example, W. H. Smith (1976) found large quantities of Na^+ in the root exudates of several hardwood trees. Nutrient uptake appears to follow a pattern typical of enzyme-mediated reactions, suggesting that specific ion carriers are present in root cell membranes (Ingestad, 1982). These carriers recognize nutrient ions in the soil solution and transport them to the internal vascular system of the root. Active uptake increases with increasing concentrations of ions in the soil solution, but eventually this process reaches a maximum rate when all carriers are saturated. The function of carrier enzymes can be described by two parameters—V_{max}, the maximum velocity of uptake, and K_m, the soil solution concentration at which

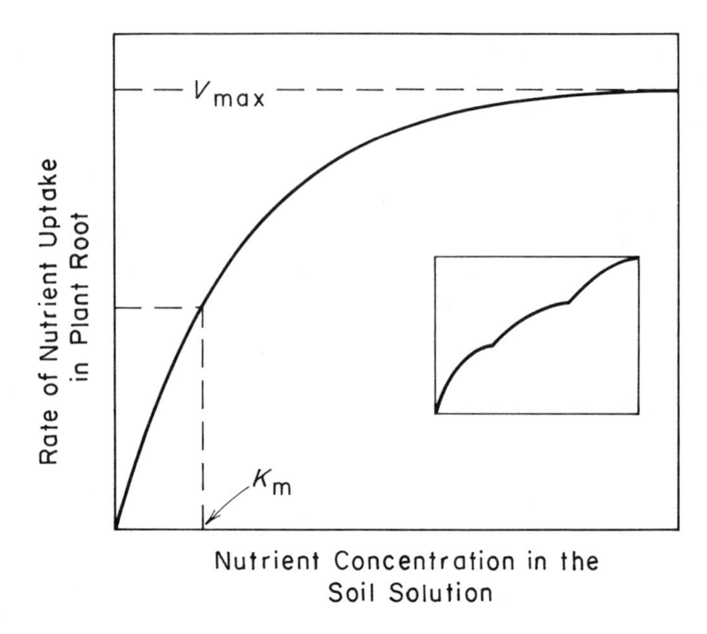

Fig. 7.3. Generalized form of the relationship between the rate of nutrient uptake and the concentration in the soil solution. Following the model for enzyme kinetics, the maximum rate of uptake (V_{max}) and the soil solution concentration which yields one-half the maximum rate of uptake (K_m) are shown. The inset shows the stepwise form of the relation, seen when the uptake of some ions is studied over a broader range of concentration. (From Luttge and Higinbotham, 1979.)

uptake is exactly one-half of the maximum level (Fig. 7.3). When examined over a broad range of solution concentrations, this simple uptake pattern often appears to result from several steplike curves of similar form (Fig. 7.3, insert), requiring more elaborate physiological models to explain uptake mechanisms at the cellular level (Luttge and Higinbotham, 1979). Further evidence of the active uptake of nutrient ions comes from the observation of competitive effects of ions in the soil solution. For example, the presence of large amounts of K^+ in the soil solution can reduce the uptake of NH_4^+ (Haynes and Goh, 1978). Such competitive interactions are seen among groups of ions with the same charge, presumably competing for the same carrier. Competitive effects are a good example of how nutrient balance is as important to plant nutrition as an adequate quantity of essential nutrients.

One can easily imagine that plants from infertile habitats might possess adaptations to enhance nutrient uptake by root enzymes. For example, species from infertile habitats might show high V_{max} when grown under fertile conditions. In fact, because diffusion limits the supply of most nutrients to the root surface, there has been little natural selection for enhanced enzymatic uptake among native species (Chapin, 1980). The most apparent belowground response of plants to low nutrient concentrations is an increase in the root/shoot ratio (Chapter 2), which increases the volume of soil exploitation and decreases diffusion distances.

The uptake of N deserves special attention because this element is so often limiting to forest production. Available forms of soil N are ammonium (NH_4^+), nitrate (NO_3^-), and dissolved organic N. Different forest species show widely differing preferences for uptake of these forms. Species occurring in sites where microbial nitrification (Chapter 6) is slow or inhibited often tend to show superior growth with ammonium (Haynes and Goh, 1978; Adams and Attiwill, 1982).

Inside the plant, both forms of inorganic N are converted to amino groups ($-NH_2$) that are attached to soluble organic compounds. In many woody species these conversions occur in the roots and N is transported as amides, amino acids, and ureide compounds through the xylem to leaf tissues (Salisbury and Ross, 1978). In some species, the reduction of NO_3^- to $-NH_2$ occurs in leaf tissues, and N is found as NO_3^- in the xylem stream. Eventually, most plant N is incorporated into protein. The conversion of NO_3^- to $-NH_2$ is a biochemical reduction reaction that requires metabolic energy and is catalyzed by the enzyme nitrate reductase, containing Mo. One might puzzle why most plants do not show a clear preference for NH_4^+, which is assimilated more easily. Several explanations have been offered. Remembering that NH_4^+ interacts with soil cation-exchange sites, whereas NO_3^- is highly mobile in temperate forest soils, the rate of delivery of NO_3^- to the root by diffusion or mass flow is much higher than that of NH_4^+ under otherwise equivalent conditions. Moreover, the uptake of NO_3^- avoids the competition that occurs between NH_4^+ and other positively charged nutrient ions. Finally, relatively low concentrations of NH_4^+ are potentially toxic to plant tissues. The advantages and disadvantages of N uptake as NH_4^+ and NO_3^- help explain the natural selection for different preferences in species from field sites with different availabilities of N in each of these forms.

Nutrient Uptake through Symbiotic Associations

In Chapter 6 we discussed how those species that form symbiotic associations with bacteria in root nodules are capable of fixing atmospheric N_2 into forms useful to plant growth. These species can result in large additions of available N to forest ecosystems. Symbiotic associations with fungi are also widespread and of utmost importance to the mineral nutrition of forest trees.

Mycorrhizal fungi include a wide variety of species that form symbiotic relations with the roots of higher plants (Harley and Smith, 1983). There are several forms of symbiosis. In temperate regions, many trees are infected by ectotrophic mycorrhizae. These form a hyphal sheath surrounding the active fine roots of forest trees and extend additional hyphae into the surrounding soil. Many familiar forest mushrooms are the fruiting bodies of Basidiomycete fungi that are ectomycorrhizal on tree roots. Most trees are infected by endotrophic mycorrhizae in which fungal hyphae actually penetrate the cells of the root cortex, but do not form a sheath around the root. By virtue of the large surface area and efficient absorption capacity of hyphae, the fungi are able to obtain soil nutrients and transfer these to the higher plant root. The fungus depends on the host for supplies of carbohydrate; thus, both members of the association benefit.

Mycorrhizal fungi are most important in the transfer of those soil nutrients with low diffusion rates in the soil. A vast scientific literature demonstrates the importance of mycorrhizae in P transfer, but mycorrhizal absorption of N and other nutrients is also known. Nutrient balance affects many experimental results, for with increased P uptake, the uptake of other nutrients may increase proportionally (Bowen and Smith, 1981). Some plants with mycorrhizal fungi show higher levels of various nutrients in foliage (Table 7.2), but frequently the enhanced uptake of nutrients results in higher rates of growth (Rose and Youngberg, 1981; Schultz and Kormanik, 1982).

The importance of mycorrhizae in infertile sites is well known. Many species of pine require ectotrophic mycorrhizae, which perhaps accounts for their success in nutrient-poor soils. Forest revegetation of mining spoils is often not effective without special efforts to establish the natural mycorrhizal associations (Reeves *et al.*, 1979). Most tropical trees appear to require endotrophic mycorrhizal associations for proper growth (Janos, 1980). Mycorrhizal fungi are also widespread among *Eucalyptus* species growing in the low-phosphorus soils of Australia.

Recent laboratory experiments with agricultural species have carefully documented the metabolic costs associated with both mycorrhizal and rhizobial (N-fixing) associations (Paul and Kucey, 1981). We know very little about the metabolic cost of mycorrhizae to tree species. Most field studies of roots in

TABLE 7.2 Some Characteristics of White Pine (*Pinus strobus*) Seedlings Grown for 1 yr with and without Mycorrhizal Infection[a]

	Seedling		Leaf (% oven dry weight)		
Treatment	Dry weight (g)	Root/shoot	N	P	K
Mycorrhizal	405	0.78	1.24	0.20	0.74
Nonmycorrhizal	321	1.14	0.85	0.07	0.43

[a] Data from Hatch (1937).

forests have not carefully separated mycorrhizal fungi from the fine root collections. Vogt *et al.* (1982) found that mycorrhizal biomass was only 1% of the ecosystem total in Pacific silver fir (*Abies amabilis*) stands in Washington. However, the annual growth of mycorrhizae accounted for about 15% of the annual net primary production. Annual turnover of fungi and fine roots dominated the nutrient return to decomposing materials in the soil of this ecosystem.

If 15% of primary production typically appears as mycorrhizal biomass and an additional, unknown amount is used in mycorrhizal respiration, then these symbiotic associations are a significant metabolic cost to forest trees. This would underscore the apparent benefits of mycorrhizae in supplying nutrients. The metabolic cost of mycorrhizae may not be expressed in reduced stem wood growth, because mycorrhizae may allow decreased allocation to the growth of absorbing roots in infected species (Table 7.2). However, Blaise and Garbaye (1983) found lower mycorrhizal infection and greater growth in fertilized beech (*Fagus silvatica*) forests.

In conditions of nutrient deficiency, plant growth usually slows whereas photosynthesis continues at relatively high rates (Chapin, 1980). In such cases, the carbohydrate content of plant tissues increases. Marx *et al.* (1977) found that high concentrations of carbohydrate in the root tissues of loblolly pine (*Pinus taeda*) stimulated mycorrhizal infections. Thus, internal plant allocation of carbohydrates to roots might result in increased nutrient uptake by mycorrhizae and an alleviation of nutrient deficiencies. Comparative observations of the occurrence of mycorrhizae along a natural or artificial fertility gradient in forest ecosystems would be a welcome addition to our understanding.

Nutrient Allocation in Trees

The nutrient requirements of forest trees are highly variable from species to species. There is also a broad range in the nutrient concentrations that may accumulate in the biomass of a species in response to variations in soil nutrient availability among sites. When nutrients are added to deficient soils, the relative growth rate of trees may increase, so that no increase in foliar nutrient concentration is observed. However, when factors such as light are limiting, nutrients may be taken up in excess of immediate metabolic requirements. This storage results in high concentrations in foliage—a condition that is called luxury consumption (Fig. 7.4). Differing leaf nutrient concentrations form the basis for the use of foliar analysis to recognize nutrient deficiencies in forest ecosystems; however, variations due to species, season, canopy position, relative growth rate, and nutrient balance must be known before the technique is of great use in forest management (van den Driessche, 1974; Linder and Rook, 1984).

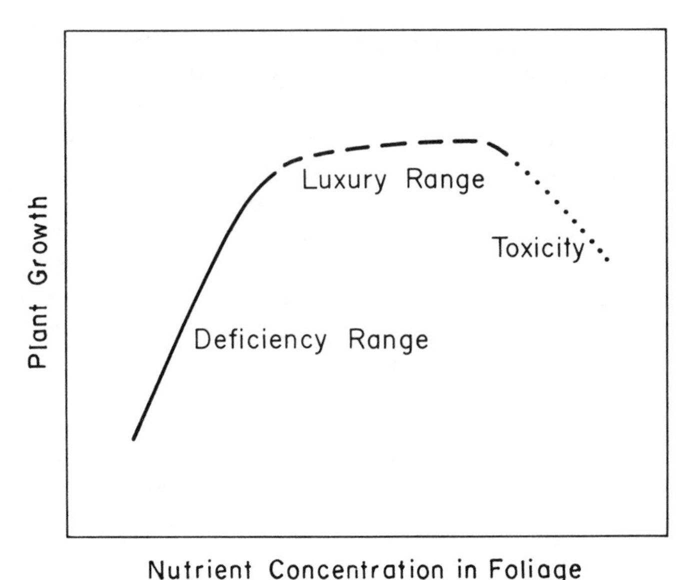

Fig. 7.4. Plant growth in relation to the nutrient content in foliage. (Reprinted by permission from *Bot. Rev.*, **40,** 347–394, Copyright 1974, van den Driessche and The New York Botanical Garden.)

One view of the long-term nutrient requirements and the balance of nutrient elements needed for forest growth may be determined by examinng the total content of nutrients in biomass. For seedlings of several tree species, Ingestad (1979b) has shown that the appropriate ratio of nutrient elements might be supplied by a solution that contains 100 parts N, 15 parts P, 50 parts K, 5 parts Ca and Mg, and 10 parts S. As the allocation of biomass changes during tree maturation so do the ratios among the nutrients accumulated in biomass, as a result of differing concentrations in leaves and wood. If we examine the ratio of nutrients in the total aboveground biomass of mature forests (Chapter 6, Table 6.5), we see that for every 100 parts N, more than 100 parts Ca have been accumulated. This increase in the relative accumulation of Ca is due to its presence in the cell walls of woody tissues and its high concentration in bark (e.g., Stone and Boonkird, 1963; Johnson and Risser, 1974).

Long-lived tissues contain most of the nutrient accumulation in forest biomass (e.g., Chapter 6, Table 6.6), but we must remember that this storage has accumulated over many years. For 20-yr-old loblolly pine (*Pinus taeda*) plantations, Switzer and Nelson (1972) found only 10–14% of the annual uptake of N, P, and K was allocated to branches and stems. Thus, total plant contents may reveal long-term nutrient requirements, but they tell us little about the annual nutrient circulation. Leaves and fine roots, on the other hand, contain only a small portion of the nutrient content in biomass, but the growth, death, and replacement of these tissues largely determine the internal nutrient cycle of a forest ecosystem.

INTRASYSTEM CYCLING

Seasonal Changes in Foliage Nutrient Content

When leaf buds break and new foliage begins to grow, the leaf tissues often have high concentrations of N, P, and K. As the foliage matures, these concentrations often decrease, while concentrations of Ca, Mg, and Fe usually increase (Chapin, 1980; van den Driessche, 1974; Fife and Nambiar, 1982; Chapin and Kedrowski, 1983). Some of these changes are due to increasing accumulation of photosynthetic products and to leaf thickening during development. Leaf weight per unit area (in milligrams per square centimeter) may increase as much as 50% during the growing season and then decline as the leaf senesces (Smith *et al.*, 1981). The initial concentrations of N and P are diluted as the leaf tissues accumulate carbohydrates and cellulose. Increases in calcium concentration with leaf age result from secondary thickening, including calcium pectate deposition in cell walls, and from increasing storage of calcium oxalate in cell vacuoles.

Once leaves are fully expanded, seasonal changes in the nutrient content per unit of leaf area indicate the actual pattern of nutrient movements between the foliage and the stem. This expression is not affected by changes in the carbohydrate content or thickness of the leaf. In scarlet oak (*Quercus coccinea*), a temperate deciduous species, leaf N rapidly accumulates during the early summer, presumably as a component of photosynthetic enzymes (Fig. 7.5). The leaf content of N, P, and K is relatively constant at high levels during the growing season, but strongly removed from leaves in autumn. Such losses often represent active withdrawal of nutrients from foliage for reuse during the next year. Similar patterns are seen in Douglas fir (*Pseudotsuga menziesii*), in which the nutrient content of developing foliage appears to be derived, in part, by withdrawals from 1-yr-old foliage still on the branches (Krueger, 1967; Waring and Youngberg, 1972, cf. Fife and Nambiar, 1982).

Leaf nutrient contents are affected by rainfall that leaches nutrients from the leaf surface (Tukey, 1970; Parker, 1983). In particular, seasonal changes in the content of K, which is highly soluble and especially concentrated in cells near the

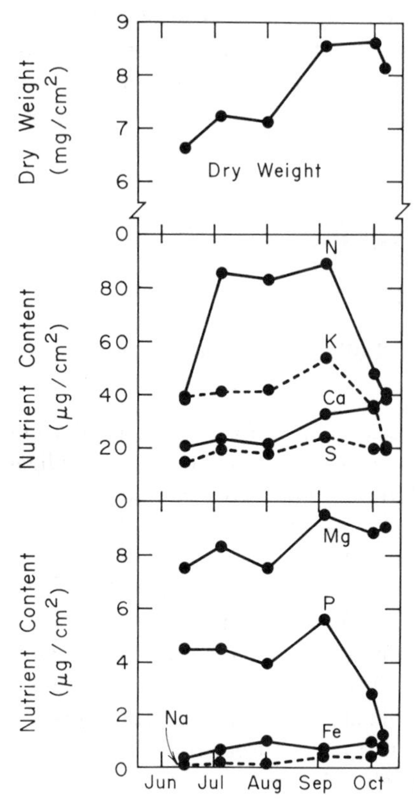

Fig. 7.5. Seasonal changes in the dry weight and nutrient content of the leaves of scarlet oak (*Quercus coccinea*) in the Brookhaven Forest, Long Island, New York. (From Woodwell, 1974.)

leaf surface, may represent losses by leaching. The losses of nutrients in leaching often follow the order:

$$K > P > N > Ca.$$

Leaching rates often increase as foliage senesces before abscission; thus, care must be taken to recognize changes due to leaching versus changes due to active withdrawals (Ostman and Weaver, 1982).

Nutrient losses by leaching differ among leaf types. Luxmoore *et al.* (1981) calculated lower rates of leaching loss from loblolly (*Pinus taeda*) and short-leaf (*P. echinata*) pines than from deciduous hardwood species in a forest in Tennessee. Such differences may be due to differences in leaf nutrient concentration, surface-area-to-volume ratio, surface texture, and leaf age. Among the trees of the humid tropics, the smooth surface of broad sclerophylls may be an adaptive response to reducing leaching by minimizing the length of time that rainwater is in contact with the leaf surface. Species-specific differences in rates of leaching may explain differences in epiphyte loads of forest species (Schlesinger and Marks, 1977).

Aboveground Nutrient Return from Vegetation

Nutrients are lost when foliage and other plant parts fall from the plant. In studies of nutrient cycling, litterfall from the aboveground vegetation is usually collected in baskets or trays, which are emptied at periodic intervals. Nutrient return in litterfall varies seasonally depending on the forest composition and abscission processes. In a typical temperate deciduous forest, Gosz *et al.* (1972) found that premature abscission of leaves by summer storms resulted in a small amount of litterfall with relatively high nutrient concentration, since nutrient reabsorptions had not occurred. In this forest, one-half of the annual nutrient return occurred as a result of the abscission of leaf tissues in autumn. Seasonal peaks in litterfall are also seen in temperate coniferous forests, whereas nutrient return by litterfall in humid tropical forests is more uniform year-round.

Rainwater that passes through the forest canopy is called throughfall, which is usually collected in funnels or troughs near the forest floor. Throughfall contains nutrients leached from leaf surfaces and is important in the cycling of nutrients such as K. Similarly, rainwater that travels down the surface of stems is called stemflow. It contains the nutrients leached from bark. The concentrations of nutrients in stemflow waters are much greater than the concentrations in throughfall, but much more water reaches the forest floor as throughfall (Chapter 5). The annual nutrient return in throughfall typically accounts for 90% of the nutrient movement by leaching. Stemflow is significant, however, to the extent that it returns highly concentrated nutrient solutions to the soil at the base of trees (Gersper and Holowaychuk, 1971). Leaching varies seasonally depending on forest type and climate. Not surprisingly, in temperate deciduous forests, the greatest losses are during the summer months. In some cases, the canopy appears to accumulate nutrients from rainfall, particularly soluble forms of N (e.g., Carlisle *et al.*, 1966b; Miller *et al.*, 1976). Of course, field measurements of leaching are also complicated by the presence of aerosols that may be rinsed off the surface of foliage. Much of the nutrient deposition in throughfall may be derived from dry deposition on foliage (Parker, 1983).

The annual transfers of nutrients from aboveground vegetation to the soil for a 20-yr-old loblolly pine ecosystem are shown as part of Table 7.3. Leaching

TABLE 7.3 Circulation of Nutrients in the Aboveground Portion of a 20-Yr-Old Loblolly Pine (*Pinus taeda*) Plantation in Mississippi[a]

Process	N	P	K	Ca	Mg
Uptake (midsummer accumulation in foliage)	59.5	5.0	27.4	7.4	4.6
Leaching	3.3	0.5	16.0	3.0	1.0
Reabsorption from senescing foliage	26.9	3.3	6.9	0.0	1.5
Litterfall	27.9	1.3	5.2	6.9	2.4

[a] All values are kilograms per hectare per year. From Switzer and Nelson, *Soil Science Society of America Proceedings,* Volume 36, 1972, pp. 143–147.

losses are greatest for K and lower for N and P. Except for K, litterfall exceeds leaching as a pathway of nutrient return to the soil. When one considers that leaf material comprises 70% of the litterfall in most forests (Meentemeyer *et al.*, 1982), the nutrient content in leaves and the seasonal changes before leaf abscission are of utmost importance in determining the annual return of nutrients to the forest floor. In the loblolly pine ecosystem, 45% of the N and 66% of the P in the leaf biomass may be derived by reabsorption from senescing foliage (Table 7.3), which greatly reduces the annual losses in litterfall. Nutrients such as Ca, which are rather immobile in the plant cell, show little reabsorption.

Belowground Nutrient Return from Vegetation

In Chapter 3 we noted that the annual production of fine root materials can exceed the production of foliage in many forests, particularly in nutrient-poor sites. The growth and death of fine roots are difficult to study, and often require sequential measurements from soil cores throughout the year. Unfortunately, few of these studies have also analyzed the fine root material for nutrient content, so we know very little about the role of fine roots in the annual nutrient circulation of forest ecosystems (McClaugherty *et al.*, 1982). For a yellow poplar (*Liriodendron tulipifera*) forest in Tennessee, the ratio of belowground litter production (fine root death) to aboveground litterfall was 2.3 (Table 7.4). Similar ratios for N and K turnover were 1.9 and 6.9, respectively. The high ratio for K was due to large amounts of K in root exudates. This forest was similar to the stands of Pacific silver fir, studied by Vogt *et al.* (1982), in that the turnover of fine root tissues, including associated mycorrhizae, comprised the largest annual return of nutrients to the soil.

TABLE 7.4 Nutrient Return from the Above- and Belowground Vegetation Components in a Yellow Poplar (*Liriodendron tulipifera*) Forest in Tennessee[a]

Process	Return (kg/ha/yr)		
	Biomass	N	K
Aboveground processes			
Litterfall	3310	42.2	10.0
Leaching	—	2.3	29.4
Total	3310	44.5	39.4
Belowground processes			
Root death	6750	76	128
Root consumption	750	9	14
Root exudation and leaching	—	—	128
Total	7500	85.0	270
Ratio of below : aboveground	2.3	1.9	6.9

[a] From Cox *et al.* (1978).

Mass Balance of the Forest Nutrient Cycle

The annual circulation of nutrients in forest vegetation, the intrasystem cycle, can be modeled using the mass-balance approach (see Chapter 6). For example, nutrient uptake can not be measured directly, but uptake must equal the annual storage in wood plus the replacement of losses in litterfall and leaching. Uptake is less than the annual requirement by the amount reabsorbed from leaf tissues before abscission. Thus the annual requirement is equal to the peak nutrient content in newly produced tissues during the growing season, plus the leaching losses through the year.[1] The requirement is the nutrient flux needed to complete a mass balance in the study of a forest ecosystem. It should not be taken as indicative of biological requirements, and in fact, it can be calculated for nonessential elements such as Na.

As an example, the mass-balance approach has been used to analyze the internal storage and the annual transfers of nutrients in an 80-yr-old deciduous forest in Great Britain (Table 7.5). These data serve to summarize many aspects of our discussion. Note that 87% of the annual requirement of N is allocated to foliage and root increment, whereas little is allocated to the bole wood increment. However, total nutrient storage in short-lived tissues is small compared to

[1]This definition is slightly different from that used by some workers. We consider leaching loss as an inherent process in forests; thus, annual replacement of these losses is a *requirement* for forest trees.

TABLE 7.5 Storage and Annual Circulation of Macronutrients in a Mixed Deciduous Forest, Meathop Woods, Great Britain[a]

	N	P	K	Ca	Mg
Storage (kg/ha)					
Foliage	85.7	4.6	38.9	41.9	9.5
Wood and branches	192.4	12.1	174.2	437.4	16.5
Roots	223.0	11.9	112.8	235.3	30.8
Understory	8.1	5.4	42.9	66.9	9.7
Total	509.2	34.0	368.8	782.5	66.5
Annual requirement (kg/ha/yr)					
Foliage	85.7	4.6	38.9	41.9	9.5
Wood increment in					
Branches	2.1	0.1	1.5	3.9	0.3
Boles	4.1	0.2	3.4	8.9	0.7
Roots	13.2	0.6	6.6	14.3	1.7
Total increment	19.4	0.9	11.5	27.1	2.7
Leaching					
Throughfall	8.8	0.7	30.3	22.8	11.9
Stemflow	0.4	0.03	4.8	5.2	3.4
Total leaching	9.2	0.73	35.1	28.0	15.3
Total requirement	114.3	6.23	85.5	97.0	27.5
Annual uptake (kg/ha/yr)					
Woody increment	19.4	0.9	11.5	27.1	2.7
Returns					
Leaching	9.2	0.73	35.1	28.0	15.3
Litterfall	63.5	2.6	19.0	83.3	9.7
Total return	72.7	3.33	54.1	111.3	25.0
Total uptake	92.1	4.23	65.6	138.4	27.7
Reabsorption (kg/ha/yr)	22.2	2.00	19.9	0	0
Comparison of annual turnovers and flux (%)					
Bole wood increment/bole wood pool	3.1	2.4	3.0	3.0	8.2
Foliage and root growth/total requirement	87	83	53	58	41
Litterfall/total return	87	79	35	75	39
Uptake/aboveground pool	18.1	12.4	17.8	17.7	41.7
Return/uptake	79	79	82	80	90
Reabsorption/requirement	19	32	23	0	0

[a] From Cole and Rapp (1981, p. 404).

storage in roots and wood that have accumulated nutrients for many years. For most nutrients in this forest, the storage in wood increases by about 3% each year. Presumably, this value has declined through time as the pool in wood has increased. Leaching losses are dominated by throughfall, but the annual return in leaching is relatively small except for K and Mg. Despite substantial reabsorption of N, P, and K before leaf abscission, litterfall is the dominant (79–87%) aboveground pathway for the return of N and P to the soil. It appears that Ca is actively exported to leaves before abscission (see also Sollins *et al.*, 1980). In this forest, annual uptake is 12–20% of the total storage in vegetation, but 80–90% of the uptake is returned each year. As in most studies, some of these calculations would be revised if belowground transfers were better understood.

Nutrient cycling changes during succession and as the allocation of net primary production changes with forest growth. Percentage turnover in vegetation declines as the mass and nutrient storage in vegetation increase. However, since the leaf area of a forest usually reaches a steady state early in development, the nutrient movements dependent on leaf area (i.e., litterfall and leaching) are quickly reestablished in forest growth after clearcutting (Marks and Bormann, 1972). In many cases the early successional species have particularly rapid growth and nutrient-rich tissues. Their high nutrient requirements tend to store available nutrients that might otherwise be lost after forest cutting or fires (Boring *et al.*, 1981; Pastor and Bockheim, 1984).

In general, understory vegetation in mature forests has a greater contribution to annual nutrient cycling than its biomass would suggest (e.g., Tappeiner and Alm, 1975; Yarie, 1980). These plants have nutrient-rich tissues and a high ratio of foliage to wood. Thomas (1969) reported that flowering dogwood (*Cornus florida*), an understory tree, contained 0.2% of aboveground biomass but 1.8% of the calcium storage in a southeastern pine forest. In most cases the importance of the understory declines when the forest canopy is fully developed, but there are exceptions. Kazimirov and Morozova (1973) found an increasing level of N uptake in the understory species of Russian spruce (*Picea abies*) forests, in which canopy openings developed at about 100 yr of age.

NUTRIENT-USE EFFICIENCY

We might expect species to differ in the rate of net primary production per unit of nutrient accumulated from the soil. We define nutrient-use efficiency in forests as the net primary production per unit of nutrient lost from vegetation (cf. Vitousek, 1982), recognizing that the annual losses must be replaced by uptake.

Over a broad range of species, photosynthesis is positively correlated to leaf N concentration (Chapter 2, Mooney *et al.*, 1978; Mooney and Gulmon, 1982; Brix, 1981); thus, there appear to be limits to nutrient-use efficiency as a result of modifications of the photosynthetic process at the biochemical level. However, nutrient-use efficiency can be increased substantially by adaptations to minimize annual losses and to increase the internal reuse of nutrients. Lower leaf turnover and higher reabsorption of nutrients before leaf abscission are two mechanisms that increase internal conservation of nutrients in trees (Small, 1972). We might ask the general question: Do these processes lead to differences in nutrient-use efficiency among forests?

Differences in nutrient-use efficiency among forests might be due to species differences among sites—forests on poor sites being dominated by species that use nutrients efficiently. Differences in nutrient-use efficiency might also appear within a species as a result of responses to differing nutrient availabilities. Both of these possibilities appear to operate in forest ecosystems. In temperate regions, the annual circulation of nutrients in coniferous stands is much lower than the circulation in deciduous stands, largely as a result of lower leaf turnover in coniferous forest species (Cole and Rapp, 1981). Leaching losses are also lower in coniferous forests (Parker, 1983). In our examples, nutrient reabsorption from loblolly pine needles (Table 7.3) appears to be more effective than reabsorption from the deciduous forest leaves (Table 7.5), especially when the amounts are expressed as a percentage of the annual requirements. These mechanisms result in greater nutrient-use efficiency in coniferous forests compared to deciduous forests of the world (Table 7.6). Higher nutrient-use efficiency in coniferous species may explain the frequent occurrence of coniferous vegetation on nutrient-poor sites and in boreal climates with slow nutrient turnover in the soil. These findings may also extend to the occurrence of broad-leafed evergreen vegetation in warm temperate climates (Monk, 1966; Beadle, 1966; Gray, 1983; Goldberg, 1982).

In general, nutrient-use efficiency for N is lower in forests with large amounts

TABLE 7.6 Net Primary Production (kg/ha/yr) per Unit of Nutrient Uptake Used as an Index of Nutrient-Use Efficiency to Compare Deciduous and Coniferous Forests[a]

	Production per unit nutrient uptake				
Forest type	N	P	K	Ca	Mg
Deciduous	143	1859	216	130	915
Coniferous	194	1519	354	217	1559

[a] From Cole and Rapp (1981).

of N circulation in litterfall and relatively high soil N availability (Vitousek *et al.*, 1982; Vitousek, 1982; Pastor *et al.*, 1984). Net primary productivity, however, is positively correlated to annual N circulation in both coniferous and deciduous forests (Cole and Rapp, 1981). Thus, a comparison of forest types suggests that abundant intrasystem cycling is associated with relatively high productivity but lower nutrient-use efficiency.

In tropical forests, the high annual rainfall and year-round growing season lead to high rates of annual production and nutrient circulation by leaching and litterfall. Thus, nutrient-use efficiency for many nutrients is relatively low in these ecosystems, despite the widespread occurrence of the evergreen leaf type. An exception is P; Vitousek (1984) found especially high nutrient-use efficiency of P in tropical forests, particularly in areas where soil P is very low. Through mycorrhizal associations and internal conservation of P, it appears that tropical trees are adapted to P-deficient soils, which are widespread in these regions.

Comparing several temperate forests in Poland, Zimka and Stachurski (1976) found that species with high rates of reabsorption of foliar nutrients tended to dominate nutrient-poor sites, which resulted in an efficient intrasystem cycle of nutrients in these ecosystems (cf. Flanagan and Van Cleve, 1983). Differences in species composition between nutrient-rich and nutrient-poor sites make it difficult to perceive the response of individual tree species to differences in nutrient availability. Among species occurring at several sites, they observed greater nutrient reabsorption before leaf fall on the poor sites (Stachurski and Zimka, 1975). Comparing beech (*Fagus sylvatica*) forests in southern Sweden, however, Staaf (1982) found that the percentage nutrient reabsorption was related to the initial content in foliage. Thus, reabsorption was greater on higher fertility sites. Other workers have also found no tendency for trees with low N and P status to reabsorb a larger proportion of leaf N and P before abscission (Lea and Ballard, 1982a; Chapin and Kedrowski, 1983).

Changes in nutrient-use efficiency have been examined in laboratory experiments that vary nutrient availability. Ingestad (1979a) found that the growth of birch (*Betula verrucosa*) seedlings increased in response to additions of N to nutrient-deficient laboratory cultures, but N-use efficiency (dry matter production per unit N) declined sharply over the same range (Fig. 7.6). At the highest levels of N, there were damaging effects and growth reductions. Similarly, fertilization often changes the nutrient-use efficiency within a forest. Miller *et al.* (1976) found greater productivity and litterfall in N-fertilized stands of Corsican pine (*Pinus nigra*). More importantly, fertilization caused a decline in the reabsorption of nutrients from senescing needles, such that the N concentration in litterfall increased (cf. Flanagan and Van Cleve, 1983). The overall result was a decline in the N-use efficiency by the vegetation. Fertilization also increases N losses in throughfall (cf., Mahendrappa and Ogden, 1973; Khanna and Ulrich, 1981; Yawney *et al.*, 1978).

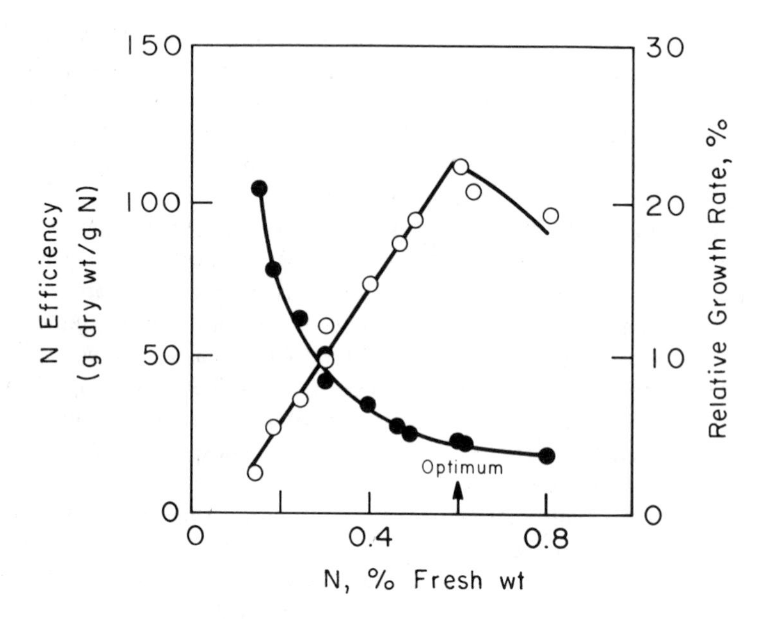

Fig. 7.6. Relationship of N-use efficiency and relative growth rate to N status of Birch (*Betula verrucosa*) seedlings. Nitrogen status is measured as N percentage of seedling fresh weight. The range in N status results from differing experimental N regimes. (Modified from Ingestad, 1979a.)

Fertilizing field plots with carbohydrates (e.g., sugar or sawdust) acts to lower soil N availability, since soil N is retained by soil microbes during the decomposition of the added materials (see Chapter 8). When fertilization experiments include sugar treatments, N concentrations often decline in fresh foliage. In stands of Douglas fir, Turner and Olson (1976) found reabsorption of N was greater on sugar-treated plots, implying greater nutrient-use efficiency under experimental conditions of lower soil nutrient availability.

Analysis of foliage can indicate conditions in which nutrients are limiting to forest growth (Fig. 7.4; van den Driessche, 1974). For example, Tilton (1978) showed strong correlations between foliar N and growth of larch (*Larix laricina*) in wetland forests of Minnesota. Upon fertilization with a specific nutrient, however, the concentrations of various leaf nutrients can show unpredictable patterns of change. Leaf N increased when Miller *et al.* (1976) fertilized the Corsican pine stands with N, but in the same samples, concentrations of P, Ca, and Mg declined. Apparently, an N fertilization of N-deficient stands stimulates photosynthesis such that the concentrations of other nutrients in foliage are diluted by accumulations of carbohydrates (Turner and Olson, 1976; Fowells and Krauss, 1959; Lea *et al.*, 1980; Binkley, 1983). In these cases, uptake of P from the soil may fall behind the rates needed for growth at the newly established level of N availability. Thus, the interpretation of foliar analysis after fertilization is often not straightforward (Timmer and Stone, 1978). We may view such single element fertilizations as disturbing the normal balance in the uptake of nutrients and their circulation in the forest ecosystem.

Natural or artificial differences in nutrient availability alter foliar concentrations and rates of nutrient reabsorption and loss in litter. The differences in the nutrient concentrations in litterfall strongly affect the rates of decomposition, as we shall see in Chapter 8. When nutrient concentrations in litter are low, as might be expected after reabsorption of nutrients, decomposition is slower. Thus, intrasystem cycling contains a positive feedback system to the extent that an increase in nutrient-use efficiency by trees may reduce the availability of soil nutrients for plant uptake.

SUMMARY

In this chapter we have considered nutrient uptake from the soil solution, nutrient allocation patterns in forest trees, and nutrient return to the soil. On an annual basis, all of these processes involve rather large movements of nutrients compared to the amounts received or lost from most forests as intersystem transfers. In an old-growth stand of Douglas fir, Sollins *et al.* (1980) found that the circulation of N and P within the forest was 6.5 and 18.2 times greater than the annual receipts from the atmosphere and rock weathering. Similarly, an analysis of rainwater at various points during its passage through a hemlock–cedar forest in British Columbia (Table 7.7) shows large nutrient transfers through the biotic portion of the ecosystem, but only small losses to streamwaters.

During growth, forests show selective uptake and retention of essential nutrients from the unrelenting biogeochemical flux that carries materials from land to the oceans. The nutrient retention is remarkably effective considering the magnitude of the annual circulation within the system and the points during the intrasystem cycle at which nutrients are exposed to potential loss. Retention of fertilizer inputs of cations emphasizes this point. In K-deficient stands of red pine (*Pinus resinosa*) in central New York State, over 70% of the K applied in fertilizer could still be found after 23 yr (Stone and Kszystyniak, 1977). Some K was retained on the soil-exchange sites and the remainder was found in the vegetation, which increased in biomass upon fertilization. Despite its rapid circulation and mobility in forest ecosystems, little K appears to have been lost to streamwaters. Melin *et al.* (1983) found 79% of N applied in fertilizer remained in a Scots pine forest after 1 yr, but retention of N is often lower than other

TABLE 7.7 Comparison of Intersystem and Intrasystem Movements of Nutrients through a Western Hemlock (*Tsuga heterophylla*)–Western Redcedar (*Thuja plicata*) Forest, as Indicated by Nutrient Concentrations in Water (mg/liter) Collected at Various points in the Ecosystem[a]

Collection	N[b]	P[c]	K	Ca	Mg
Precipitation	0.22	0.003	0.06	0.20	0.06
Throughfall	0.26	0.042	1.11	1.00	0.30
Lysimeters					
Forest floor	0.08	0.033	1.65	2.90	1.82
Mineral soil	0.11	0.010	0.35	1.80	0.65
Streamwater	0.08	0.003	0.09	1.50	0.31

[a] Calculated from "Nutrient movement through western hemlock–western redcedar ecosystems in southwestern British Columbia" by M. C. Feller, *Ecology*, 1977, **58**, 1269–1283. Copyright © 1977 by the Ecological Society of America. Reprinted by permission.

[b] N available as NH_4 and NO_3.

[c] P available as H_2PO_4.

nutrients as a result of microbial reactions (denitrification) that cause losses to the atmosphere (Chapter 6).

In a growing forest, the annual retention of nutrients in perennial tissues should approximate the annual input of limiting nutrients to the ecosystem, i.e., losses from the system will be small. As a percentage of inputs, the long-term retention of essential nutrients is, of course, lower when forest biomass is not accumulating or when the nutrients are not in short supply. As we saw in Chapter 6, nonessential elements are freely lost in streamwaters. Disruption of the intrasystem cycle, as in clearing forest land for agriculture, leads to greater nutrient losses and lower retention of applied nutrients. This conversion usually removes the potential for long-term nutrient storage in biomass and for nutrient-use efficiency by internal conservation of nutrients in vegetation.

Atmospheric inputs supply only a small percentage of the annual requirement of nutrients such as N. Cole and Rapp (1981) found that internal recycling of N can supply as much as one-third of the requirement in many deciduous forests. The remainder is derived from the decomposition of plant parts, which releases nutrients to the soil solution for plant uptake. Decomposition is affected by litter quality and climatic factors, but the rate of decomposition by soil microbes is a critical factor that determines the productivity and intrasystem nutrient circulation of a forest. The biological processes that occur in the forest floor, including litter decay, are the subject of the next chapter.

8

Decomposition and Forest Soil Development

INTRODUCTION

The general term *decomposition* is used to describe a large number of interrelated processes by which organic matter is broken down to smaller particles and to soluble forms of nutrients that are available for plant uptake. In forest ecosystems, various soil animals are important in the decomposition process, but the vast majority of the chemical transformations are performed by bacteria and fungi in the upper soil layers. In the course of decomposition, resistant organic substances are produced by microbes and accumulate in the soil as humus. Thus, many ecologists prefer to use the term *mineralization* for those decomposition events that actually release inorganic compounds (e.g., CO_2, H_2O, NH_4^+, Ca^{+2}) from organic matter. Remember that this usage does not imply the formation of secondary minerals as in rock weathering.

In Chapter 7 we saw that large quantities of nutrients circulate within a forest ecosystem. Part of the annual nutrient requirements for forest growth can be met by nutrient reabsorption before the loss of tissues. The remaining requirements must be supplied by uptake from the soil. The majority of the nutrient pool that is available for uptake is derived from the decomposition of dead organic matter. Thus, decomposition is a critical link that makes nutrients available to circulate in the intrasystem cycle of a forest ecosystem. When decomposition is slow, as in boreal forests, soil organic matter accumulates and forest productivity may be very low.

Soil organic matter consists of fresh litter, partially decomposed material, and humus. The total dead organic matter in an ecosystem is sometimes called *detritus* (e.g., Schlesinger, 1977), though some ecologists prefer to use this term only for the layer of litter on the soil surface. Since most fresh litter from aboveground vegetation and dead roots enters near the surface of the soil, the upper layers are often easily recognized and separated as the forest floor.

The distribution and transformations of organic matter through the lower soil profile are important processes in the development of forest soils. Humic substances produced during decomposition tend to accumulate in the mineral soil. While humus typically comprises only 2% of the dry weight of upland forest soils, this content has a major role in determining the soil's physical structure, aeration, and moisture-holding capacity. Humic substances also provide cation exchange sites (Chapter 7) for the soil solution.

In this chapter we review the major processes of decomposition and the factors that determine them. We also draw on material from previous chapters to provide a synthesis of soil profile development in forest ecosystems.

SOURCES OF DETRITUS

The annual loss of leaves, twigs, flowers, fruits, and bark fragments are the obvious forms of litterfall in forest ecosystems. Leaf litter typically comprises 70% of the total litter from aboveground (O'Neill and DeAngelis, 1981; Meentemeyer *et al.*, 1982). The composition and quantity of litterfall are often variable between years, which requires long-term studies to obtain good data. Some recent studies have carefully documented subtle forms of organic detritus in forests, including substances in aphid exudates and dissolved organic compounds in throughfall (Gosz *et al.*, 1976). Such materials are minor components of litterfall, except during insect outbreaks when frass may account for 7.8% of the aboveground deposition (Carlisle *et al.*, 1966a,b). Windstorms are important in treefall; deposition of this component of litterfall is highly variable in time and in spatial distribution. An accurate estimate of the nutrient return in large wood requires many years of study (Grier, 1978; Sollins, 1982).

In most cases, the mass of detritus in standing dead trees is not large. Reiners (1972) reported 4,000–11,000 kg/ha in three deciduous forests in Minnesota (cf. Chapter 6, Table 6.5). Wood litterfall increases through succession (Long, 1982). Decomposition of wood can be relatively rapid in tropical forests, but in many ecosystems fallen logs are a conspicuous component of the forest floor in old-growth stands. As a result of their relatively large size and low nutrient concentrations, fallen logs decompose slowly in temperate forests and require many years to become fragmented and mixed into the forest floor. Lang and Forman (1978) found that 71% of the forest floor mass consisted of fallen trees and large branches in a 250-yr-old oak forest in New Jersey. Even higher values are reported for old-growth Douglas fir in Oregon (Grier and Logan, 1977). Such detritus can contain a large portion of the nutrient pool in the forest floor, despite the low concentrations in wood.

The mean annual litterfall from aboveground vegetation increases from boreal forests to the tropics (Fig. 8.1), following the gradient of productivity (Bray and Gorham, 1964). As we have seen, there is also a large amount of litter produced in the death of fine roots in the upper layers of forest soils. Unfortunately we know very little about fine root turnover from various forests, so it is not possible to generalize about worldwide trends in the total production of detritus from both above- and belowground sources.

The chemical composition of litter determines the quality of the material as a food and substrate for decomposers. Most leaf litterfall has lower concentrations of nitrogen (N), phosphorus (P), and soluble carbohydrates than fresh leaf material as a result of nutrient reabsorptions before abscission. At the same time, concentrations of calcium (Ca), tannin, and structural leaf materials such as cellulose and lignin may be higher (Chapter 7). Often litter is exposed to leaching, fragmentation, and fungal attack while it is still attached to the tree (Jensen,

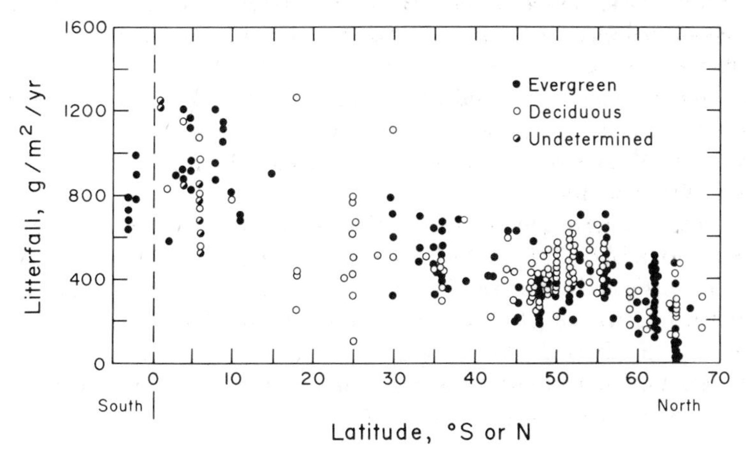

Fig. 8.1. Annual litterfall in forest ecosystems of the world (From Van Cleve *et al.*, 1983, with permission from *Can. J. For. Res.* **13,** 747–766.

1974). This litter arrives at the forest floor in a stage of partial decomposition. Wood decomposition may be well advanced before trees fall to the forest floor (Boddy and Rayner, 1983). Thus, forest litter is a heterogeneous mixture derived from the seasonal loss of various tissues of different species.

DECOMPOSITION

Fragmentation and Mixing

In many forests, leaf litter is fragmented and mixed into the lower layers of the soil within a year after abscission. The physical breakage of logs, after fungal attack has weakened the original cellular structure of wood, may largely determine the rate of their disappearance from the forest floor. The physical reduction and mixing of litter is largely carried out by an abundance of soil animals, ranging from microscopic nematodes to large earthworms (Hole, 1981). The result is an increased surface area for microbial attack and the movement of litter to more constant conditions of temperature and moisture in the lower profile.

There is an enormous diversity of soil invertebrates involved in the fragmentation or comminution of litter (Fig. 8.2). The functional roles of each are poorly known, partly because of the interactions between these organisms. For example, destruction of wood cellulose by termites is due to the action of symbiotic protozoa and bacteria contained in their lower digestive tract. In the forest floor, some litter microarthropods are primarily predators on species that feed directly on fresh litter (Moulder and Reichle, 1972; Bauer, 1982). Others feed on fungal hyphae that invade the forest floor layers (Mitchell and Parkinson, 1976). Dead soil animals and their fecal materials may be eaten by other animals, which, in turn, may be consumed by predators. Such interactions are an important aspect of the animal community in soil detritus in contrast to the more linear trophic relationships among the animals that feed aboveground (Swift *et al.*, 1979). The result is a rapid reduction in the size of litter material and an increased surface area for microbial attack.

Nematodes are among the smallest soil animals. Their populations range from 1,700,000 to 6,300,000/m^2 among various forest types (Sohlenius, 1980), and tend to be particularly abundant in coniferous forests. Springtails (Collembola) and mites (Acari) are also prevalent in coniferous forests, ranging up to 400,000/m^2 (Swift *et al.*, 1979; Hole, 1981). Earthworms, on the other hand, are intolerant of coniferous litter and are more abundant in temperate deciduous and tropical forests (Swift *et al.*, 1979; Phillipson *et al.*, 1978; Romell, 1935).

Fig. 8.2. Classification of some of the important groups of small soil animals important in processing forest litter. (From Hole, 1981.)

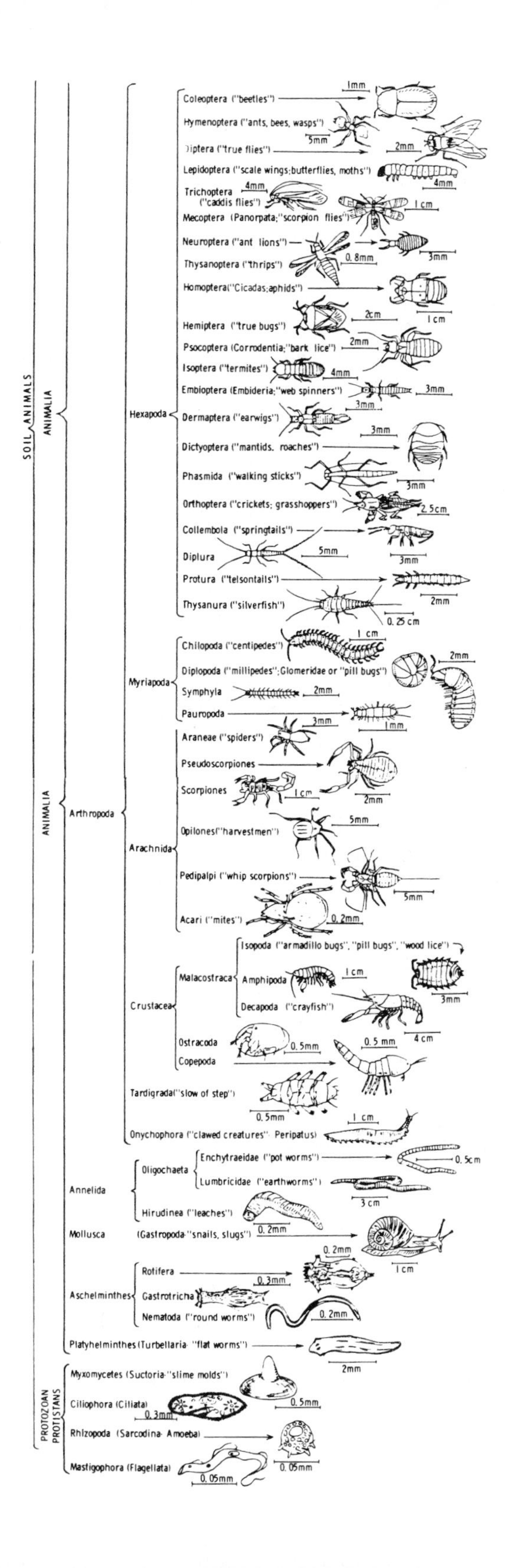

SOIL ANIMALS
ANIMALIA
Arthropoda
Hexapoda
Coleoptera ("beetles")
Hymenoptera ("ants, bees, wasps")
Diptera ("true flies")
Lepidoptera ("scale wings; butterflies, moths")
Trichoptera ("caddis flies")
Mecoptera (Panorpata; "scorpion flies")
Neuroptera ("ant lions")
Thysanoptera ("thrips")
Homoptera ("Cicadas; aphids")
Hemiptera ("true bugs")
Psocoptera (Corrodentia; "bark lice")
Isoptera ("termites")
Embioptera (Embideria; "web spinners")
Dermaptera ("earwigs")
Dictyoptera ("mantids, roaches")
Phasmida ("walking sticks")
Orthoptera ("crickets; grasshoppers")
Collembola ("springtails")
Diplura
Protura ("telsontails")
Thysanura ("silverfish")
Myriapoda
Chilopoda ("centipedes")
Diplopoda ("millipedes"; Glomeridae or "pill bugs")
Symphyla
Pauropoda
Arachnida
Araneae ("spiders")
Pseudoscorpiones
Scorpiones
Opiliones ("harvestmen")
Pedipalpi ("whip scorpions")
Acari ("mites")
Crustacea
Malacostraca
Isopoda ("armadillo bugs", "pill bugs", "wood lice")
Amphipoda
Decapoda ("crayfish")
Ostracoda
Copepoda
Tardigrada ("slow of step")
Onychophora ("clawed creatures" Peripatus)
Annelida
Oligochaeta
Enchytraeidae ("pot worms")
Lumbricidae ("earthworms")
Hirudinea ("leaches")
Mollusca
(Gastropoda "snails, slugs")
Aschelminthes
Rotifera
Gastrotricha
Nematoda ("round worms")
Platyhelminthes (Turbellaria "flat worms")
PROTOZOAN PROTISTANS
Myxomycetes (Suctoria "slime molds")
Ciliophora (Ciliata)
Rhizopoda (Sarcodina Amoeba)
Mastigophora (Flagellata)
1mm
5mm
2mm
4mm
4mm
1 cm
0.8mm
3mm
2cm
1 cm
2mm
4mm
3mm
3mm
3mm
3mm
2.5cm
5mm
3mm
2mm
0.25 cm
1 cm
2mm
2mm
3mm
1mm
1 cm
2mm
5mm
5mm
0.2mm
1 cm
3mm
4 cm
0.5mm
0.5 mm
0.5mm
1 cm
0.5cm
3 cm
0.2mm
0.2mm
1 cm
0.3mm
0.2mm
2mm
0.5mm
0.3mm
0.05mm
0.05mm

Despite the large number of soil organisms in boreal and coniferous forests, the total biomass of soil animals increases by a factor of six from boreal to tropical forests (Witkamp and Ausmus, 1976; Swift *et al.*, 1979). Biomass of earthworms is sometimes as high as 250 g fr. wt./m^2 (Witkamp, 1971). Earthworms are especially effective in mixing the litter material through the soil profile, and their movements leave open channels for air and water transport in the soil (Aina, 1984). Termites are also most abundant in warm temperate and tropical forests, where their wood-tunneling activity causes a relatively rapid disappearance of fallen logs (Gentry and Whitford, 1982; Lang and Knight, 1979). Wood (1976) calculated that 3500 termites with a biomass of 2.4 g consumed 168 g/m^2/yr of detritus in a savanna woodland in Nigeria.

Soil organisms have been classified by body size and trophic structure in attempts to understand their function. For the ecosystem ecologist, however, the most practical approach often is to look for the net effect of these organisms on the rates of energy flow and nutrient cycling in the detritus layers. In such studies, the decomposition of fresh abscissed litter has been monitored in experimental conditions that either allow or exclude the normal activity of soil animals (e.g., Witkamp and Crossley, 1966; Anderson, 1973; Seastedt and Crossley, 1980). The presence of soil organisms often allows a significantly greater overall rate of decomposition, largely through a more rapid fragmentation of litter (Witkamp and Ausmus, 1976).

Soil animals utilize the carbon (C)-containing compounds of detritus as an energy source, respiring C as CO_2. Collectively these animals can respire up to 7% of the mass of litterfall in forest ecosystems (Witkamp and Ausmus, 1976). Since the biomass of soil animals is never a significant fraction of the forest floor, nutrient retention in these organisms does not slow nutrient cycling through the ecosystem (Seastedt and Tate, 1981). Compared to fresh litter, their fecal matter is often slightly higher in concentrations of N and P, as well as in lignified compounds that are difficult to digest. If litter material is first processed by soil animals, bacteria and fungi find a substrate that is different both physically and chemically from fresh litter.

Microbial Decomposition and Mineralization

The movements and feeding of soil animals are important to the inoculation of litter materials with bacteria and fungi (Swift *et al.*, 1979). Thus, microbial decomposition that may have started in the forest canopy continues concurrently and interactively with the comminution of litter on the forest floor. Fungal hyphae penetrate the cell structure of plant tissues while bacteria are predominantly surface colonists.

Soil microbiologists have been frustrated by attempts to enumerate microbial populations or to measure their live biomass. Traditional microbiological approaches of isolation and culture may result in microbial populations that are very different from those inhabiting substrates in the soil (Witkamp, 1974). A large portion of the bacteria and fungal hyphae may be inactive in forest soil (Söderström, 1979). It is difficult to distinguish between live and dead fungal hyphae, and there is often strong seasonal variation in activity in relation to temperature and moisture (e.g., Bååth and Söderström, 1982).

Some workers have attempted to measure the content of ATP in detrital layers, assuming that this labile molecule is contained only in the cells of living microbes (Jenkinson *et al.*, 1979). This and similar approaches have usually indicated that from 1 to 5% of the organic carbon in soils is contained in microbial biomass (Anderson and Domsch, 1980). In most soils, fungal biomass exceeds bacterial biomass by a factor of three, but this varies among major soil groups, as we shall see in a later section. The energy flow through microbial populations has been estimated by measuring changes in the release of respired CO_2 following the complete or selective inhibition of particular groups of decomposers (Jenkinson and Powlson, 1976; Parkinson *et al.*, 1978; Anderson and Domsch, 1975, 1978; Voroney and Paul, 1984). Although fungi may exceed bacteria in biomass, a higher turnover rate and energy flow in bacterial populations may dominate soil metabolism.

Soil bacteria and fungi obtain metabolic energy from the breakdown of dead organic matter to CO_2 and H_2O. The decomposition processes proceed by the release of extracellular degradative enzymes (Burns, 1982). Often a sequence of decomposers is involved. In the degradation of plant protein to NO_3:

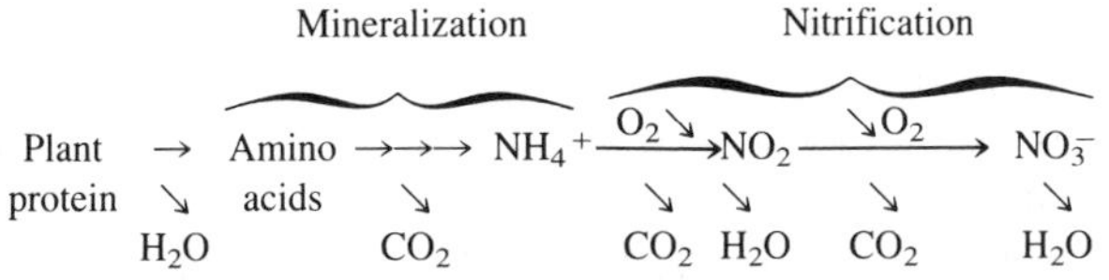

$$\begin{array}{ccccccccc} & & \multicolumn{3}{c}{\overbrace{\hspace{6em}}^{\text{Mineralization}}} & \multicolumn{4}{c}{\overbrace{\hspace{8em}}^{\text{Nitrification}}} \\ \text{Plant protein} & \rightarrow & \text{Amino acids} & \rightarrow\rightarrow\rightarrow & NH_4^+ & \xrightarrow{O_2 \searrow} & NO_2 & \xrightarrow{\searrow O_2} & NO_3^- \\ \searrow & & & \searrow & & \searrow \; \searrow & & \searrow & \searrow \\ H_2O & & & CO_2 & & CO_2 \; H_2O & & CO_2 & H_2O \end{array}$$

many forms of bacteria and fungi are involved, including the nitrifying bacteria, *Nitrosomonas* and *Nitrobacter,* which we discussed in Chapter 6. Note that the C and N are converted from reduced forms in the initial substrate to oxidized forms, CO_2 and NO_3^-.

Bacteria and fungi typically have high nutrient requirements; some nutrient ions from the detrital materials are retained for the synthesis of protein and other components of the decomposer biomass. Retention of nutrients in soil microbial biomass is known as immobilization. When nutrient concentrations in litter are very low, microbes may also accumulate and immobilize soluble nutrients from throughfall and available nutrients from the soil solution. (Remember in Chapter 7, how experimental additions of sugar reduced available N for plant uptake from the soil solution.)

For many years, soil scientists and ecologists have used ratios between the concentration of organic C and various nutrients to follow nutrient immobilization by microbes during the decomposition of litter. Table 8.1 illustrates this

TABLE 8.1 Ratios of Nutrient Elements to Carbon in the Litter of Scots Pine (*Pinus sylvestris*) at Sequential Stages of Decomposition[a]

	C/N	C/P	C/K	C/S	C/Ca	C/Mg	C/Mn
				Needle litter			
Initial	134	2630	705	1210	79	1350	330
After incubation of:							
1 yr	85	1330	735	864	101	1870	576
2 yr	66	912	867	ND	107	2360	800
3 yr	53	948	1970	ND	132	1710	1110
4 yr	46	869	1360	496	104	704	988
5 yr	41	656	591	497	231	1600	1120
				Fungal biomass			
Scots pine forest	12	64	41	ND	ND	ND	ND

[a] Some values for fungal tissues are also given. Note that C/N and C/P ratios decline, which indicates retention of these nutrients as C is lost, whereas C/Ca and C/K ratios increase, which indicates that these nutrients are lost more rapidly than carbon. From Staff and Berg (1982).

approach. Fresh litter from Scots pine (*Pinus sylvestris*) contains 0.37% N and 0.02% P, for C/N and C/P ratios of 134 and 2630, respectively. Fungi in the same forest contain 4.17% N and 0.78% P; C/N is 12 and C/P is 64. During the growth of fungal biomass, which metabolizes the organic C contained in litter, N and P are retained in microbial tissue while C is respired as CO_2. Thus, C/N and C/P ratios decrease during the first 5 yr of decomposition. Similarly when the decomposition of deciduous leaves is monitored, the N concentration in the residue increases as the mass of the original leaf material declines (Fig. 8.3). This linear relationship is found with a great many leaf and woody litter materials (Aber and Melillo, 1980). Toward the end of the decomposition process, much of the residue consists of microbial tissue that has replaced the original substrate.

Immobilization occurs more rapidly when decay is rapid, although the total amount immobilized per unit weight may be greater for litter that decays slowly, such as wood (Bosatta and Staaf, 1982; Aber and Melillo, 1982a). Fallen logs are extremely low in N content and immobilization of N is especially evident during log decay (Allison *et al.*, 1963; Grier, 1978; Lambert *et al.*, 1980; Graham and Cromack, 1982; Foster and Lang, 1982; Fahey, 1983). Asymbiotic N fixation is often found in bacteria that are important in the decomposition of fallen logs, presumably because there are such low N concentrations in wood (Roskoski, 1980; Larsen *et al.*, 1982; Silvester *et al.*, 1982). N fixation is also reported for the symbiotic bacteria in the hindgut of termites that digest wood cellulose (Breznak *et al.*, 1973).

The biomass of bacteria and fungi does not contain a large percentage of the nutrient pool in most soils (Anderson and Domsch, 1980), but the growth of these organisms frequently immobilizes N and P that might otherwise be made available from decomposition. A few studies have observed immobilizations of sulfur (S) (e.g., Staaf and Berg, 1982; Gosz *et al.*, 1973; Saggar *et al.*, 1981). In contrast, there are relatively rapid losses of K, Ca, magnesium (Mg), and manganese (Mn) from litter in most instances (Parmentier and Remacle, 1981; Lousier and Parkinson, 1976; Edmonds, 1980; Reiners and Reiners, 1970; Schlesinger and Hasey, 1981; Anderson *et al.*, 1983b; Fig. 8.4). These are not usually limiting for microbial growth and are released as soluble ions to the soil solution. Much K is removed by leaching well in advance of microbial decay, whereas Ca, a cell wall constituent, is often lost concurrently with the disappearance of litter mass. In some cases, Ca may be immobilized as Ca-oxalate in fungal tissue.

The transformations of litter often occur sequentially as organic matter is fragmented and mixed through the forest floor. Eventually enough C is lost that the C/N ratio of the residue is equivalent to that of bacterial and fungal tissues. Jorgensen and Wells (1973) found a progressive decrease in soil respiration (= microbial activity) and C/N ratio through the soil layers of a loblolly pine forest.

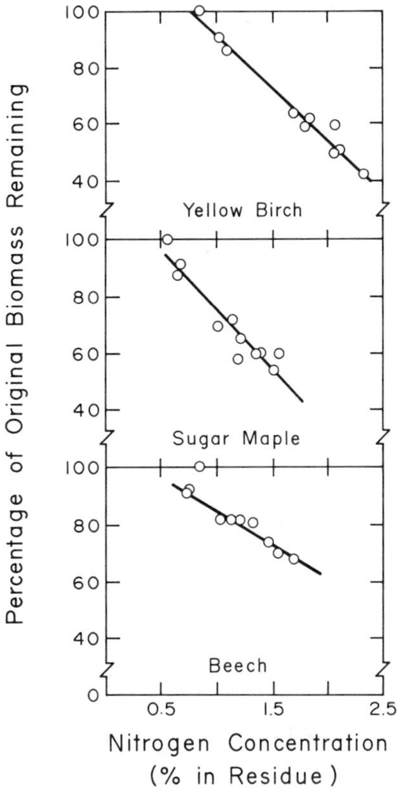

Fig. 8.3. Percentage of the original mass of leaf litter remaining as a function of its N concentration during the first 12 months of litter decomposition. (Data are from Gosz *et al.*, 1973, in the Hubbard Brook Experimental Forest, as replotted by Aber and Melillo, 1980.)

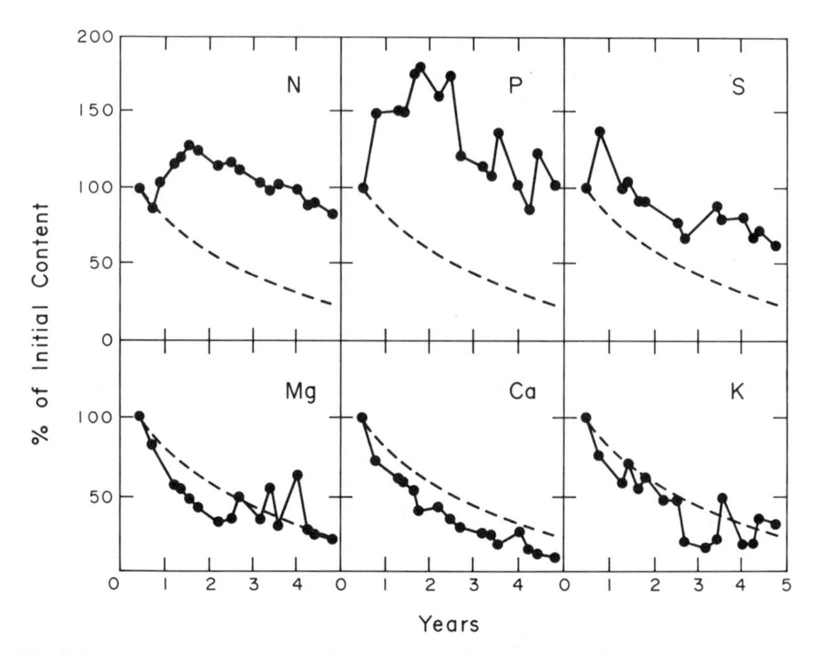

Fig. 8.4. Loss of nutrient elements from the litter of Scots Pine (*Pinus sylvestris*) during the first 5 yr of decomposition in Sweden. For each nutrient, the solid line indicates the percentage of the initial content remaining at various intervals. Note that K, Ca, and Mg are lost more rapidly than the disappearance of the organic mass of litter (dashed line), whereas N, P, and S are retained during the period of litter decay. (From Staaf and Berg, 1982.)

When microbial activity slows, there is little further nutrient immobilization. As the microbial populations die, available N is released as NH_4^+ from dead microbial tissue. This mineralization of N often commences with C/N ratios near 30 : 1 (Lutz and Chandler, 1946; Berg and Staaf, 1981), but this can vary depending on substrate and the assimilation efficiency of the decomposer (Rosswall, 1982). Using ^{15}N as a tracer, Marumoto *et al.* (1982) have shown that much of the N mineralized in soil is released from dead microbes and not directly from soil organic matter. The presence of soil animals that feed on bacteria and fungi can increase the rates of release of N and P from microbial tissues (Cole *et al.*, 1978; Bååth *et al.*, 1981; Anderson *et al.*, 1983a).

Mineralization of N, P, and S is usually greatest in the lower forest floor (Federer, 1983). Release of N, P, and S from soil organic matter is likely to occur at different rates (McGill and Cole, 1981). Nitrogen is largely bound directly to C in amino groups ($—C—NH_2$). Thus, N is mineralized as a result of the balance between the degradation of organic substances for energy and protein synthesis by microbes. While some S is also bound to C, much of the S and P is bound in ester linkages (i.e., C—O—S and C—O—P). These organic groups may be mineralized by the release of extracellular enzymes (e.g., phosphatases) in response to microbial demand for nutrients. For P, organic transformations are increasingly important through primary succession, since inorganic P is complexed into secondary minerals (Chapter 6, Fig. 6.3).

The relative content of organic substances is important in determination of comparative rates of litter decomposition. Soluble carbohydrates are rapidly leached from fresh litter and quickly metabolized by microbes. Proteins in litter and dead microbial tissues are also quickly degraded. Plant cell wall constituents, cellulose and lignin, are more resistant to decomposition. Lignin is a complex polymer of aromatic rings that is deposited in the secondary thickening of cell walls, particularly in woody tissues (Chapter 2). Only certain fungi can attack and decompose the intact lignin molecule; thus, lignin content is a major factor that determines rates of decomposition of different litter materials (Fogel and Cromack, 1977; Meentemeyer, 1978a). Phenolic compounds, often loosely called tannins, also retard decomposition (Benoit and Starkey, 1968; Harrison, 1971). These are easily leached from fresh litter by rainfall, but they bind with proteins to form a highly resistant complex (Suberkropp *et al.*, 1976; Handley, 1961; Schlesinger and Hasey, 1981). This effect of tannins is, of course, why they are useful in the preservation of leather. As decomposition proceeds, labile materials are lost and resistant materials increase in relative content in the remaining soil organic materials (Minderman, 1968; Berg *et al.*, 1982).

In many cases, nutrient-rich litter tends to decay more rapidly than nutrient-poor litter in the same forest environment. The initial C/N ratio is often a good predictor of the rate of decomposition. Experimental additions of N and P can increase the rate of decay of litter on the forest floor. Litter materials that have high contents of cell wall materials are often low in concentrations of plant nutrients. Knowledge of both organic and nutrient concentrations can be combined to predict decomposition rates. Melillo *et al.* (1982) have shown that the rate of decomposition is inversely proportional to the initial lignin/nitrogen ratio in leaf litterfall (Fig. 8.5).

Humus Formation

Plant litter and soil microbes constitute the cellular fraction of soil organic matter. As decomposition proceeds, there is an increasing content of amorphous organic matter, humus, that appears to result from microbial activity. The structure of humus is poorly known, but appears to contain aromatic rings with phenolic (—OH) and organic acid (—COOH) groups (Flaig *et al.*, 1975). The molecule appears to have no consistent molecular weight or repeating units in its structure. It has been suggested that no two humus molecules are alike! Humus

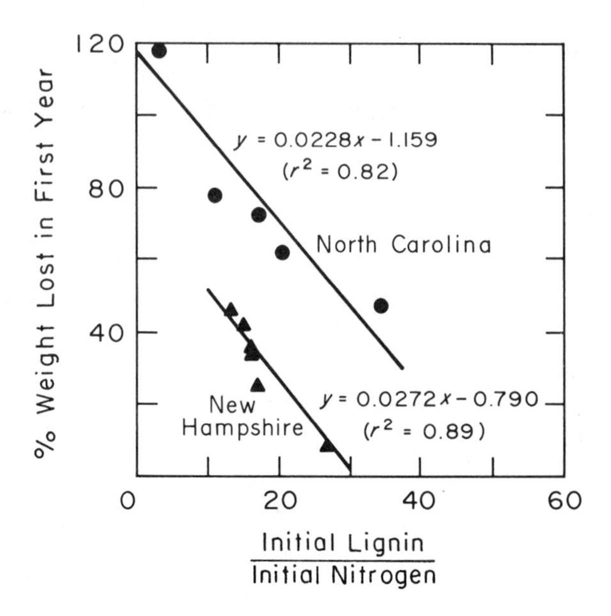

Fig. 8.5. Decomposition of forest leaf litter as a function of the lignin/nitrogen ratio in fresh litterfall of various species in New Hampshire and North Carolina Forests. (From "Nitrogen and lignin control of hardwood leaf litter decomposition dynamics" by J. M. Melillo, J. D. Aber, and J. F. Muratore, *Ecology,* 1982, **63,** 621–626. Copyright by the Ecological Society of America. Reprinted by permission.)

accumulates in the lower horizons of forest soils and is often complexed with clay mineral components. Humus accumulation appears to be associated with lignin degradation. Most humus substances also have a high concentration of N in amino ($—NH_2$) groups. Some workers have suggested that humus results from the formation of stable lignoprotein complexes. Despite a high N content, it is very resistant to microbial attack. Humic materials from a forest soil in Saskatchewan showed a weighted mean ^{14}C age of 250 yr (Campbell *et al.*, 1967). Some forms of humus are water soluble (e.g., fulvic acids), and the downward movement of humic acids controls the movement of plant nutrients in the soil solution (Dawson *et al.*, 1978; see also Chapter 6, Table 6.4). Chemical characterizations of humus are often based on the solubility of humic and fulvic acid components in alkaline and acid solutions, respectively.

While the structure and formation of humus substances remain elusive, the importance of humus in forest ecosystems is clear. In most forests, humus in the soil profile exceeds the combined content of organic matter in the forest floor and aboveground biomass (Schlesinger, 1977). Annual increments in the humus pool are small but continuous during soil profile development. It is unlikely that the humus pool is ever in steady state. Humus accumulations in glacial till soils, 12–15 kg C/m^2, imply accumulation rates of 1 g C/m^2/yr over the entire interval since the end of the Pleistocene. If similar rates of accumulation are found in most forests, then about 0.1% of the annual net primary production is stored as net ecosystem production in humus.

As a result of high concentrations of N, P, and S, humus contains an overwhelming proportion of the pool of these nutrients in forest soils (Chapter 6, Table 6.6). The stability of humus substances in the mineral soil means that this large nutrient pool turns over very slowly. Active transport by plant roots results in rapid uptake of limiting nutrients and low concentrations of available forms in the soil solution. The *rate* of mineralization, not the underlying pool size, determines the availability of these ions for forest growth (Ellenberg, 1977). Thus, one-time measurements of quantities in the soil solution are of limited value in determining site fertility. We will examine some techniques for measuring mineralization of these nutrients as an index to site fertility in a later section.

In addition to storing the majority of the pool of N, P, and S, humus substances are a major source of cation-exchange sites for ions in the soil solution. These exchange sites hold a pool of cationic nutrients such as K and Ca, which are rapidly released from forest litter; *rates* of mineralization do not usually determine the available quantities of these nutrients in the soil over short intervals. Consequently, direct measurements of available soil contents of these nutrients are useful in comparative studies of site fertility.

Effects of Temperature and Moisture

The quality of litter influences the comparative rate of disappearance of individual plant materials, but the overall rate of decomposition in a forest ecosystem is largely determined by temperature and moisture. Temperature is of primary importance. Microbial activity increases exponentially with increasing temperatures (e.g., Edwards, 1975). This relation often shows a Q_{10} of two, i.e., a doubling in activity per 10°C increase in temperature (Singh and Gupta, 1977). High temperatures result in rapid decomposition in tropical forests versus slow decomposition in boreal forests. Van Cleve *et al.* (1981) found that the thickness of forest floor layers in black spruce (*Picea mariana*) forests in Alaska was inversely related to the cumulative days favorable to decomposition each year (Fig. 8.6). In this region the leaf area and forest growth are directly related to N mineralization and are greatest on sites with warmer soils (Van Cleve and Alexander, 1981). When snowfall insulates the forest floor, microbial respiration may continue even at temperatures close to 0°C.

Most forests grow in mesic regions; thus, soil moisture does not usually limit

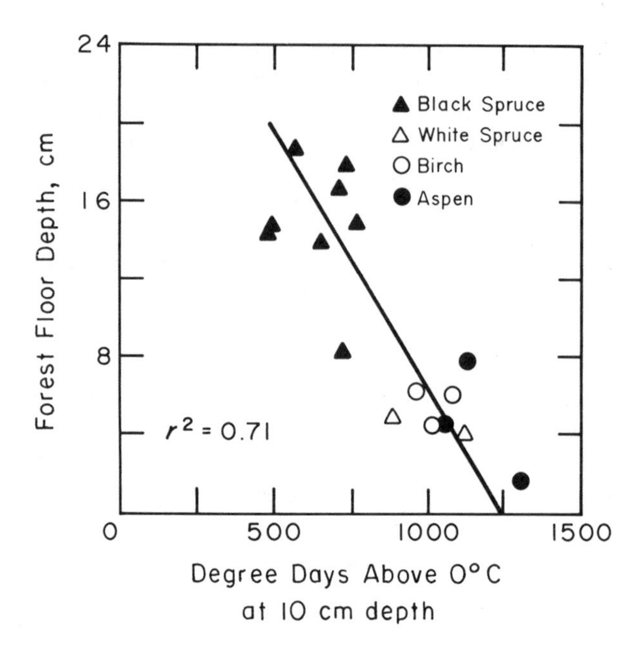

Fig. 8.6. Depth of the forest floor in boreal forest stands in Alaska as a function of the duration of warm soil temperatures during the summer. Degree days are the sum of the mean soil temperature for each day that soil temperature exceeded 0°C during the growing season. (From Van Cleve *et al.*, 1981.)

microbial activity. However, when precipitation is not evenly distributed throughout the year, soil moisture can limit decomposition during the dry season (Waring and Franklin, 1979; Schlesinger and Hasey, 1981; Ellis, 1969; Swift *et al.*, 1981). Microbes obtain moisture from metabolic degradation of their substrate, so their activity may continue even after plants are limited by low soil water potential. Limited moisture affects the upper forest floor more frequently than the humus and mineral soil layers. Microbial response to soil moisture is parabolic (Fig. 8.7). When soils are saturated, oxygen diffusion is negligible and anaerobic conditions develop. In these conditions decomposition is inhibited and in extreme conditions peat accumulates (Reiners and Reiners, 1970; Schlesinger, 1978). Thus, large accumulations of soil organic matter are found in wetland forests, even in tropical regions (Brown and Lugo, 1982). In temperate climates, wet soils are slow to warm in the springtime. Decomposition in the waterlogged soils of boreal regions is inhibited by high moisture as well as low temperatures (Van Cleve *et al.*, 1981).

Fluctuations of soil temperature and moisture can result in greater microbial activity than in constant favorable conditions with the same mean values (Soulides and Allison, 1961; Biederbeck and Campbell, 1973). Apparently these variations cause greater turnover of the microbial populations and mobilization of nutrients. Similarly, the periodic addition of fresh litter materials can stimulate the degradation of resistant substrates in the soil (Fig. 8.8). This "priming effect" is widely known from agricultural studies and can be observed in forests with seasonal litterfall.

Accumulations of soil organic matter reflect the effects of moisture and temperature on the balance between primary production and decomposition. Accumulations are greatest in high latitude and high elevation forests and least in lowland tropical forests (Schlesinger, 1977; Post *et al.*, 1982; Franz, 1976). Since net primary production shows the opposite trend, the accumulation of soil organic matter is largely due to differences in decomposition. Thus, compared to the process of primary production, microbial processes are more sensitive to regional differences in temperature and moisture. Meentemeyer (1978b) used published data from various decomposition studies to relate decomposition to calculated actual evapotranspiration. This parameter is useful because it incorpo-

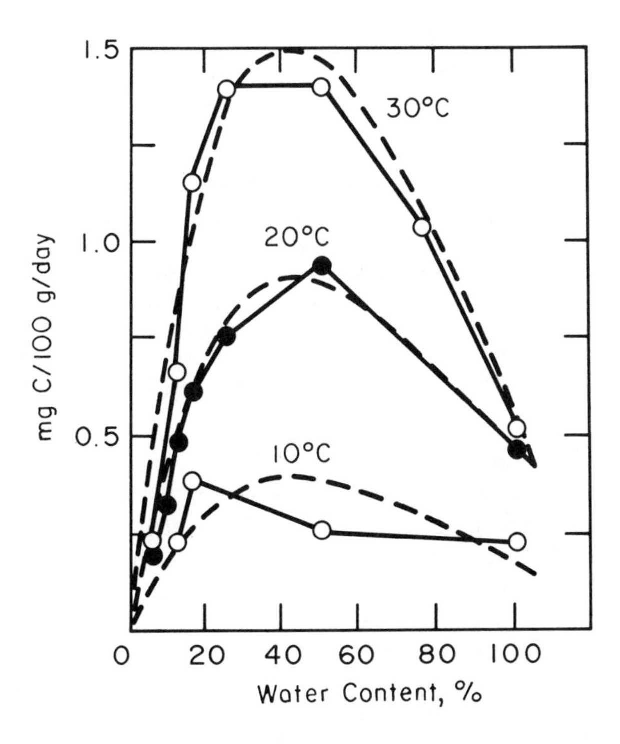

Fig. 8.7. Release of C as CO_2 due to the activity of soil microorganisms at varying temperature and soil moisture content. (From Ino and Monsi, 1969.)

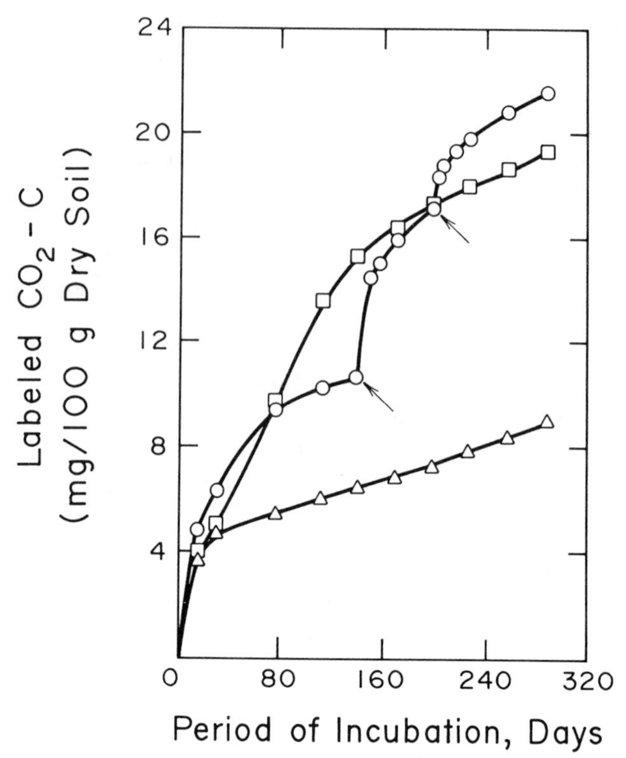

Fig. 8.8. Release of radioactively labeled C from organic matter decomposing in the soil. Three experimental conditions are imposed: △ = soil kept moist continuously; □ = soil dried and rewetted every month; ○ = soil to which unlabeled organic matter was added at points indicated by arrows. Note that the addition of the new substrate stimulated the decomposition of the remaining residues of the original organic matter. [Reprinted with permission from (*Soil Biol. Biochem.* **6,** Sørensen, Rate of decomposition of organic matter in soil as influenced by repeated air drying-rewetting and repeated additions of organic material), Copyright (1974), Pergamon Press Ltd.]

rates both temperature and moisture conditions for a given site. The resulting equation could be used to predict regional patterns of decomposition. His map for the United States (Fig. 8.9) shows a gradient of increasing decomposition from north to south in the eastern United States. Decomposition in western forests is relatively slow. (Compare this map to that for forest production in Chapter 3, Fig. 3.9.) These observations are consistent with observations on accumulations of soil organic matter in the eastern United States (Lang and Forman, 1978). Even better predictions of decomposition are possible when the mean lignin content of litterfall is included (Meentemeyer, 1978a), but extrapolation of these relations to nonforested regions is suspect (Whitford *et al.*, 1981; Elkins *et al.*, 1982).

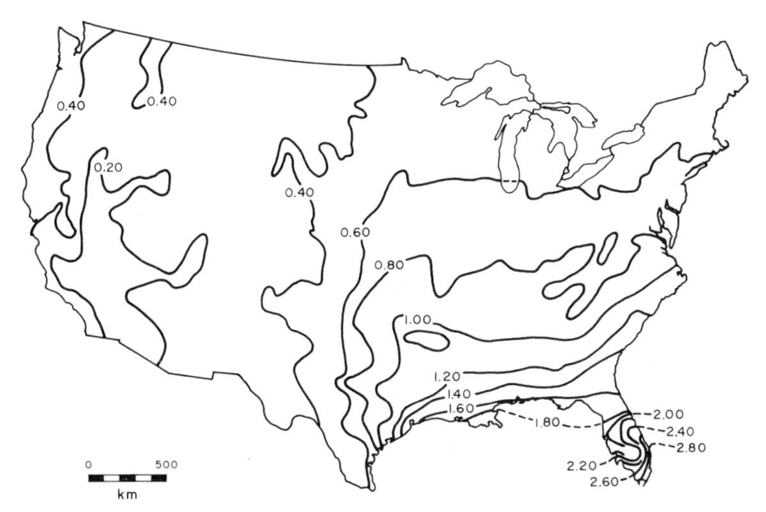

Fig. 8.9. Rates of decomposition of fresh litter in the United States predicted by a simulation model using actual evapotranspiration as a predictive variable. Isopleth values are the fractional loss rate (k) of mass from fresh litter during the first year of decay after abscission. (From Meentemeyer, 1978b. Reproduced by permission of the U.S. Department of Commerce, National Technical Information Service.)

Measurements of Turnover in Detrital Layers

During primary succession, the mass and nutrient content of the forest floor and humus layers increase (Chapter 6). In peatlands, the accumulation may continue for thousands of years, but in most upland forests the storage of soil organic matter eventually achieves an apparent steady-state level characteristic of the climatic conditions. At this stage, the large amount of soil organic matter masks any small, continuing accumulations of stable humus substances in the lower soil profile. Losses of organic C to surface- and groundwater are small (Schlesinger and Melack, 1981; Edwards and Harris, 1977); thus, most C that enters the soil as plant detritus must be respired as CO_2. Accurate measurements of soil respiration, the loss of CO_2 at the soil surface, are very difficult to obtain (Singh and Gupta, 1977; Schlesinger, 1977), but a compilation of such data from world forests shows an increase from boreal forests to the tropics (Fig. 8.10). Part of this gradient is due to a greater deposition of aboveground litter in the tropics (cf. Fig. 8.1), but the carbon lost as CO_2 averages 2.5 times greater than that received in aboveground litterfall in world forests. The additional C is derived from the decomposition of root detritus and from the respiration of live roots and mycorrhizal fungi. As a result of root respiration, it is difficult to use

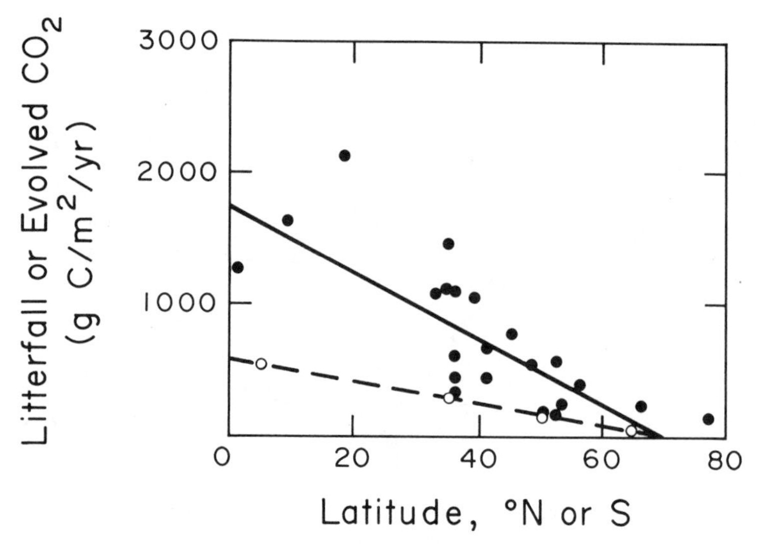

Fig. 8.10. Solid line is the best fit to solid points which are measurements of the annual release of C as CO_2 from the soil surface of forest and woodland ecosystems arranged as a function of latitude (From Schlesinger, 1977, reproduced, with permission, from the *Annual Review of Ecology and Systematics,* Vol. 8, © 1977 by Annual Reviews Inc.) Dashed line is the C deposited in aboveground litterfall (mean of values in latitudinal regions, derived from Bray and Gorham, 1964, see also Fig. 8.1).

measurements of CO_2 evolution to indicate the rate of decomposition in a forest ecosystem.

Alternatively, a mass-balance approach suggests that the annual decomposition should equal the annual input of fresh detritus in forests in which the detrital storage is in steady state. If decomposition is a fraction, k, of detrital mass, then at steady state:

$$\text{Litterfall} = k \text{ (detrital mass)}$$

and

$$\frac{\text{Litterfall}}{\text{Detrital mass}} = k$$

This turnover coefficient is frequently calculated as a comparative measure of decomposition in forest ecosystems (e.g., Lang and Forman, 1978). Most values are calculated for the forest floor. Where decomposition rates are fast, there is little surface accumulation and coefficients are greater than 1.0 (e.g., tropical rain forests). In such systems, decomposition processes have the potential to respire more than the annual input of C in litterfall (cf. Fig. 8.9). In contrast, in boreal peatlands, values for k are very small (e.g., 0.001, Olson, 1963). One might prefer turnover coefficients that are calculated for the detritus in the entire soil profile, but these values can be misleading. The large mass in the lower profile decomposes very slowly, so that a weighted average constant for the entire profile will suggest much slower decomposition rates than what actually occurs in the forest floor, where most CO_2 is released. More serious is our ignorance of the input of fresh litter from the death of fine roots. Most published turnover coefficients are too low since fine root litter has often been ignored as an input (Vogt *et al.*, 1983).

There is, of course, no reason to expect that the turnover of organic mass in the forest floor is equal to the proportional mineralization of the nutrients that it contains (McGill and Cole, 1981). Table 8.2 shows the mean residence or turnover time ($= 1/k$) for plant macronutrients in the forest floor of various forest ecosystems. Some nutrients such as K are easily leached from litter and may show mineralization rates in excess of the loss of litter mass. Others such as N turn over more slowly due to the immobilization in microbial tissue. Much N is released in the lower forest floor layers. For nutrients that are immobilized during decomposition, calculation of nutrient mineralization using the steady-state approach must consider the entire soil profile. Temporary immobilization in fresh litter should be balanced by equivalent mineralizations in the lower profile. If the total nutrient pool in detritus is not changing significantly, then annual mineralization must approximate the nutrient returns (litterfall plus leaching) from vegetation.

There are many instances in which the use of a mass-balance approach is questionable. Steady-state forest floor mass is not found in successional forests

TABLE 8.2 Mean Residence Time (yr) for Organic Matter and Its Nutrient Content Comprising the Forest Floor in World Forests[a]

		Mean residence time (yr)					
Forest region	No. of sites	Organic matter	N	K	Ca	Mg	P
Boreal coniferous	3	353	230	94	149	455	324
Boreal deciduous	1	26	27.1	10.0	13.8	14.2	15.2
Temperate coniferous	13	17	17.9	2.2	5.9	12.9	15.3
Temperate deciduous	14	4.0	5.5	1.3	3.0	3.4	5.8
Tropical rain forest	4	0.4	2.0	0.7	1.5	1.1	1.6

[a] Values are calculated by dividing the forest floor mass by the annual litterfall input. Assuming a steady state, mean residence time equals $1/k$. Boreal and temperate values are from Cole and Rapp (1981), tropical values are calculated from Edwards and Grubb (1982) and Edwards (1977, 1982).

or forests with frequent fire (Birk and Simpson, 1980). In some forests, the forest floor is difficult to define and separate from the mineral soil, whereas in other cases movement of fallen litter results in a nonuniform distribution of forest floor mass (Orndorff and Lang, 1981; Welbourn *et al.*, 1981). Where most litter gathers in pits and depressions, the actual litter decomposition and nutrient release may be very different from that implied by turnover coefficients that are calculated from mean values for litterfall and forest floor mass (Dwyer and Merriam, 1981). In these instances we are faced with the need for alternative ways to calculate k for litter mass and its nutrient content.

Many forest ecologists use the litterbag approach to characterize decomposition rates of different materials in different environments. Fresh litter is confined in mesh bags that are placed on the forest floor and collected for measurements at periodic intervals (Singh and Gupta, 1977). The litterbag is somewhat artificial since soil animals may be excluded. The loss in nutrient content is usually taken as evidence of mineralization from litterfall. Simple models are based on an exponential pattern of loss. The fraction remaining after 1 yr is given by:

$$\frac{X}{X_o} = e^{-k}$$

Here, k is the turnover coefficient and shows high values for leaf litter in tropical forests and lower values in cool temperate and boreal forests (Table 8.3). Decomposition coefficients for log decay show the same regional pattern, ranging from -0.006 to -0.506 between northern coniferous and tropical forests, respectively (Foster and Lang, 1982). As in Table 8.2, mean residence time for litter is $1/k$, and the time to achieve 95% of the steady-state forest floor accumulation is $3/k$ (Olson, 1963). Calculation of the latter statistic is particularly useful in as much as litterbag measurements can validate the assumption of steady-state detrital accumulations.

In theory, mineralization for an entire forest ecosystem might be calculated with knowledge of the proportional composition of fresh litter and the nutrient loss coefficients from each, e.g.,

$$\text{Ecosystem Ca mineralization} = \sum_{i=1}^{N} k_{\text{leaf-Ca}_i}\,(\text{leaf litter}_i) + k_{\text{bark-Ca}_i}\,(\text{bark litter}_i) + \ldots$$

where i represents each species in the forest. This approach is impractical, even in single-species stands. Thus, values for k are most useful in studies that compare different litter materials. For example, turnover coefficients are inversely correlated to the lignin/nitrogen ratio in litter, but the values are uniformly higher

TABLE 8.3 Some Litter Decomposition Constants (*k*) for Deciduous Leaves in Various Forest Environments

Biome and species	k (yr^{-1})	Reference
Boreal		
Betula papyrifera	−0.46	Van Cleve (1971)
Populus tremuloides	−0.39	
Cool-temperate		
Populus tremuloides	−0.30	Lousier and Parkinson (1976)
Populus balsamifera	−0.28	
Acer saccharum	−0.51	Gosz *et al.* (1973)
Fagus grandifolia	−0.37	
Betula allegheniensis	−0.85	
Warm-temperate		
Quercus prinus	−0.61	Cromack and Monk (1975)
Quercus alba	−0.72	
Acer rubrum	−0.77	
Cornus florida	−1.26	
Mediterranean woodland		
Salvia mellifera	−0.40	Schlesinger and Hasey (1981)
Eucalyptus species	−0.30 to −0.94	Specht (1981); Birk (1979)
Tropical rain forests		
Ficus dubium	−0.65 to −0.77	Anderson *et al.* (1983b)
Parashorea macrophylla	−0.62 to −0.69	
Pentaclethra macroloba	−2.77	Gessel *et al.* (1981)
Simaruba amara	−4.16	

in climatic regions more favorable to decomposition (Fig. 8.5, where 100% loss rate is the same as $k = -1.00$).

The simple exponential assumption is often reasonable for the pattern of disappearance of total organic mass, cellulose, and Ca. However, for some constituents such as K, leaching may remove most of the initial content very rapidly, whereas the rate of loss of the remaining content may be rather slow. Two-phase models have been applied in these instances (Bunnell and Tait, 1974) and other models are also available (Olson, 1963; Wieder and Lang, 1982).

Turnover coefficients for N and P are often difficult to obtain from litterbag data. During the first year of decomposition, microbial immobilizations often result in increases in content (Fig. 8.4). Net mineralization of these nutrients is seen in long-term litterbag studies, but the complex pattern of loss is poorly modeled by the exponential model. Moreover, by the time these mineralizations occur, the litter is thoroughly fragmented and difficult to retain in mesh bags.

Since it is the rate of mineralization of N, P, and S that determines their availability for plant growth, an ideal index of site fertility and detrital turnover for these elements would express the mineralization of available forms for plant uptake through the growing season (Ellenberg, 1977). A variety of indices are available (Keeney, 1980; Powers, 1980; Lea and Ballard, 1982b). Forest ecologists have measured potential mineralization and nitrification rates by incubating forest floor and humus samples in the absence of plant uptake or potential leaching losses. In laboratory incubations, changes in soil NH_4^+ and NO_3^- concentrations over 7- to 30-day intervals are indicative of these microbial transformations. The results often correlate reasonably well with forest characteristics (Fig. 8.11), including responses to fertilization. These mineralization rates can be measured under varying conditions of temperature, moisture, microbial inoculation, and substrate quality.

Field measurements can be performed in trenched plots; a block of soil (often 1 m^2) is isolated on all sides by trenching and the trenches are lined with plastic

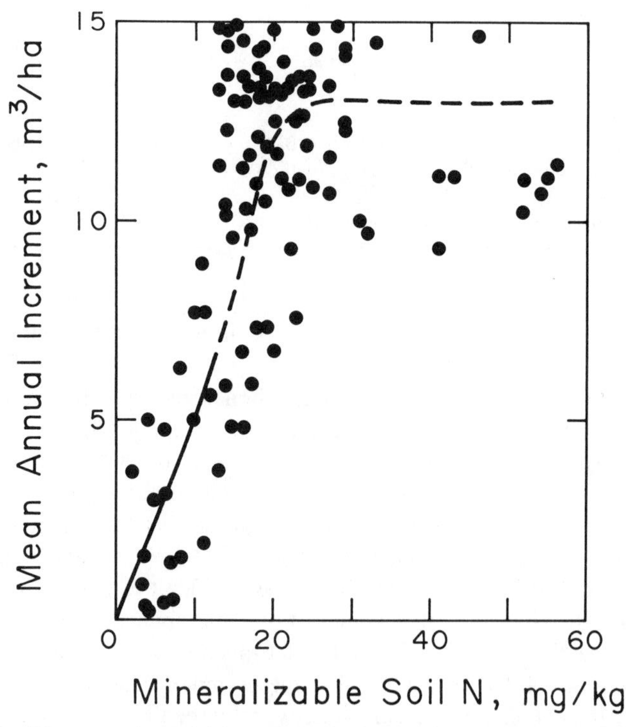

Fig. 8.11. Mean annual wood increment in Ponderosa pine (*Pinus ponderosa*) forests as a function of the mineralizable soil N measured in laboratory incubations. At values above 20 mg/kg, N does not limit the growth of these forests. (From Powers, *Soil Science Society of America Journal,* Volume 44, 1980, pages 1307–1320 by permission of the Soil Science Society of America.)

to prevent the ingrowth of roots. Plants rooted in this plot are removed but the area is not otherwise disturbed. Periodic measurements of NH_4^+ and NO_3^- may indicate rates of mineralization and nitrification in the absence of plant uptake (e.g., Vitousek *et al.*, 1982). Since trenching also eliminates plant uptake of water, this approach measures microbial activity at artificially high soil moisture and with potential losses from the ecosystem due to leaching and denitrification. To avoid the leaching of NO_3^- from the soil profile, some workers, especially in Europe, have enclosed small soil samples in polyethylene bags that are permeable to O_2 and CO_2 but not to water (Ellenberg, 1977). When these bags are used for field incubations, microbial activity continues, but soluble NO_3^- is not lost. The bags, however, are an artificial environment because the samples are not mixed by soil animals, and during the incubation interval soil moisture cannot fluctuate as it may in undisturbed forest soil. Such incubations have also been performed for studies of P and S mineralization rates (e.g., Johnson *et al.*, 1982a; Pastor *et al.*, 1984). As an index to availability, the flux of ions through the soil can be captured in lysimeters (Chapter 6) or by ion-exchange resins placed in nylon bags in forest soil horizons (Smith, 1979). While the resin-bag approach does not measure actual mineralization, it correlates well to traditional mineralization indices in many instances (Binkley and Matson, 1983). Other approaches to mineralization rates include the use of isotopic labeling of plant detritus or of the available pool in the soil (^{15}N, ^{32}P, ^{34}S). The soil solution is then monitored for subsequent changes in isotope content (Harrison, 1982; Saggar *et al.*, 1981; Van Cleve and White, 1980).

FOREST SOILS

Horizons

The soil in a forest ecosystem usually consists of a number of layers, or horizons, that collectively comprise the complete soil profile (Fig. 8.12). Recognition of the processes that occur in these horizons is an essential part of understanding nutrient cycling in forest ecosystems. Conversely, our knowledge of processes such as rock weathering, water movement, and decomposition is essential in understanding the development of forest soils (Jenny, 1980). In this section, we first describe patterns of soil development in temperate-zone forests, and then briefly contrast these to boreal and tropical regions.

It is often easy to separate the forest floor from the underlying layers of mineral soil, but these two major categories can be further subdivided. The forest

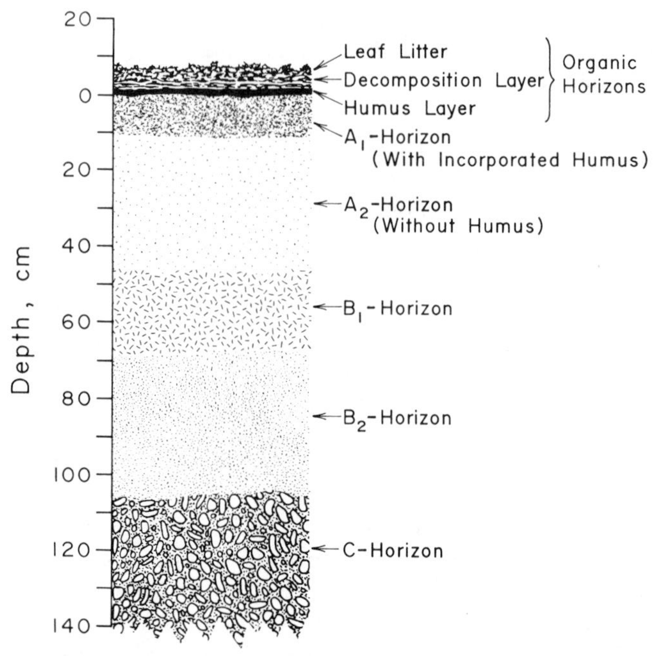

Fig. 8.12. Diagram of the profile of a podzolic soil under an old shortleaf pine (*P. echinata*) stand in North Carolina showing the three organic horizons over the A-, B-, and C-horizons of the mineral soil. (From "Plants and the Ecosystem," Third Edition, by W. D. Billings. © 1978 by Wadsworth Publishing Company, Inc. Reprinted by permission of Wadsworth Publishing Company, Belmont, California 94002.)

floor often consists of L-, F-, and H-layers (Fig. 8.12). The L-layer consists of fresh, undecomposed litter, easily recognized by species. The F-layer lies immediately below the L-layer and consists of fragmented organic matter in a stage of partial decomposition. This layer is dominated by organic materials in cellular form, and fungi and bacteria are common. The designation F-layer is derived from "fermentation," but this does not imply that the environment for microbial processes is anaerobic. Beneath the F-layer lies the H- or humus layer, primarily consisting of amorphous, resistant products of decomposition and with lower proportions of organic matter in cellular form. The lower portion of the H-horizon often shows an increasing proportion of inorganic mineral soil constituents. Thus, the differentiation of the H-layer from the uppermost layer of mineral soil is sometimes difficult, but a greater predominance of organic content versus mineral content is a useful criterion.

The upper mineral soil is designated as the A-horizon. It may vary in thickness from several centimeters to 1 m. The A-horizon is recognized as a zone of removal or eluvial processes. Soil water percolating through the forest floor contains organic acids derived from the humic materials. These waters remove iron (Fe), aluminum (Al), and other cations by weathering of the mineral components of the A-horizon (see Chapter 6, Table 6.4). Iron and aluminum are complexed with the water-soluble fulvic acids in the soil solution and percolate to the lower horizons. Clay minerals are also removed from the A-horizon. When the removal is very strong, a whitish A_2-horizon is easily recognized and may consist of nearly pure Si, which is relatively insoluble in acid conditions (Pedro *et al.*, 1978).

Substances leached from the A-horizons are deposited in the underlying B-horizon. This is defined as the zone of deposition or illuvial horizon. Iron and aluminum precipitate in the B-horizon and secondary clay minerals are formed (Ugolini *et al.*, 1977a). Soluble humic materials are complexed with the clay mineral components of soil. The removal of iron and soluble organic compounds from the A-horizon and their deposition in the B-horizon is known as podzolization. When the illuvial horizon is strongly differentiated, it is known as a spodic horizon and is designated B_{hir}. Soils with spodic horizons are Spodosols, and in extreme conditions the downward-moving humic acids are precipitated just below the A_2-horizon, forming a dark spodic B_h horizon.

Below the B-horizon, the C-horizon consists of coarsely fragmented soil material with little organic content. When the soil has developed from local materials, the C-horizon shows mineralogical similarity to the underlying parent rock, but when parent materials have been deposited by transport (Chapter 6), there may be little resemblance between the C-horizon and the underlying bedrock. In either case, carbonation weathering tends to predominate in the C-horizon (Ugolini *et al.*, 1977b).

Soil profile development on steep slopes is often incomplete as a result of landslides and other mechanical weathering events. Not all forest soils show the differentiation of all horizons. The thickness and presence of the L-, F-, and H-layers varies seasonally, especially in regions where litterfall is strongly seasonal. Strongly eroded soils of the piedmont of the southeastern United States may show a forest floor directly on top of B-horizons. Here the A-horizon has been lost through erosion during past agricultural practices. Soil profiles in floodplain areas and those exposed to deposits of volcanic ash may have ''buried'' horizons. These observations indicate that soil profile development is slow compared to the changes in vegetation. Nevertheless, in Sweden where spruce plantations have replaced beech forests, the forest floor shows changes toward lower pH and nitrification rates, which are characteristic in many coniferous forests (Nihlgard, 1971).

Geographic Patterns

Within the general description of temperate forest soils, ecologists have long differentiated between mor and mull forest floors. In broad geographic terms, mors develop in cooler climates, often characterized by coniferous vegetation. Decomposition in the forest floor is slow and incomplete, resulting in a thick organic layer. Moreover, the litter of coniferous species contains high concentrations of phenolic substances and lignin that yield acid decomposition residues. The pH of the soil solution is often as low as 4.0 (Bertram and Schleser, 1982). In these conditions, fungi predominate over bacteria. As early as 1935, Romell pointed out that the conditions of the mor forest floor tend to reinforce its own development, since fungal decomposition of lignin may yield large amounts of acid humus substances. Earthworm populations are low in mor forest floors (Phillipson *et al.*, 1978) and, in the absence of rapid fragmentation and mixing by earthworms, the forest floor is often sharply differentiated from the whitish A_2-horizon of the underlying soil. Mor forest floor is often associated with well-developed Spodosols, which are called podzols by some workers.

Mull forest floors are typically found under deciduous forests in warm-temperate climates. Most of the characteristics of mulls are in contrast to those of mors. Decomposition is more rapid, residues are less acidic, and earthworms are more abundant. Bacteria play a greater role in decomposition processes in mull forest floors, and nitrification is often more rapid in soils of near-neutral pH. Fragmentation and mixing often make differentiation of the forest floor difficult and obscure sharp boundaries between the mineral soil horizons. Under conditions of pH 5.0–7.0, which are typical of these soils, Si is relatively soluble. Thus, Si, Fe, and Al are removed in relatively equal proportions from the A-horizon minerals and there is no sharply defined A_2 horizon (Pedro *et al.*, 1978). The presence of even a weak spodic horizon classifies many warm-temperate soils with mull forest floors as Spodosols; however, in other cases these soils are often classified as Alfisols on the basis of a horizon with illuvial clay (B_t) and relatively high base saturation.

The distribution of forest floor and soil groups forms a continuous gradient over broad geographic regions, in response to parent materials, topography, climate, vegetation, and time (Jenny, 1941, 1980). However, the transition from mull to mor forest floors is gradual and patchy as one passes from warm-temperate to cool-temperate forests. In the large region mapped as Spodosol soils, one may find deciduous forests with mull forest floors at lower elevations and well-developed mor forest floors in high elevation sites under coniferous forests (Stanley and Ciolkosz, 1981). In Great Britain, Phillipson *et al.* (1978) found that either type of forest floor could develop under beech forests, depending on the underlying parent material. Mull forest floors are more likely to develop in deciduous forests dominated by species that produce nutrient-rich litterfall (e.g., *Acer, Carya, Liriodendron*) than in forests of beech (*Fagus*) and oak (*Quercus*), which produce litter with a relatively high content of phenolic substances, high C/N ratio, and low Ca (Davies, 1971).

In the coastal states of the southeastern United States (e.g., Georgia and

Alabama), the warm humid climate results in rapid decomposition. Sandy soils of the coastal plain are often dominated by conifers and have acid mors with a spodic horizon near the level of groundwater. On upland sites, weathering and soil development have produced secondary clay minerals dominated by kaolinite, as a result of the removal of Si from the soil profile (Chapter 6). In these areas, iron oxides precipitate in the lower profile, which gives a yellowish to deep-reddish coloration to the B-horizon. These soils are known as Ultisols (yellow-brown podzolics), and represent transitional intermediates to the highly weathered soils of tropical forests.

We can extend our gradient of soil development and soil formation processes to both boreal and tropical regions. To the north, decomposition in the forest floor is very slow due to cold and waterlogged conditions. Thick forest floor accumulations may comprise the entire rooting zone. In many areas of the far North, Spodosols gradually give way to peatland soils (Histosols) underlain by permafrost. In these areas the underlying mineral soil is effectively isolated from the nutrient cycling processes in the forest ecosystem. In much of the boreal region, the mineral soil is derived from glacial deposits. Weathering of primary minerals is slow in cold conditions, and there is little leaching of the soil profile when it is frozen for much of the year. Thus, the A- and B-horizons are poorly differentiated; the dominant soils are Inceptisols.

As one travels from the temperate zone to the lowland tropics, decomposition is progressively more rapid and complete. There is little forest floor mass (Chapter 6, Table 6.5) and there are lower concentrations of humic acids in percolating soil waters. In this environment, Si is more soluble than Fe and Al. Long periods of weathering under high rainfall have removed Si and cations from the entire soil profile. Resistant soil materials are hydroxides of Fe and Al (see Chapter 6). Recall that aluminum hydroxides may produce acidity (H^+) in soil waters, depending on the state of hydration (Chapter 7), and that P forms highly stable complexes with Fe and is unavailable (Chapter 6). Thus, over large portions of the lowland tropical rain forest region, soils are acid and infertile with P deficiency (Sanchez *et al.*, 1982a). Most lowland tropical soils are classified as Oxisols or Ultisols, but in extreme conditions these soils are known informally as laterite.

Soils in tropical forests may be many meters in depth, since in many areas they have developed over millions of years without disturbances such as glaciation. In the absence of clear zones of eluviation and illuviation, distinction of the A- and B-horizons is difficult. The profiles are also well mixed by earthworms. The absence of a thick forest floor does not imply that these soils are low in soil organic content. Throughout the lower profile, light-colored fulvic acids are complexed with the mineral soil materials and yield a significant storage of soil C in organic form (Sanchez *et al.*, 1982b).

A useful index of the regime of soil formation and the degree of weathering is seen in the ratio of Si to sesquioxides (Fe and Al) in the soil profile (Table 8.4).

TABLE 8.4 Silicon/Sesquioxide (Al_2O_3 + Fe_2O_3) Ratios for the A- and B-Horizons of Some Soils in Different Climatic Regions[a]

Region	Number of sites	Mean Si/sesquioxide ratio		Reference
		A-horizon	B-horizon	
Boreal	1	9.3	6.7	Leahey (1947)
Cool-temperate	4	4.07	2.28	MacKney (1961)
Warm-temperate	6	3.77	3.15	Tan and Troth (1982)
Tropical	5	1.47	1.61	Tan and Troth (1982)

[a] Note that the removal of Al and Fe results in high values in boreal and cool temperate soils, especially in the A-horizon. Lower values characterize tropical soils and there is little differentiation between horizons as a result of the removal of Si from the entire profile in long periods of weathering.

In boreal forest soils, Si is relatively immobile and Fe and Al are removed, which results in high values for this ratio in the A-horizon. The accumulation of the secondary mineral montmorillonite in the moderately weathered soils of the glaciated portion of the United States yields Si/sesquioxide ratios of two to four, as a result of the ratio of Si to Al in the crystal lattice (Chapter 6). Silicon/sesquioxide ratios are lower in more highly weathered soils. In the southeastern United States, lower ratios characterize soils in which kaolinite has accumulated as a secondary mineral. Tropical soils show very low values for this ratio in all horizons; they are dominated by iron and aluminum hydroxides and in highly weathered conditions relatively little Si remains.

One should note that our discussion of forest soil development has touched on patterns that result from differences in weathering, climate, vegetation, and time. The accumulation of secondary minerals (Chapter 6), cation-exchange capacity and base saturation (Chapter 7), and horizon differentiation all interact in soil profile development and in determining site fertility for forest growth. Our treatment, however, considers only general trends. One might expect very different soil formation processes on sandy parent materials, volcanic ash, and wetland sites.

SUMMARY

The accumulation of nutrients within a forest represents many years of input from atmospheric and rock weathering sources. However, the annual intrasystem circulation of nutrients, for example, in plant uptake, is much larger than the delivery of new quantities of available nutrients from these sources (Table 8.5).

Mineralization of nutrients from dead organic matter is the critical process that determines supplies and site fertility.

We have seen how nutrient immobilization by soil microbes may temporarily retard the release of available nutrients from detritus. Nutrients incorporated in humus may be stored for centuries. With this in mind, one might view the nutrient reabsorption by trees (Chapter 7) as a means of "short-circuiting" the intrasystem cycle in a forest ecosystem, as well as a means of increasing nutrient-use efficiency in conditions of limited supply. Note, however, that the abscission of litter with low nutrient concentrations, such as after reabsorptions, would be expected to retard decomposition and mineralization rates (Fig. 8.13). Thus, efficient nutrient use by plants is likely to be balanced by a lower rate of mineralization through microbial processes in the soil (Vitousek *et al.*, 1982; Pastor *et al.*, 1984).

Streamwater losses of nutrients represent the balance between nutrient mineralizations and plant demands in the context of the soil environment in which available ions are retained on exchange sites. At the Hubbard Brook Experimental Forest, losses of mobile forms of limiting nutrients such as NO_3^- are greatest during spring snowmelt, when mineralization most strongly exceeds plant demand (Likens *et al.*, 1977). Loss of anions must be balanced by an equivalent removal of cations (Chapter 6); thus, Gosz *et al.* (1973) found that maximum streamwater losses of Ca, Mg, and K were not during the season of greatest mineralization from fresh litterfall, but instead, were recorded during the springtime periods of high streamflow.

TABLE 8.5 Percentage of the Annual Requirement of Nutrients for Growth in the Northern Hardwoods Forest at Hubbard Brook, New Hampshire That Could Be Supplied by Various Sources of Available Nutrients[a]

Process	N	P	K	Ca	Mg
Growth requirement (kg/ha/yr)	115.6	12.3	67.3	62.2	9.5
Percentage of the requirement that could be supplied by:					
Intersystem inputs					
Atmospheric	18	0	1	4	6
Rock weathering	0	13	11	34	37
Intrasystem transfers					
Reabsorptions	31	28	4	0	2
Detritus turnover (includes return in throughfall and stemflow)	69	81	86	85	87

[a] Nutrient requirements for growth follow the definitions of Chapter 7. Reabsorption data are from Ryan and Bormann (1982); Copyright © 1982 by the American Institute of Biological Sciences. All other data are from Likens *et al.* (1977) and Wood *et al.* (1984).

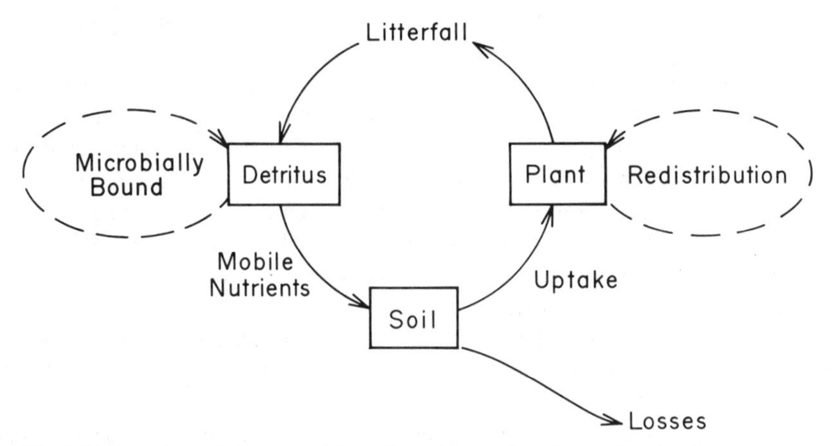

Fig. 8.13. A schematic diagram of the major pathways of nutrient cycling in a forest ecosystem. When available nutrients are limited, nutrient reabsorption by vegetation and nutrient immobilization by soil microbes increase. High nutrient availability may lead to losses of nutrients from the ecosystem, but also to high net primary productivity.

Understanding processes in both the vegetation and the soil allows us to predict the effects of forest management. In most instances, it appears that nitrifying bacteria are less efficient than plant roots in utilizing NH_4^+; thus, where forest growth is rapid, available NH_4^+ is rapidly removed from the soil solution (Chapter 6, Fig. 6.10). When experimental fertilizations with NH_4^+ exceed both plant and microbial demand, the excess will increase rates of nitrification, and NO_3^- will increase in streamwater (Johnson and Edwards, 1979). Similarly, in the absence of plant uptake and microbial immobilization, NO_3^- losses increase upon forest harvest (Vitousek and Melillo, 1979; Vitousek and Matson, 1984). Nitrification and denitrification losses and nutrient immobilizations in microbial tissue and humus explain why such small percentages of fertilizer applications of N are usually found in the living biomass, even after many years of growth (Keeney, 1980).

The humus substances in forest soils represent a large, stable pool of organic C and nutrient elements. The nutrient pool in humus acts to buffer the effects of natural and human disturbance of the living biomass in forest ecosystems. Worldwide, the mass of soil organic matter exceeds living biomass on land by a factor of two to three times. This pool represents a continuing storage of net ecosystem production during forest development. The production of stable humus substances in the soil is a good example of how biotic processes act to retard the expected geochemical flux of some chemical elements from land to the sea. In this sense, the study of soil microbial activity is an essential part of global biogeochemistry.

9

Susceptibility and Response of Forests to Natural Agents of Disturbance

INTRODUCTION

Natural agents of disturbance such as windstorms, fire, and insect and disease outbreaks are usually viewed as unpredictable. Where natural disturbances are prevalent, efforts are often mounted to control their spread or to salvage resources before further damage occurs. In cases where wild fires have been prevented or their spread curtailed, there is often increased mortality of trees attributed to insects or disease. The question arises as to whether forests may not experience times when many of the trees are highly susceptible to agents of mortality, followed by periods of relatively high resistance. If so, preventing mortality of one kind could increase risk from another. A disturbance tends to prevent forests from maintaining maximum canopy leaf area. In that way, it is a thinning agent, often irregular in extent and sometimes very selective.

In Chapters 2 and 3 the effects of reducing forest canopy and fertilizing the soil were shown generally to improve the growth efficiency of individual trees and stands. Absence of disturbance fosters increased competition among trees and reduced growth efficiency (Chapter 3). Individual trees or entire stands may become so limited in resources that few reserves are available for protective responses or for maintaining beneficial associations with symbiotic microorganisms. Under such circumstances, a major disturbance often modifies the entire forest structure. When major disturbances occur, the rate of revegetation may depend on the delivery of propagules by wide-ranging vertebrates.

The chapter is organized so that we first address the impact of natural disturbances upon normal rates of tree mortality. Then the general effects of disturbance upon ecosystem structure and function are analyzed as they influence the availability of carbon (C), nutrients, and water to primary producers. Subtle changes in the chemistry of leaves and other organs may affect the susceptibility of vegetation to herbivory and pathogenic attack. In the last part of the chapter, we review three experimental manipulations that altered the susceptibility of forests to insects, disease, and other agents of disturbance. Observed changes in forest susceptibility are interpreted as predictable responses to linked changes in ecosystem processes and properties.

PATTERNS OF MORTALITY

Self-Thinning and Disturbance-Induced Mortality

In forests where natural thinning operates, the number of trees decreases steadily as stands age (Fig. 9.1). Although the actual number of trees present decreases exponentially, the rate of death each year is rather consistent, often between 1 and 2% (Harcombe and Marks, 1983; Mohler *et al.*, 1978; White, 1981). The number of living trees in fully stocked stands is related to average tree size or biomass (Yoda *et al.*, 1963). In general, the number of trees present in a stand decreases exponentially as the size of individuals increases. This relationship rests on the fact that once a forest reaches maximum leaf area, further growth requires a corresponding reduction in the number of surviving individuals (Mohler *et al.*, 1978).

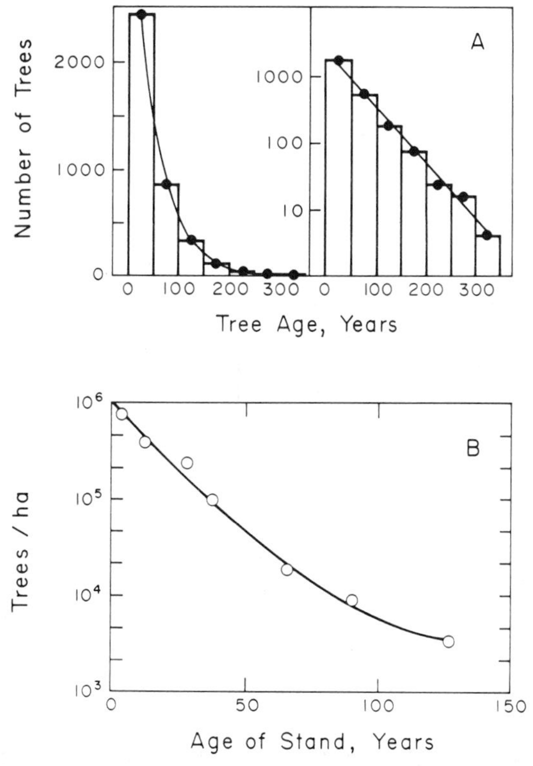

Fig. 9.1. Self-thinning in fully stocked stands tends to cause a relatively constant rate of mortality. (A) age distribution of white oaks (*Quercus alba*) in mature oak–hickory forests. Based on data from some 1500 measurements of 74 ha, cited by Whittaker (1975). (B) from age distribution in pure stands of *Abies* in Japan. (After Tadaki *et al.*, 1977.)

When a natural disturbance such as a windstorm passes through a forest, the short-term pattern of mortality deviates from the pattern expected as a result of competition. For example, in 50 yr, three major windstorms each caused high rates of mortality in forest of Sitka spruce and western hemlock in Oregon (Harcombe, 1985). Yet, when mortality data were averaged over 5-yr periods, the rates were amazingly consistent, averaging 1% per year as a result of compensatory decreases in mortality following disturbance (Fig. 9.2).

Frequency versus Intensity of Disturbance

Forests differ in their susceptibility to disturbance, but all are subject to some agent. When long periods pass with little disturbance, an outbreak of insects, a

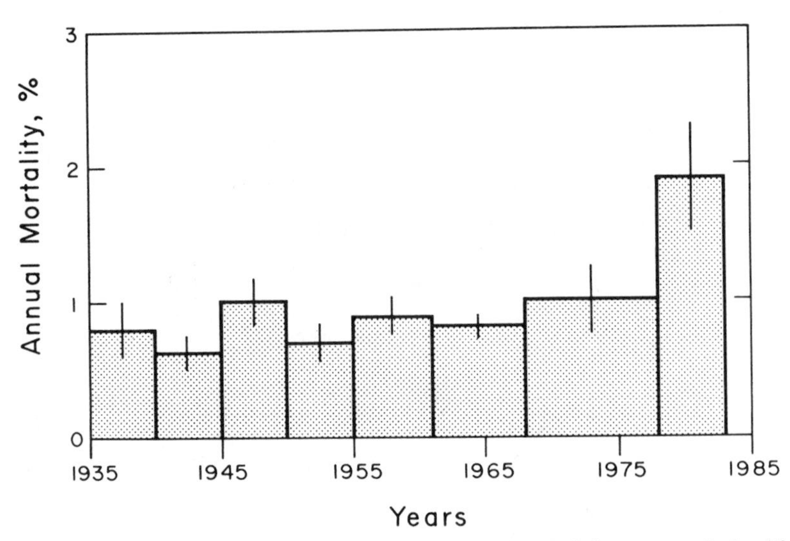

Fig. 9.2. Annual mortality rates in a Sitka spruce–western hemlock forest were calculated from surveys made at approximately 5-yr intervals over a period of 50 yr (Harcombe, 1986). Values are averages of nine plots, each 0.4 ha. Although three major storms increased mortality rates in single years, the effects were unnotable over a decade. Annual mortality for the entire period averaged about 1% (0.9% ± 0.1 SE) of the trees present.

fire, or a windstorm is likely to damage a large portion of the stand. Of various disturbances, the frequency of fire is the best documented because when fire passes through a forest, surviving trees are often scarred and the wounds may be precisely dated.

When fires are frequent, they are confined to the forest floor, killing only very small trees or those with particularly thin bark. Some individual large trees show scars dating as many as 20 separate fires (Arno, 1976). Before the era of fire protection, the probability of fire was closely related to the frequency of thunderstorms (Coulson *et al.*, 1983). Forests in the Gulf Coastal states, the central Rocky Mountains, and the Sierra Nevada of California are particularly prone to lightning strikes in the United States (Brown and Davis, 1973). In the pine forests characterizing the drier, most fire-prone environments in each of these regions, ground fires occurred at intervals of less than 10 yr (Kilgore, 1973; Wagener, 1961; Weaver, 1951; Christensen, 1981). Fires burned at similar frequency through the pine forests along the prairie border in Minnesota (Frissell, 1973). The high frequency of fire prevented fuel from accumulating and prepared a seedbed favorable for perpetuating pines (Cary, 1936; U.S. Department of Agriculture, 1965).

Where fires burn at less frequent intervals, sufficient fuel accumulates to permit fires to kill nearly all trees. With fire frequencies of 25–100 yr, many species regenerate from sprouts, from seeds released from resinous cones, or from seeds derived from long-distance dispersal. The diversity of the original forest vegetation in the Great Lakes–St. Lawrence region depended on such periodic destructive fires (Heinselman, 1973, 1981). Periodic fires at intervals of less than a century have been important in maintaining diversity in other regions (Yarie, 1981; Romme, 1982; Forman and Boerner, 1981; Zackrisson, 1977; Christensen, 1981). Douglas fir forests in many parts of the Pacific Northwest and the Rocky Mountains were regenerated following destructive fires at intervals between 125 and 400 yr (Arno, 1976; Burke, 1979; Romme, 1982). Even the giant sequoia (*Sequoiadendron*) that may live for more than 1000 yr is fire-dependent for perpetuation in its native habitat (Kilgore and Taylor, 1979).

In some areas, such as tropical rain forests and hardwood forests in the northeastern United States, forest fires were historically relatively unimportant (Wein and Moore, 1977; Fahey and Reiners, 1981; but see Sanford *et al.*, 1985). In these areas, however, windstorms and periodic insect outbreaks often serve as agents of disturbance (Bormann and Likens, 1979b; Morrow and LaMarche, 1978; Worrell, 1983; Janzen, 1975).

With modern human activity, the area burned annually is likely to decrease as protection of property investments becomes important. In most areas of the western United States, wildfires are now contained within a day, or at most a few weeks, where previously some burned across the landscape for months (Burke, 1979). Throughout the United States, even in dry ponderosa pine forests, extensive fires have not burned for the last 50–70 yr (Mitchell and Martin, 1980; Martin, 1982; McCune, 1983; Fahey and Reiners, 1981). Where fire frequencies have been drastically reduced, the short-lived fire-dependent species have been replaced (Heinselman, 1981) and the long-lived species are threatened by an abnormally high accumulation of dead material and by the growth of understory species that transmit fire to the upper canopy (Romme, 1982; Parsons and DeBenedetti, 1979).

An analysis of fire histories suggests that frequent, moderate burns were typical in fire-adapted forests and that it was the infrequent fire that was damaging. This principle has been tested by accident when wildfires have entered forests in which previous controlled burning was conducted. Crown fires often drop to the ground and burn harmlessly through the treated areas (Davis and Cooper, 1963; Biswell *et al.*, 1973). Prescribed fires are now used to protect and perpetuate pine plantations in the southeastern United States (Davis and Cooper, 1963; Sackett, 1975) and in Brazil (Goldammer, 1983).

High-intensity windstorms as associated with hurricanes, typhoons, and tornadoes can destroy whole stands. Usually the destruction is incomplete. For example, in 1938 a hurricane struck 4.5 million ha of forest in the northeastern United States (Smith, 1946) but only about 5% of the stands experienced heavy damage. Otherwise, damage was restricted to defective trees or to those occupying particularly wind-prone sites (reported in Bormann and Likens, 1979a; Henry and Swan, 1974).

Windstorms in tropical rain forests (Lugo *et al.*, 1983) cause frequent treefall, averaging about one tree per hectare per year with an impact area of about 120 m^2 (Hartshorn, 1978). The regularity of these disturbances constrains the maximum age of trees in forests of Costa Rica to a range of 80–140 yr. Comparable rates of wind-throw are found elsewhere in the tropics (Putz and Milton, 1982) and in northern hardwood forests (Henry and Swan, 1974; Bormann and Likens, 1979b). The latter authors emphasized that many of the wind-thrown trees appeared susceptible because of previous decay or general senescence.

The comparative resistance of trees to blowdown is somewhat similar to the situation with fire. Certain trees are more susceptible than others. When leaning and defective trees are removed early in stand development, the open canopy encourages more wood to be produced near the base of the remaining trees, which strengthens the bole against the forces of wind or heavy snow (Petty and Worrell, 1981). Trees that have grown in dense stands all their life, on the other hand, are particularly susceptible to wind-throw when surrounding trees are removed.

Wind-thrown and lightning-killed trees provide habitat for populations of bark beetles and some disease organisms (Coulson *et al.*, 1983). Often, apparently healthy trees are infected or attacked when they are adjacent to dead trees (Bakke, 1983). After a few years, however, mortality rates in the stand decrease, often to below normal rates.

Where fire- or wind-caused mortality is absent or very low, the probability of insect and disease outbreaks increases. In the Rocky Mountain region, where fire frequency has been reduced by two orders of magnitude in the last 70 yr, outbreaks of defoliating insects have become increasingly prevalent (McCune, 1983; Johnson and Denton, 1975; Fellin, 1980). Similarly, the extensive forests of lodgepole and ponderosa pine that once experienced large, infrequent forest fires are now widely attacked by bark beetles and other insects. Damage from root rots and mistletoe has also increased during recent decades (Wicker and Leaphart, 1976; Fellin, 1980). Where insect outbreaks and windstorms are rare, pathogens often kill large patches of forest and start a regeneration cycle (McCauley and Cook, 1980; Barklund and Rowe, 1981; Setliff *et al.*, 1975). Normally, pathogens are more likely to attack overtopped trees or those sustaining injury from other causes. Even where pathogens kill dominant trees as well as suppressed individuals, stand growth rates may be maintained over a decade while stocking declines 30% (Oren *et al.*, 1985).

Biological agents of disturbance are usually native components of ecosystems. Where they have been recently introduced into an area, however, extensive tree mortality may result (Mattson and Addy, 1975; Manion, 1981). This situation is observed in the United States as a result of the introduction of Gypsy moth

(*Porthetria dispar*), white pine blister-rust (*Cronartium ribocola*), Dutch elm disease (*Ceratocystis ulmi*), and chestnut-blight (*Endothia parasitica*). Although these biological agents of disturbance were recently introduced, the total impact upon stand growth is now relatively minor as other species or more resistant varieties of trees become dominant and the disturbing agents become less virulent (Doane and McManus, 1981; Campbell and Sloan, 1977; Jaynes and Elliston, 1982). Tropical rain forests provide an extreme case in which the variety of herbivores and host trees is so high that specific mortality in one group of primary producers has little total effect on ecosystem function or structural diversity (Janzen, 1981). Moreover, insects, which are the major tropical forest herbivores, consume less than 1% per day of the new growth and a maximum of 0.5–0.25% per day of older foliage on climax and pioneer tree species, respectively (Coley, 1983).

We conclude this section with the observation that frequent but moderate disturbance initiated by fire, insects, wind, or disease tends to have little long-term effect on forest structure, growth rates, or other ecosystem properties. Infrequent disturbance, on the other hand, changes forest structure and alters the rates of various processes, often favoring the perpetuation of certain types of forests. Attempts to reduce the frequency of disturbance by fire protection or widespread application of insecticides may lead to situations where large areas of otherwise unmanaged forests are more susceptible than normal to catastrophic disturbances.

EFFECTS OF DISTURBANCE ON ECOSYSTEM STRUCTURE AND FUNCTION

Selective Effects of Animals

We have mentioned how wind, fire, and certain introduced pathogens can selectively remove certain types of trees from a forest. Herbivores, both large and small, may also have very selective impact on forest growth and composition. Grazing and browsing vertebrates can have a profound impact on forest composition when trees are seedlings and saplings. For example, browsing in the winter by deer and small mammals can completely remove some tree species from European forests (Peterken, 1966). Similar examples are documented from other regions (Ross *et al.*, 1970; Crawley, 1983).

In general, large herbivores such as vertebrates exert little direct effect on forest canopies once trees are well established. However, when numerous species of vertebrate herbivores were introduced into New Zealand, a land previously without native mammals (except for bats), large areas of native forests were greatly affected by overbrowsing of understory trees and shrubs and by compaction of the soil by large herds of animals uncontrolled by predators (Howard, 1964, 1967).

The most significant effect that vertebrates have on forest composition is not through browsing, however, but through selective dispersal of propagules across inhospitable environments. Vertebrates, particularly large vertebrates and birds, move long distances carrying seeds in their digestive tracts or on their bodies. In extreme cases, some tree species have seeds that only certain kinds of vertebrates can disseminate successfully. The tree *Calvaria major,* a native on the island of Maritius in the western Indian Ocean, had no natural regeneration for a period of 300 yr because its seeds needed to be processed through the extinct dodo bird, *Raphus cucullatus* (Temple, 1977). In Africa, elephants consume great quantities of the fruits of *Balanites willsoniana.* When elephants are absent, the fruits, which are generally toxic to other animals, rot. Janzen (1979) observed that *Simaba cedron* trees growing in tropical forests of Central America share fruit characteristics with *Balanities* and concluded that the present restricted distribution of *Simaba* is related to the extinction of mastodons in the last 10,000 yr. Some seed coats are so hard that only a few animals can crack them. The reintroduction of the horse into Central America by Spaniards increased dispersal of the seeds of such species (Janzen and Martin, 1982).

Rapid recolonization of forest sites following fire or other major disturbance may well depend on the presence of certain vertebrates. The truffles, fungi that produce potato-shaped fruiting bodies underground, form important symbiotic mycorrhizal associations with tree roots (Chapter 7). These fungi are almost exclusively hunted for, stored, and dispersed by small mammals (Maser *et al.*, 1978). Loss of small mammals from an ecosystem would reduce the likelihood of barren areas being revegetated with mycorrhizae of the truffle group.

Besides humans, birds have the greatest impact on long distance seed dispersal (Cain, 1944; Good, 1953; Livingstone, 1972). Blue jays (*Cyanocitta cristata*) cached more than 50% of the entire mast crop of an oak forest an average distance of more than 1 km from the source (Darley-Hill and Johnson, 1981; Chettleburgh, 1952). The ability of birds and small mammals to cache large nuts may help explain how hazelnut (*Corylus*) was able to colonize the landscape after the continental glaciation in Europe as fast and sometimes faster than species with wind dispersed seeds, birch and pine (Walter, 1954). Insects, birds, and bats also play important roles as pollinators (Crawley, 1983). Animals, through their eating habits and movement, may concentrate nutrients within one system or distribute them between ecosystems. For example, egrets and pelicans distribute resources from the sea to mangrove ecosystems (Onuf *et al.*, 1977); salmon move upstream to spawn and die, providing food to eagles and nutrients to the nearby forest (McClelland, 1973); and hippopotami move resources from the land into the water (Viner and Smith, 1973). Large vertebrates also affect forest ecosystems by changing the microhabitat through trampling, burrowing, or in the case of animals such as beavers, by felling trees (Chapter 10). The absence of large animals can, therefore, indirectly limit the populations of other animals and some plants. For example, when 10-ha segments of Amazon forests were isolated, piccaries no longer provided wallows and three species of frogs disappeared, along with other types of animals and plants (Lewin, 1984).

The role that large or small predators play in controlling herbivore populations is still not critically established, although in forests and other ecosystems where pesticides have been widely applied, rare insect herbivores sometimes become more abundant (Crawley, 1983; Matteson *et al.*, 1984; Metcalf and Luckmann, 1982). There is some consensus that herbivores at endemic population levels may be held in check by predators, but once herbivore populations increase substantially, only limited food supply, unfavorable weather, or disease are likely to be significant to restore populations to normal levels (May, 1981; Crawley, 1983). For example, when birds were prevented from foraging for insects on an understory deciduous shrub in a northeastern hardwood forest, populations of Lepidoptera larvae increased substantially (Holmes *et al.*, 1979). On the other hand, in conditions when Gypsy moth larvae defoliated more than 90% of all plants in these forests, bird populations were unable to exert any measureable control (Doane and McManus, 1981).

There is an important and often unappreciated link between food quality and predation. Reduced palatability or scarcity of food requires herbivores to spend more time exposed to predators while satisfying their nutritional requirements (Price *et al.*, 1980). Only a fraction of a percent change in the growth or reproductive rates of a herbivore may shift the balance from an endemic to an epidemic population (Van Emden, 1966). Similarly, small changes in the digestibility and nutrient balance of forage can alter dramatically the susceptibility of herbivores to parasites and pathogens (Price *et al.*, 1980).

In summary, complex food webs exist in most ecosystems; there are close interactions among populations of herbivores, their predators, parasites, and host plants. Normally, herbivores consume much less than 10% of aboveground primary production (Crawley, 1983). Selective herbivory, however, can affect forest composition significantly. At times certain invertebrates, even those native to a forest, may increase to epidemic levels sufficient to alter the rates of many processes and to affect the general structure and composition of the forest. Usually vertebrates play a much smaller role in forests than invertebrates. Following destruction of the forest cover, however, vertebrates, particularly birds, small mammals, and humans, may speed the rate of tree reestablishment by transport and preparation of seeds and spores and by favorable modifications in the microhabitat conducive to seed germination and seedling establishment. Selective browsing on young seedlings and saplings may also strongly affect the composition of forests.

General Effects of Reducing Canopy Leaf Area

The influence of moderate and frequent disturbance on forest ecosystems may be similar to a silvicultural system in which the forest canopy is reduced and small amounts of fertilizer are spread upon the forest floor. This sequence happens following fire (Wright and Bailey, 1982), insect (Mattson and Addy, 1975), and disease-induced mortality (Matson and Boone, 1984). Table 9.1 summarizes some anticipated responses in the ecosystem.

A reduction in canopy leaf area increases the penetration of radiation (Chapter 2) and precipitation (Chapter 5) to the forest floor. The soil temperature increases and the reduction in transpiration temporarily increases the available water supply (Chapters 4 and 5). Increasing the soil temperature and moisture stimulates microbial activity and mineralization (Chapter 8), and an additional supply of fresh litter, insect frass, or ash should further enhance these processes (Owen, 1980). During the recovery of the forest, nutrient and water uptake will increase per unit of leaf area (Chapters 3, 5, and 7). Additional light penetration generally increases photosynthetic rates in the lower canopy (Chapter 2), and additional access to water and essential minerals means plants allocate proportionally less carbohydrate to roots. For these reasons, the rate of wood production per unit of

TABLE 9.1 Ecosystem Responses Predicted from Moderately Increased Mortality in Forests at Maximum Leaf Area

A. Structural modifications	
1. Canopy leaf area index	(−)
2. Quality of detrital biomass	(+)
3. Total nutrient supply in soils	(+)
4. Water supply in soils	(+)
B. Environmental modifications	
1. Radiation through canopy	(+)
2. Precipitation reaching ground	(+)
3. Soil temperature	(+)
C. Functional modifications	
1. Mineralization	(+)
2. Water and nutrient uptake	(+)
3. Photosynthetic rates	(+)
4. Growth efficiency	(+)
5. Susceptibility to attack	(−)

leaf area should increase (Mattson and Addy, 1975; Chapters 2 and 3). Associated with a more open canopy, the foliage develops structural and biochemical characteristics that decrease the efficiency of herbivores and pathogens (Schultz and Baldwin, 1982; Piene and Percy, 1984). We return to these predicted responses in later sections as we interpret the effects of specific manipulations on the functioning of the trees and the disturbing biological agents.

General Changes in Host Susceptibility

In a seminal paper, Mattson and Addy (1975) emphasize that outbreaks of defoliating insects may sometimes consume up to 40% of the foliage without necessarily reducing stand growth (also see Kulman, 1971). There is increased growth efficiency by the remaining trees because fewer leaves are involved in the production process. Although nitrogen (N) content in the remaining foliage usually increases (Heichel and Turner, 1983; Piene, 1980), a primary change involves improved water relations (Chapter 4), permitting stomata to open more widely at similar levels of irradiance. This results in increasing net photosynthetic rates by 50% in hardwoods such as oak and maple (Heichel and Turner, 1983). At the same time, foliage exposed to higher irradiance is likely to be thicker (Chapter 2), tougher due to higher fiber content, more covered with hairs and waxes, and lower in water content per unit mass than in the original canopy (Levitt, 1980, Chapters 2 and 4). These characteristics alone are likely to reduce the efficiency of herbivores and pathogens (Schoeneweiss, 1975; Gilbert, 1979; Rhoades and Cates, 1976; Crawley, 1983; Coley, 1983).

Mattson and Addy (1975) further observed that insect outbreaks rarely persist in a given stand for more than 2–3 yr. During that time, considerable tree mortality may occur (Kulman, 1971) and populations of insect predators and parasites may increase (Crawley, 1983). It is possible that surviving trees develop leaves and other organs anatomically and chemically less susceptible to insect herbivores. In a few long-term studies, surviving trees have, after a delay, shown improved growth rates that continued for some decades following extensive defoliation (Wickman, 1980).

Host Biochemistry and Susceptibility

Generally, plant tissue is only marginally nutritious to most insects and pathogens. A variety of changes in biochemistry can directly increase plant susceptibility to attack from insects and pathogens by making the tissue more nutritious and accessible. Increasing concentrations of soluble N and carbohydrates in the cells or exudates tends to favor colonization by microbes and increasing activity by herbivores. The kind of sugar or amino acid can be as significant as the amount (Hare, 1966; Li and Bollen, 1975; Schoeneweiss, 1975; Van Emden and Bashford, 1971; Hale *et al.*, 1982). Nitrogen in the form of amino acids is usually more readily available to herbivores and pathogens than N in protein (McNeil and Southwood, 1978; Fox and McCauley, 1977; House, 1971).

Predisposition of plants to attack may often be traced to climatically induced stresses and to deficiencies or excesses in available water and nutrients (Lambert and Turner, 1977; Onuf *et al.*, 1977; Mattson, 1980; Hesterberg and Jurgensen, 1972; White, 1969, 1974; Worrell, 1983). Concentrations of sugars and amino acids within plant tissue may be unusually high when plants are stressed or supplied with excess nutrients or water. On the other hand, carbohydrate reserves, such as starch, and certain proteins may be in short supply when plants are growing either rapidly or extremely slowly (Levitt, 1980; Chapter 2).

Along with nutritional compounds in leaves and other tissues, we must consider a large variety of secondary metabolites suspected to have a function in defense against herbivory and pathogenic infection. Two broad categories may be identified, based on whether N is included in their structure. Compounds that contain N include cyanogenic glucosides, alkaloids, and nonprotein amino acids. Defensive compounds without N include tannins, terpenes, phytoalexins, steroids, and phenolic acids. Each kind of compound may serve in a variety of ways against various organisms (Table 9.2).

Some plants produce fungistatic and bacteriostatic compounds that prevent colonization by pathogens. Other compounds act as physical barriers, such as waxes on the leaf surface or resins or lignin in cell walls. Increasing fiber content decreases digestibility of plant tissue and reduces herbivore growth rates and survival (Van Soest, 1967; Crawley, 1983). Tannins have the ability to precipitate protein, which inhibits most enzyme reactions and makes the proteins present in plant tissue nutritionally unavailable to most animals and microbes (Zucker, 1983). Phytoalexins are lipid-soluble compounds, activated only following an attack by pathogens, and exhibiting antibiotic properties (Harborne, 1982). The alkaloids are found in angiosperms and are particularly toxic to a variety of mammals (Freeland and Janzen, 1974; Swain, 1977).

Changes in host biochemistry may also reduce the colonization by organisms that are helpful to the host plant. These include protective ant colonies, microbes that graze on bacteria, and symbiotic associations with N-fixing bacteria and mycorrhizal root fungi. These beneficial organisms may directly infect or prey upon attacking organisms, release antibiotics, or provide nutrients (Tilman, 1978; Clarholm, 1981; Coleman *et al.*, 1983; Marx, 1969). Plants provide resources for the growth of various beneficial organisms in the form of litterfall and various leaf or root exudates, including a variety of polysaccarides, organic

TABLE 9.2 Major groups of Secondary Plant Metabolites Known to Contain Products Important for Defense[a]

Class	Number known	Contains N	Protection against:
Alkaloids	1000	Yes	Mammals
Amino acids	250	Yes	Insects
Ligans	50	No	Insects
Lipids	100	No	Fungi
Phenolic acids	100	No	Plants
Phytoalexins	100	No	Fungi
Quinones	200	No	Plants
Terpenes	1100	No	Insects
Steroids	600	No	Insects

[a] From Swain (1977).

acids, and amino acids. Many of these latter compounds are also used by herbivores and pathogens, so the loss of beneficial associates may favor attack. Some compounds produced by plants are detrimental to the germination or growth of associated vegetation. These include various phenolic acids, quinones, and a host of other chemicals known collectively as allelopathic (MacClaren, 1983).

In general, defensive compounds that lack N reach rather high concentration in cells, often 10–15% by weight, whereas N-containing compounds are usually at levels below 1%. Plants use less energy in the synthesis of N-containing defensive compounds, but compounds without N tend to be more stable (Fox, 1981). Nitrogen-containing compounds are most frequently found in deciduous, fast-growing vegetation, whereas defensive compounds without N are more characteristic of slow-growing plants, particularly evergreens (Gartlan *et al.*, 1980; Mooney and Gulmon, 1982; Bryant *et al.*, 1983). Regardless of the defensive compound synthesized, no plant is completely immune to attack. In fact, there are specialized insects and pathogens that not only detoxify toxic compounds but actually require them for optimal growth (Bernays and Woodhead, 1982). These organisms are restricted to a few host species, but they may attack even vigorous individuals (McLaughlin and Shriner, 1980). Other less-specialized organisms accommodate a wider range of biochemical challenges and attack a wider variety of plants. Tropical rain forests characteristically contain many specialized insects and other animals that feed on selected species with certain chemical compositions. The year-round presence of insects in the tropics can result in heavy herbivory upon some trees (Walda and Foster, 1978; Coley, 1982) and may help account for the floristic richness and relative isolation of tree species (Gillett, 1962; Connell, 1971; Leigh, 1982). Experimental evidence supporting rarity as an adaptation against herbivory, however, is very limited and not conclusive (Coley, 1983).

The selection of host plants by attacking organisms probably involves a large number of compounds volatilized or exuded by plants. Adult insects seeking to lay eggs on a suitable host may use their antennae to sense volatile compounds at levels as low as 10^{-12} g/cm^3. By direct tasting, insects may discriminate non-volatile compounds at concentrations of 1 mg/1000 cm^3 in tissue, which is far below toxic levels (Swain, 1977). To meet this latter challenge, many plants are able to produce toxic compounds quickly and to construct barriers that consist of dead, resin, or gum-filled tissue almost immediately following attack (Schultz and Baldwin, 1982; Haukioja and Niemelä, 1979; Raffa and Berryman, 1982, 1983; Hare, 1966). In response to localized insect activity, foliage throughout an entire tree may become less palatable (Haukioja and Niemelä, 1979). Morphological responses such as stiffer thorns on *Acacia* trees may also be induced by grazing (Seif el Din and Obeid, 1971). Why the woolly aphid (*Adelges piceae*) only reaches epidemic populations on an introduced fir is due to the ability of the native host to slough bark following attack (Kloft, 1957).

Induced responses to attack are only effective if sufficient resources can be quickly mobilized (Croteau *et al.*, 1972; Firmage, 1981; Lagenheim *et al.*, 1981). The rate at which stored carbohydrates or proteins may be converted into mobile forms (sugars and amino acids) and transported to sites of attack may be limiting (McLaughlin and Shriner, 1980). This may not affect canopy responses to partial defoliation if photosynthetic rates are high. However, changes in the allocation of current photosynthate to remote organs such as the lower bole or roots cannot be accomplished rapidly because of the distance involved and limitations in phloem transport (Chapter 2). For this reason, concentrations of stored reserves in roots, stems, and twigs are a good indicator of a tree's potential to survive localized attack by insects or pathogens (Ostrofsky and Shigo, 1984). Wargo *et al.* (1972) demonstrated that defoliation of sugar maple (*Acer saccharum*) greatly reduced starch content in roots at the end of the growing season. Low starch reserves in the roots predisposed trees to attack by root rot (Wargo, 1972).

A variety of environmental stresses may reduce a tree's ability to respond to attack. Limitations in sulfur (S) (Lambert and Turner, 1977), phosphorus (P) (Bowen, 1969), and micronutrients (Lambert and Turner, 1977), reduced photosynthesis from shading (Durzan, 1974; Li and Bollen, 1975), or excess metabolic expenditures (Wargo *et al.*, 1972) have all been documented to increase amino acid levels and make plants susceptible to a variety of root diseases. Drought, likewise, can initiate a breakdown in leaf protein and eventually reduce starch reserves (Bradford and Hsiao, 1982; Hsiao, 1973). Drought makes the foliage of Douglas fir and white fir (*Abies concolor*) more palatable to spruce budworm and tussock moth larvae (McMurray, 1980).

Plants depending on defensive compounds rich in N are at a competitive disadvantage whenever N is in short supply. On the other hand, plants producing C-rich defensive compounds are at a disadvantage when growing in shade with an abundant supply of N (Bryant *et al.*, 1983; Cates *et al.*, 1983). Host plants must maintain a balance between carbohydrate and N resources that can be mobilized quickly. Excess starch cannot be mobilized without N to construct critical enzymes; excess N cannot be combined into protein without adequate carbohydrates. Those plants native to fertile sites may be expected to build a variety of defensive compounds from N. Thus, alkaloids predominate in the foliage of trees in many lowland tropical forests where N is relatively abundant (McKey *et al.*, 1978). Plants growing in areas where N is scarce will generally form C-enriched defensive compounds, as do tropical forests growing on extremely sterile soils (Gartlan *et al.*, 1980). This pattern has been widely observed in boreal and temperate forests as well (Rhoades and Cates, 1976; Bryant *et al.*, 1983). At the time of foliage elongation, when N is relatively available even for plants growing in nutrient-poor habitats, N-based defensive compounds may also be produced (Prudhomme, 1983; Dement and Mooney, 1974).

Tissue Quality and Growth Efficiency

Having developed a case for the importance of a balance between available carbohydrates and N compounds to inhibit certain types of herbivores and pathogens, we now seek a more explicit link between these resources and wood production per unit of leaf area (Chapters 2 and 3).

The availability of a moderate pool of amino acids in the leaf or other tissue is of critical importance in permitting rapid and continuous growth (Bradford and Hsiao, 1982). Without such a source, protein must be broken down in older tissue, which results in a reduction in photosynthesis and carbohydrate assimilation. Beyond a certain level, however, added N is not converted into enzymes or pigments but rather is stored as amino acids or in extreme cases as nitrate (Pate, 1980; Titus and Kang, 1982).

When imbalances in C and N occur, resources are shifted to their most critical use, usually favoring some adjustment of the imbalance. Reduced N uptake results in lower enzyme activity and photosynthetic assimilation (Chapter 2). The relative excess of carbohydrates may initially go into construction of thicker cell walls, increased phenolic or tannin production, or starch. Eventually, however, carbohydrate allocation shifts disproportionately to the roots, favoring uptake of more N and water per unit of leaf area (see review by Waring, 1983, and Chapter 2).

Reduced irradiance can lead to an excess of N accumulation. Excess N may be expected first to accumulate in tissue. Because amino acid synthesis is energy consuming, resource allocation to root growth will be reduced. Reduced root growth permits shoots to expand, which in forests results in more access to light and enhanced C uptake. A reduction in allocation to roots can be harmful if other minerals are limiting (Chapter 7) or if protective or symbiotic associates are lost (Same *et al.*, 1983).

By providing trees with a balanced source of nutrients, water, and light, the relative proportion of carbohydrates allocated to roots may be reduced by more

TABLE 9.3 Chemical Composition and Performance of Willow Leaves (*Salix aquatica*) under Specified Environments Permitting Constant Relative Growth Rates[a]

	High light, high nutrients	Low light, high nutrients	High light, moderate nutrients
	A. Compound		
Phenolics (relative units)	100[1]	31[2]	79[1]
Tannins (relative units)	65[1]	64[1]	100[2]
Leaf nitrogen (mg N/dm^2)	21.5[1]	13.4[2]	14.0[2]
Amino acids (mg N/dm^2)	2.4[1]	2.3[1]	0.9[2]
Nitrate (mg N/dm^2)	1.0[1]	1.7[2]	0.0[3]
Starch (% dry weight)	5.1[1]	5.3[1]	20.7[2]
Lignin (% dry weight)	20.8[1]	13.4[2]	24.5[3]
	B. Variable		
Photosynthesis (mg $CO_2/dm^2/h$)	12.5[1]	1.9[2]	13.1[1]
Dark respiration (mg $CO_2/dm^2/h$)	7.4[1]	5.5[2]	2.9[3]
Relative growth rate (% per day)	16.1[1]	6.8[2]	5.5[2]
Unit leaf rate (mg shoot/dm^2 leaf/day)	105[1]	28[2]	67[3]
Specific leaf weight (g dry weight/dm^2)	0.48[1]	0.28[2]	0.80[3]
Herbivory (mg consumed/g leaf/day)	0.18[1]	1.00[2]	0.23[1]

[a] Values with different superscripts differ significantly at $p = 0.5$. From Waring *et al.* (1985); Larsson *et al.* (1986).

than 50% and growth efficiency improved by more than threefold compared to unfertilized and unirrigated trees (Axelsson, 1981; Waring, 1983; Chapter 2). The implications of such changes in allocation may extend to protecting forest vegetation in general and to increasing the amount of harvestable material in particular.

Modifying the relative availability of carbohydrates and amino acids affects the production of defensive compounds such as tannins and phenolics in willow clones growing at constant relative growth rates under stable environments (Table 9.3). Plants grown under high light had similar photosynthetic rates but those with unlimited N and other minerals had significantly higher unit leaf rates (closely related to growth efficiency index) than plants with moderate nutrient supply. With moderate nutrient supply, leaves of plants growing under high light converted more carbohydrates into tannins, starch, and lignin as compared to plants with higher supplies of N and other nutrients. Increasing the supply of N greatly increased amino acid content and led to the appearance of nitrate in leaf tissue. Finally, when leaf-defoliating beetles (*Galerucella lineola*) were provided free access to all willows, the insects fed selectively on willow leaves with the highest amino acid and lowest levels of available carbohydrates and defensive compounds (low-light, high-nutrient treatment). High levels of herbivory were associated with particularly low rates of photosynthesis as well as with qualitative and quantitative reductions in phenolics (Larsson *et al.*, 1986).

EXPERIMENTAL MANIPULATIONS

To illustrate principles that determine the susceptibility of primary producers to agents of disturbance and to link these principles to ecosystem level responses (Table 9.1), we have chosen two simple examples in pure stands of conifers and one example of a disturbance that at present affects a variety of temperate forests. In each case a manipulation of the system leads to series of predicted responses culminating in increasing or decreasing susceptibility of the trees to disturbance agents. Regrettably, few such experiments have been conducted, so we draw heavily on local experience. The three manipulations are an experiment with bark beetles, another with root rot, and a third with air pollution.

Pine Forest and Bark Beetles

Epidemic outbreaks of mountain pine beetle (*Dendroctonus ponderosae*) in widely distributed forests of lodgepole pine (*Pinus contorta*) offered an ideal situation for experimentation in eastern Oregon. Waring and Pitman (1985) established plots in 120-yr-old stands to examine whether limited N or light led to low growth efficiency and insect attack. Available N was known to be extremely low from soil incubation tests (Table 9.4). Treatments included (1) N fertilizer; (2) fertilizer combined with a reduction in the canopy of about 80%; (3) added sugar and sawdust to limit mineralization by microorganisms; and (4) untreated plots. Uniform dispersal of insects throughout treatments was encouraged by baiting each plot with equal amounts of synthetic attracting pheromones. In plots with canopy reduction, the larger-diameter trees were purposely left because they were assumed to be the most susceptible to attack (Cole and Amman, 1969; Amman and Baker, 1972; Berryman, 1976). In fact, many trees showed scars of attack. Changes in various ecosystem properties were followed over a 3-yr period.

Fertilizing initially increased available N in the soil by 10-fold. Foliar N paralleled increases in mineralizable soil N. To a lesser, but significant amount, N levels also increased in leaf litter, fine roots, and inner bark following fertilization. No changes in the availability of water were noted, in part because grass grew luxuriantly on the fertilized treatments with thinning. Additions of carbohydrates prevented significant increases in available N over the entire experimental period (cf. Chapter 8).

TABLE 9.4 Changes in Carbon and Nitrogen Indices after Thinning and Fertilization of a 120-Yr-Old Lodgepole Pine Forest[a]

Index	Untreated	Sugar and sawdust	N added	Thinned + N
Carbon				
Initial LAI	4.7[1]	4.9[1]	4.5[1]	1.0[2]
Growth efficiency (g wood/m^2 leaf/yr, after 3 yr)	73[1]	88[1]	108[2]	120[2]
Nitrogen				
Current leaves (after 1 yr, g/m^2)	5.1[1]	5.2[1]	6.8[1]	6.7[2]
Inner bark (after 2 yr, %)	0.2[1]	0.2[1]	0.3[2]	0.4[2]
Fresh litter (g/m^2)	1.6[1]	1.4[1]	2.1[2]	2.0[2]
Soil mineralizable (N, ppm, after 1 yr)	4.4[1]	3.8[1]	48.8[2]	40.7[2]

[a] Treatment mean values not differing at the 5% level of significance share a common superscript, 1 or 2. From "Modifying lodgepole pine stands to change susceptibility to mountain pine beetle attack" by R. H. Waring and G. B. Pitman, *Ecology*, 1985, **66**, 889–897. Copyright © 1985 by the Ecological Society of America. Reprinted by permission.

TABLE 9.5 Growth Efficiency (Wood Production per Unit Leaf Area) under Various Treatments (n = 12)[a]

Treatment	Growth efficiency (g wood/m² foliage/yr)			
	1979	1980	1981	1982
Control	51.0	59.0	73.0	65.6
Sugar and sawdust	76.0	80.8	88.3	72.2
Fertilized	66.9	83.8	108.0	87.1
Fertilized and thinned	77.0	95.0	120.0	115.5

[a] Means connected by brackets are significantly different at $p = 0.05$. From Waring and Pitman (1983).

Insects attacked the treatments rather broadly the first year, when growth efficiencies averaged 50–80 g wood production per square meter of foliage annually (Table 9.5). After 2 yr, the surviving trees in both fertilized treatments, thinned by man or beetles, were able to survive all insect attacks, and growth efficiencies were above 100 g of annual wood production per square meter of foliage (Waring and Pitman, 1983). In Fig. 9.3, a precise relationship is shown between growth efficiency and the susceptibility of single trees to beetle attack. We note that the number of insects required to kill lodgepole pine increased linearly with improving growth efficiency up to the threshold value. Above a growth efficiency of 100 g wood per square meter of foliage, beetles did not attack at sufficient densities to overcome defensive responses of the more vigorous trees.

In the stands that received additions of carbohydrates, insect attacks continued unabaited for 3 yr (Waring and Pitman, 1985) with surviving trees continuing to grow at susceptible growth efficiencies. Tree mortality rates in the untreated stands began to drop notably by the third year as a result of a beetle-induced thinning response that provided significant increases in available soil N as compared to carbohydrate-enriched plots (Waring and Pitman, 1985). Growth efficiency values also indicated increased resistance to beetle attack by the third year in untreated stands (Table 9.5).

The association between growth efficiency thresholds, attack densities, and resulting mortality matched patterns described in thinning studies of lodgepole pine (Mitchell *et al.*, 1983) and ponderosa pine (Larsson *et al.*, 1983). Starch reserves in the sapwood initially were below 1%, but following the conclusion of the experiment, starch was notably higher in surviving trees. This experiment

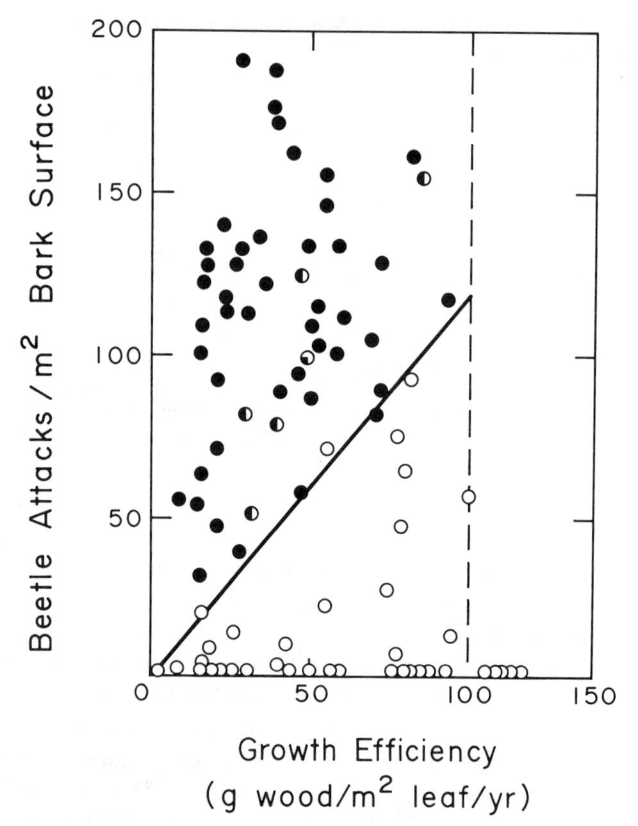

Fig. 9.3. Growth efficiency provides an index to the density of bark beetle attack required to kill lodgepole pine trees. Filled or partly filled circles represent the proportion of conducting tissue killed on attacked trees. Open circles represent trees able to halt all beetle attacks before any conducting tissue was killed. The dotted vertical line indicates the boundary above which beetle attacks are unlikely to cause tree mortality. (From Waring, 1983; Waring and Pitman, 1983.)

indicates that management of overmature forests can rapidly stimulate changes in the system. Removal of trees for harvest may increase growth efficiency and reduce the probability of disturbance to the remaining trees (Sartwell and Stevens, 1975; Mitchell *et al.*, 1983; Larsson *et al.*, 1983). In the absence of harvest, insect attack may produce similar results.

From a wide range of western coniferous forests, Schroeder *et al.* (1982) reported that nearly all unmanaged stands have average growth efficiencies below apparent safe values. Many of these forests are now experiencing defoliation by tussock moth and spruce budworm. Forests prone to massive outbreaks of native insects may be managed to maintain growth efficiencies above an empirically defined susceptible level, as suggested by Larsson *et al.* (1983) and Pitman *et al.* (1982).

Mountain Hemlock Forest and Root Rot

A few studies with pathogens have involved controlled inoculation of trees in various susceptibility classes (Rishbeth, 1951; Gibbs, 1967; Raffa and Berryman, 1982), but in most studies growth efficiency and other ecosystem processes have not been evaluated. We selected an experiment in subalpine forests of mountain hemlock (*Tsuga mertensiana*) where laminated root rot (*Phellinus weirii*) causes mortality. McCauley and Cook (1980) described wavelike patterns of mortality in the Oregon Cascade Mountains and noted that when older stands are killed, reinfection (or more likely successful infection) does not occur for 85 yr or more (Hadfield and Johnson, 1977). Mineralizable N in the undisturbed forest soils is very low but increases significantly in the recently disturbed zones (Fig. 9.4). Growth efficiency of trees in the zone that experiences release or regeneration improves measureably following death of the overstory (Fig. 9.4). The pathogen is known to remain viable in dead roots and stumps for over 50 yr, and it may reinfect trees of all ages and sizes. However, successful reinfection is unlikely until growth efficiency decreases, approaching values similar to those observed in the mature forest.

Similar examples of wave mortality have been reported in subalpine forests of *Abies* in Japan (Kohyama and Fujita, 1981) and in the northeastern United States (Sprugel, 1976), where high winds and ice accumulation may contribute to tree mortality. However, in both cases, analyses of trees along transects indicated extremely low growth efficiencies in mature forests and relatively high levels where regrowth has recently been established (Kimura, 1982; Marchand, 1984).

An experiment was designed to test whether plants limited by nutrients and/or light were more susceptible to the root pathogen. Seedlings of mountain hemlock were planted in native soil and introduced into controlled environments. In autumn, following completion of all shoot growth, treatments affecting the level of irradiance and the availability of N commenced (Matson and Waring, 1984). After 2 months under the various experimental conditions, seedlings were inoculated at their root collar with the fungus or with a sterile inoculum. Environmental conditions were maintained for an additional 9 weeks at which time seedlings were carefully assessed for discoloration and loss of foliage, and then harvested for analysis.

Seedlings growing without added nutrients suffered significantly greater foliage damage as a result of pathogen inoculation (Fig. 9.5A) than did those growing with nutrients. Unfertilized treatments, while showing adequate levels

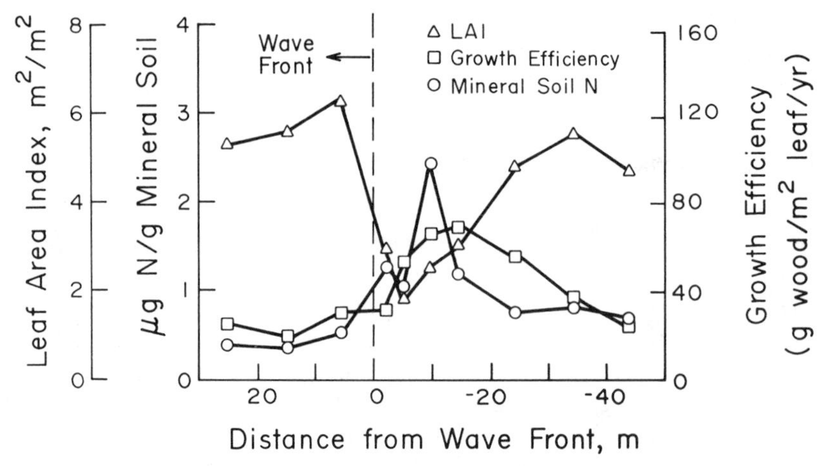

Fig. 9.4. In a subalpine forest of mountain hemlock, 220-yr-old trees are killed by a root disease approaching along a broad front. As older trees are killed, younger ones replace them and eventually regrow into forests. The availability of both mineralizable N (○) and light peak in the bare zone where young trees are becoming reestablished. Growth efficiency of trees also peaks in the area just behind the bare zone (0–10). The disease is present in decaying roots but only successfully reinfects trees with low vigor and limited N supply (R. H. Waring, unpublished). (From "Natural disturbance and nitrogen mineralization: Wave form dieback of mountain hemlock in the Oregon Cascades" by P. A. Matson and R. Boone, *Ecology,* 1984, **65,** 1511–1516. Copyright © by the Ecological Society of America. Reprinted by permission.)

of starch reserves, were extremely deficient in N. Total concentrations in their roots were generally about 0.3% of dry weight. Fertilized seedlings had values threefold higher. Shading, which reduced photosynthetic rates by 40%, significantly increased susceptibility whether or not nutrients were added. Under low light and high nitrogen, starch levels were reduced and the seedlings were more susceptible to root rot. Thus, susceptibility increased when either carbohydrate or N reserves dropped below certain values.

Natural resistance to laminated root rot is associated with improvement in the rates of mineralization following death and collapse of the overstory (Matson and Boone, 1984). Moreover, the exposed seedlings and saplings receive direct solar radiation in the open zone and display their maximum growth efficiencies (Fig. 9.4). As young saplings grow, however, the available N is again reduced and increased canopy competition reduces photosynthetic efficiency (Chapter 2). Growth efficiencies decrease in association with changes in both carbohydrate and N reserves. At a certain level, the pathogen, which still survives in decaying roots and stumps, successfully reinoculates the trees, initiating another wave of dieback. As an alternative, thinning and fertilization appear likely to create conditions favorable for maintaining a relatively high growth efficiency and favorable internal balance of C/N. Following such a prescription, we should be able to evaluate critically whether losses from at least some types of root rot may be reduced substantially.

Acid Rain

We now present a rather speculative analysis of how acid rain might extensively affect forest growth, susceptibility to insects and disease, and a variety of

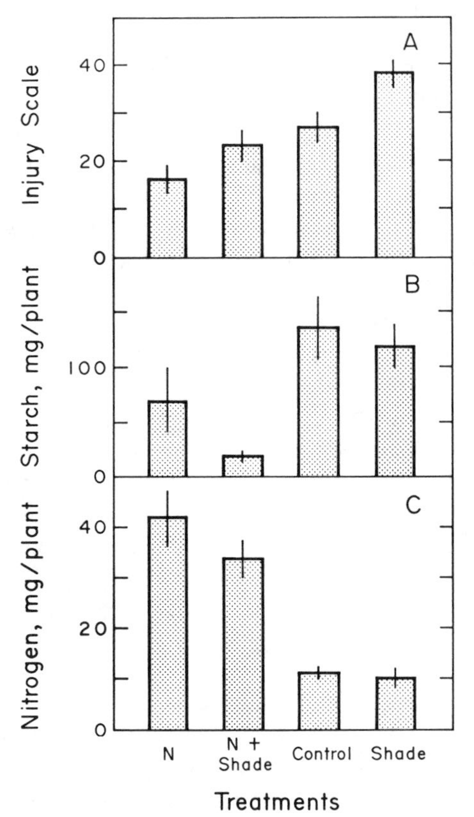

Fig. 9.5. Mountain hemlock seedlings maintained under controlled environments were inoculated with a root disease that caused foliage discoloration and loss (injury scale in A). Injury was highest when seedlings were shaded and photosynthesis was reduced by 40% compared to unshaded conditions. Addition of N decreased injury as long as seedlings were not shaded. The most resistant plants had a balance of carbohydrate reserves and N (B and C). (After "Effects of nutrient and light limitation on mountain hemlock: Susceptibility to laminated root-rot" by P. A. Matson and R. H. Waring, *Ecology,* 1984, **65,** 1517–1524. Copyright © 1984 by the Ecological Society of America. Reprinted by permission.)

other ecosystem processes. In this analysis, we first look at the problem to suggest a possible series of subtle changes that may eventually limit forest productivity. We end with a series of predictions for what might happen following certain experimental manipulations. The problem, we believe, may be very serious; this example, however, is only illustrative of an approach. Other interpretations are possible, but whatever the hypothesis, we stress the need for process-linked experiments.

In the last 30–40 yr, industrialized nations have increased the concentration of gases such as sulfur dioxide, oxides of nitrogen, and ozone in the atmosphere (Stoker and Seager, 1976; Tyler, 1972; Rühling and Skärby, 1979). Gorham *et al.* (1984) have shown that acid deposition is caused mainly by the conversion of SO_2 and oxides of nitrogen to acids, and that the resultant H^+ is partially neutralized by atmospheric NH_3 and calcium carbonate dust from various sources. They suggest that SO_2 is probably the major cause of acidity because the sum of base cations (NH_4^+ + Ca^{+2}) is usually sufficient to balance the NO_3^- resulting from conversion of oxides of nitrogen (Gorham *et al.*, 1984). These ions may enter the forest as a result of precipitation or aerosol impaction (Jacobson, 1984). In solution, the ions may be taken up directly through leaves or by roots. Ammonia is converted quickly into amino acids, but both SO_4^{-2} and NO_3^- may be converted to organic form or remain in inorganic forms in cell vacuoles (Johnson *et al.*, 1982b). Acid rain on foliage can cause the breakdown of lipids and thus injure membranes (Chia *et al.*, 1984), and excess NO_3^- in leaves may damage chloroplasts (Mohr, 1983).

In addition to acid rain, gaseous pollutants are able to diffuse into leaves through open stomata (Abrahamsen and Tveite, 1983). Gaseous pollutants and dryfall of other pollutants such as heavy metals are most concentrated near industrial centers, but wetfall of sulfuric and nitric acid may occur over forests far from sources of emission such as in the northeastern United States and in the Nordic countries of Europe (Fig. 9.6). The acidity of precipitation has increased accordingly but pH is generally still above 3.0 and therefore not directly harmful to most forest trees (Wood and Bormann, 1975; Haines *et al.*, 1980; Abrahamsen and Tveite, 1983).

The indirect effects of acid rain, however, may be considerable. The availability of mobile anions such as nitrate and sulfate tends to accelerate replacement of calcium (Ca), magnesium (Mg), and other cations in the soil with hydrogen ions (Chapter 7; Hovland *et al.*, 1980). Initially, this mobilization process increases the availability of scarce elements and uptake by plants is enhanced (Abrahamsen, 1980), as shown by analysis of pine needle litter collected in Finland over the period from 1959 to 1979 (Raunemaa *et al.*, 1982). Forest productivity in Nordic countries such as Sweden has actually increased substantially in the last 2 decades for a variety of reasons and even the most sensitive conifer species (*Abies alba*) has yet to show direct signs of damage from air pollution (Eriksson, 1984).

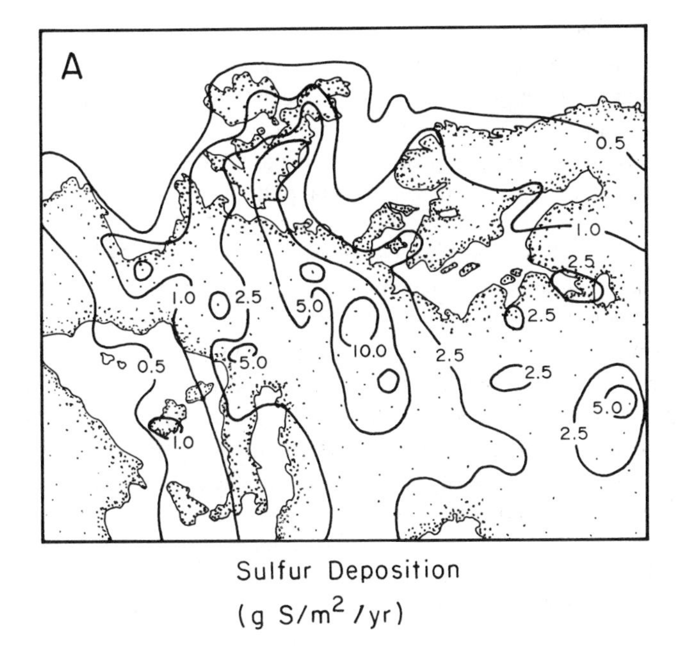

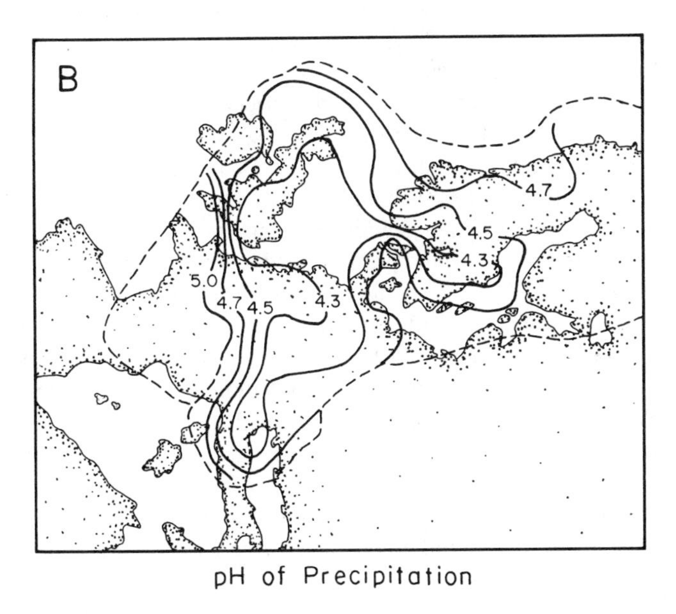

Fig. 9.6. Sulfur deposition (A) and pH of precipitation (B) over Europe in 1974 showed some areas of heavy deposition in relatively unindustrialized areas such as southern Norway. Nitrogen deposition followed the general pattern illustrated by sulfur. (From Swedish Ministry of Agriculture, 1982).

After a point, the essential cations may be largely replaced by hydrogen or by various metal ions in the soil. This is associated with increasing acid conditions that make some elements such as P less soluble and mobilize other elements such as aluminum, iron, copper, and zinc (Chapter 7, Fig. 7.2). Zinc and other heavy metals such as lead, nickel, and cadmium often are deposited in precipitation and dust that has been contaminated by industrial pollution. Concentrations of metals may occasionally reach toxic levels to roots. The heavy metals are toxic to mosses, lichens, and fungi at low concentrations (Tyler, 1972). The main effect of heavy metals, which tend to accumulate in the leaf litter, is to reduce the rates of decomposition and mineralization (O'Neill *et al.*, 1977).

Acidic conditions may result in aluminum (Al) ions being leached into rivers and lakes where small concentrations are toxic to fish (Baker and Schofield, 1980). Under normal soil conditions in boreal forests, Al ions are mobilized so the impact of acid rain is often difficult to prove (Johnson *et al.*, 1982b). Coniferous forests composed of spruce are among the more sensitive to acid rain in Europe and the United States, yet this genus is particularly well adapted to acidic soils. It has been speculated that Al concentrations are sufficient to be toxic to roots and that this is the underlying cause of dieback in many forests of central Europe (Ulrich *et al.*, 1980; but see Richter, 1983; Abrahamsen, 1983). Alternatively, a scarcity of Mg resulting from increased leaching from injured leaves and from the soil has also been noted in some forest areas (Prinz *et al.*, 1982; Zech and Popp, 1983; Zöttl and Mies, 1983). Deficiencies in Mg are, however, by no means universally associated with acid rain (Abrahamsen, 1983) and damage to forests extends across a wide variety of soils, some rich in Ca and Mg (Schröter, 1983; Reichelt, 1983).

Deposition of nitric acid, ammonia, or direct conversion of gaseous NO_2 to NO_3 within leaves (Mohr, 1983) should increase the availability of N. Together with enhanced mobility of cations and abundant sources of S, this N may temporarily improve the aboveground primary production by forests (Johnson *et al.*, 1982b; Hari *et al.*, 1984). This should be particularly true for trees in the boreal and temperate zone where N is often limiting. If forest growth responds to added nutrients, the maximum canopy leaf area initially should increase, at least on infertile soils (Chapter 3). With continued deposition of pollutants, major cations and P are likely to become less available. If this happens, forest growth should decrease (Baes and McLaughlin, 1984) and trees may show various symptoms of physiological stress. Eventually, canopy leaf area should be reduced, probably in association with outbreaks of insects and diseases associated with nutrient imbalance (Johnson *et al.*, 1982b; Lambert and Turner, 1977) and reduced supplies of storage carbohydrates.

Excess N together with reduced availability of carbohydrates to roots may result in loss of mycorrhizal fungi (Bowen, 1982; Ratnayake *et al.*, 1978; Hütterman, 1983) and favor bacteria over free-living fungi (Bååth *et al.*, 1978). Recently, Littke *et al.* (1984) has shown that high soil NO_3^- may even be toxic to mycorrhizal fungi. This would immediately reduce P uptake by trees and lead to an imbalance, where available N was in excess compared to available P (Gill and Lavender, 1983). Along three air pollution gradients in southern California, Zinke (1980) showed that N content in the foliage of *Pseudotsuga* increased from 1% to more than 2% while P content decreased abruptly, changing the ratios of N/P from about 7 in pollution-free areas to 20–30 in the most polluted areas. The presumed results of this sequence of events, as summarized in Table 9.6, might explain the general increase in susceptibility of central European forests to disease (Schütt, 1981; Barklund and Rowe, 1981; Binns and Redfern, 1983). In addition, excess N may also make trees more susceptible to winter desiccation and frost (Johnson and Siccama, 1983).

Assuming our diagnosis is valid, what might be done to restore a balance to such forest ecosystems, at least temporarily? Clearly we should strive to decrease the availability of N and possibly S while increasing that of P. Widespread fertilization programs would be prohibitively expensive, and, under acid conditions, not likely to increase the available pool of P. Heavy thinning of the forest coupled with shorter rotations might increase the incorporation of N into protein, while still assuring that sufficient reserves of carbohydrates were available to keep roots functioning through the dormant season and during periods of rapid aboveground growth. Burning of the slash, although creating additional pollution, would reduce both N and S levels in the soil while increasing the relative availability of other nutrients. Favoring deciduous species, including conifers

TABLE 9.6 How Shifts in Carbohydrates and Nutrient Balance Associated with Acid Rain Might Alter Tree Susceptibility

I. Primary production
   A. Increasing N and cation availability may initially enhance shoot growth
   B. Increasing canopy leaf area reduces photosynthetic efficiency
   C. Amino acid and nitrate content should increase while starch reserves decrease
   D. Trees as they grow larger and increase maintenance respiration become more susceptible to attack from a variety of insects and diseases
II. Decomposition
   A. Litter becomes enriched in N and S, but reduced in P and possibly other essential minerals
   B. Decomposition may eventually slow, but N should be mineralized at a faster rate than P
   C. Bacteria/free-living fungi ratio increased
III. Root-related processes
   A. Reduced availability of carbohydrates and less uptake of P may limit establishment of mycorrhizal fungi on roots
   B. Roots systems should be less extensive and less efficient
   C. Cation uptake should initially improve, but later could prove inadequate
   D. Roots may become more susceptible to pathogens

such as *Larix*, would reduce exposure to gaseous pollutants, aerosols, and fog drip (White and Turner, 1970; Kellman *et al.*, 1982; Lovett *et al.*, 1982).

Although this diagnosis and prescription may be incorrect, it forces us to concentrate on the basic processes involved rather than on indirect problems like disease or insect outbreaks. Most importantly, we suggest an experiment that includes measurements of a chain of interlinked predicted responses. This experimental approach is an important alternative to observational correlations and will quickly clarify weaknesses in any of the underlying assumptions.

SUMMARY

Historical analysis of wind, fire, insect, and disease-induced disturbances suggests that these factors do not usually alter the long-term mortality rates associated with intense competition. Certain factors, however, may affect the composition of the forest significantly without greatly modifying major ecosystem processes or the general forest structure. Vertebrates may play a particularly important role in selective dissemination of propagules for germination following loss of the forest canopy and seed supply in the litter. Once a forest canopy is established, frequent disturbances tend to remove stressed trees. When disturbances are infrequent, most trees exhibit low growth efficiency and can be expected to succumb to some injury. Under a full canopy, intense competition results in reduced photosynthetic efficiency and inadequate reserves of carbohydrates or N to produce various morphological, anatomical, or biochemical defenses. Whenever the balance between carbohydrate and N supply in tissue is adversely altered, the susceptibility of a tree is likely to increase because the imbalance affects growth rates and allocation of resources between roots and shoots. Mineral imbalances also adversely affect growth rates and tree resistance. Knowledge of growth efficiency and the relative balance of available carbohydrates and N may provide a general index useful in diagnosing the susceptibility of forests to agents of disturbance. With insight into basic ecosystem processes, it may be possible to recommend remedial actions. As an example of the approach, three experimental manipulations that involved insects, disease, and the impact of acid rain were analyzed within this general frame of reference. Prescriptions were suggested that should improve stand conditions and reduce susceptibility to a broad array of disturbing agents.

10

Linkage of Terrestrial and Aquatic Ecosystems

INTRODUCTION

In this chapter we expand our perspective from the small watersheds discussed in Chapter 5 to large drainage basins. This enlarged perspective is necessary to see how modifications of forest ecosystems may affect (1) sediment transport and deposition in rivers; (2) enrichment or pollution of downstream waters; and (3) the maintenance of important groups of aquatic animals.

Modifying the forest canopy, forest floor, and forest soil conditions affects the amount of water available, its quality, and the path of its flow from the system (Chapter 5). As water flows, it transports both organic and inorganic compounds in sediment and in solution. The kind and amount of material transported influence not only small streams but also rivers and estuaries.

In Chapters 6 and 8 we discussed some important processes that affect nutrient export such as rock weathering, plant uptake, and decomposition. In this chapter we identify major geomorphic processes that affect sediment and solution transport from the land and through aquatic ecosystems. Further, we consider how these processes may be altered by various forest practices or natural events. The kind of disturbance and the frequency of disturbance are both important. Normally, a river and its tributary streams are in steady state. Change in the steady state, particularly permanent change, results in predictable modifications in sediment budgets, channel morphology, and biotic life. The latter part of this chapter deals with recognizing biological signs of changing stream conditions. Throughout the chapter we suggest various ways of minimizing instability in order to maintain certain desired stream attributes.

GEOMORPHIC PROCESSES

Definitions

Material transport processes operating in drainage basins are broadly grouped into those that affect hillslopes and those that are active in stream channels. Hillslope processes supply dissolved and particulate material, of both organic and inorganic origin, to the channel. In the channel, other processes operate that modify terrestrially derived materials, and eventually transport them, sometimes in an altered form, further downstream.

Major kinds of geomorphic processes that occur in steep forested landscapes are identified in Fig. 10.1 and described in detail by Swanson *et al.* (1982a). Transport of dissolved material begins when rainfall passes through vegetation, soil, and fragmented bedrock as described in Chapter 6. Surface erosion is the particle-by-particle transfer of material on the surface of the ground by overland flow, raindrop impact, frost heaving, and ravel that occurs during dry periods. Creep is the slow downslope movement of soil and weathered bedrock by plastic deformation. Debris avalanches are rapid, shallow (1–4 m deep) mass movements of soil. Slumps and earthflows are slow, deep-seated (5–10 m) displacements of soil, rock, and covering vegetation associated with slippage along a particular subsurface layer, commonly where the conductivity to water flow decreases abruptly (Swanston and Swanson, 1976).

Solution transport occurs in the streamwater. In addition, fine sediment may be suspended in flowing water and larger particles may be moved along the stream bottom as bedload. Debris torrents are rapid, turbulent movements of

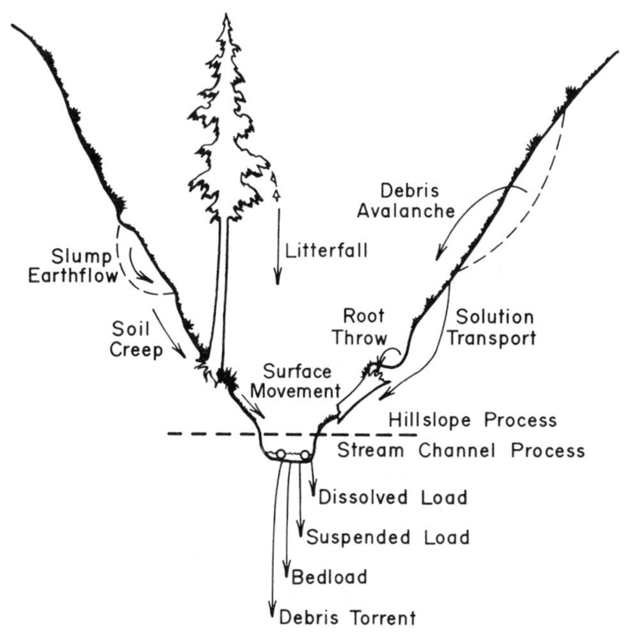

Fig. 10.1. Geomorphic processes that transfer organic and inorganic material from hillslopes and stream channels. (From Swanson *et al.*, 1982a.)

sediment and organic matter down stream channels. Entire trees may be included. Streambank erosion occurs as water, sediment, and organic debris entrain materials delivered to the streamside by various processes.

Linkages

In a single watershed, many geomorphic processes may be operating at the same time. Moreover, the processes are interlinked; one may initiate another. For example, root-throw initiated by wind might also cause surface erosion and could, under some conditions, trigger a debris avalanche that would affect many channel processes. Examples of this and other related kinds of interactions are identified in Fig. 10.2.

From an ecosystem standpoint, it is important to identify those combinations of factors that are associated with a disturbance. The probability of a debris avalanche, slump, or earthflow differs, for example, depending upon the likelihood of intense storms or earthquakes (Garwood *et al.*, 1979), the geomorphology of the slope, and the time since roads were constructed or vegetation was removed by logging or fire (Swantson and Swanson, 1976).

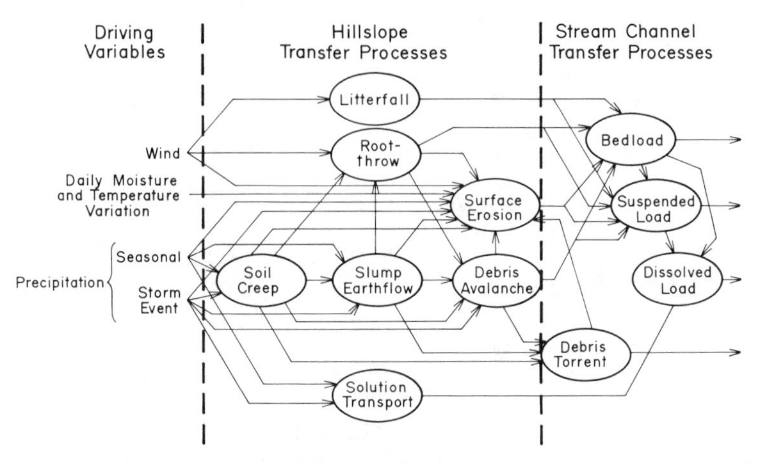

Fig. 10.2. Various environmental factors can trigger one or more geomorphic processes because of cascading interactions that cause material to move from one place to another. (From Swanson *et al.*, 1982a.)

Major events that lead to massive erosion are typically infrequent. For example, in old-growth Douglas fir forests of the Pacific Northwest, a careful assessment of nutrient losses from a 10-ha drainage basin indicated that the leaching process accounted for approximately 65% of all inorganic material and for more than 40% of the organics leaving the basin, excluding mass erosion events (Table 10.1). From an historical evaluation of the same drainages, Swanson *et al.* (1982a) estimated that a small debris torrent was likely to occur about every 580 yr under forested conditions. In such an event, the amount of inorganic and organic material lost from the basin would more than equal (200% and 133%, respectively) that transported in nearly 6 centuries of leaching.

Influence of Vegetation

Certain features are harbingers of potential instability in drainage basins. Trees leaning in various directions or exhibiting asymmetrical boles are often indicative of recent soil movement. Slumps and earthflows leave distinctive scars on the landscape that can be identified for decades to millenia (Fig. 10.3). Where depressions exist in the bedrock, subsurface flow of water is concentrated and the likelihood of landslides is increased (Dietrich and Dunne, 1978; Pierson, 1980). Placing roads across such unstable topography concentrates the flow of water and increases the risk of landslides. Deep-seated processes such as slumps and earthflow are relatively unaffected, except hydrologically, by vegetation because the shear zone is normally located below the maximum depth of rooting and the

TABLE 10.1 Transfer of Organic and Inorganic Material to the Channel by Hillslope Processes and Export by Channel Processes from a 10-ha Forested Basin in Oregon[a]

Process	Inorganic (t/ha/yr)	Organic (t/ha/yr)
Hillslope processes		
Solution transfer	3	0.3
Litterfall	0	0.3
Surface erosion	0.5	0.3
Creep	1.1	0.04
Root-throw	0.1	0.1
Debris avalanche	6.0	0.4
Slump/earthflow	0.0	0.0
Total	10.7	1.4
Total particulate		
Including debris avalanche	7.7	1.1
Excluding debris avalanche	1.7	0.7
Channel processes		
Solution transfer	3.0	0.3
Net suspended sediment	0.6	0.1
Bedload	0.6	0.3
Debris torrent	4.6	0.3
Total	8.9	1.0
Total particulate		
Including debris torrent	5.9	0.7
Excluding debris torrent	1.3	0.4

[a] From Swanson *et al.* (1982a).

vegetation is only a minor component compared to the weight of soil on the slope (the latter weighs at least 1 t/m^3).

The kind of vegetation on hillslopes has a significant influence on many surface geomorphic processes. In the absence of vegetation, of course, there is neither litterfall nor root-throw. As discussed in Chapter 5, increasing the canopy generally decreases the amount of water that reaches the soil but increases the potential rate of infiltration. In most cases, the erosive potential of water is reduced with vegetative cover because (1) the soil is saturated less often; (2) snowmelt intensity and frost heaving are reduced; (3) roots form a network of reinforcing links across the slope; and (4) associated dead organic matter protects the soil surface from rain splashing and serves as a barrier to soil movement downslope (Bormann *et al.*, 1974). Unlike logging or road construction, a fire does not remove large organic debris or cause compaction. Nevertheless, a

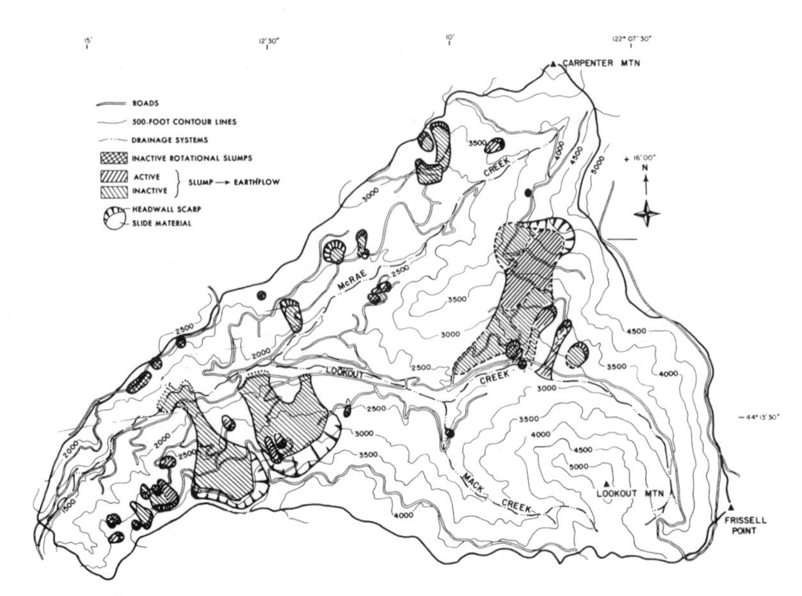

Fig. 10.3. Unstable areas such as slumps and earthflows can be identified and mapped so that roads and harvesting activities can be planned to minimize risks. In this map of an Oregon drainage, some roads unfortunately were placed across active slumps and subsequently failed. (From Swanson and James, 1975.)

destructive fire dramatically increases the rates of many geomorphic processes on hillslopes and in stream channels.

Living vegetation absorbs water through roots and transpires it from leaf surfaces. A reduction in soil moisture content attributed to vegetation can delay the period when soils approach saturation and are most prone to erosion. This is seen in areas such as the Pacific Northwest where summer drought is followed by heavy autumn rains but most major erosional events occur in midwinter when soils finally reach saturation (Day and Megaham, 1975; Harr, 1977).

Roots stabilize the soil mantle by anchoring it both vertically and laterally across potential failure zones (Swanston, 1970; Nakano, 1971). Roots are only effective if a significant part of the failure plane falls within rooting depth. The depth of rooting varies with the age of the forest, the soil profile, and the kind of vegetation. In addition, the strength of roots in anchoring soil is a function of their diameter, density, and spatial volume of distribution. Often, shrub species contribute disproportionately to maintaining slope stability because their root systems remain alive and they resprout following logging or fire. Hardwoods have dense wood with greater tensile strength than many species of conifers. For example, *Nothofagus* and *Metrosideros,* hardwoods native to New Zealand, exhibit nearly twice to three times the tensile strength of roots of the introduced conifer, *Pinus radiata* (O'Loughlin and Watson, 1979, 1981).

Once roots die and begin to decay, their tensile strength declines exponentially. Depending on soil temperature and moisture content, the anchoring capabilities of dead roots may be reduced to half that of live roots after 2–5 yr (O'Loughlin, 1974; Ziemer, 1981; O'Loughlin and Ziemer, 1982). The strength of the total network of both living and dead roots reaches a minimum some years following destruction of a forest (Bormann *et al.*, 1969; Nakano, 1971; Ziemer, 1981). It is then that hillslopes are most sensitive to mass failures (Fig. 10.4). In cases where new species of shrubs quickly become established following disturbance, root anchoring strength may be almost equivalent to that provided by the previous forest (Ziemer, 1981). If shrub cover is controlled or allowed to grow to maximum age without the reestablishment of a dense cover of trees, net root strength will again decrease, as in the example shown in Fig. 10.4 about 20 yr following logging.

Forest Practices

The impact of forest practices upon erosion depends on a complex of factors: climate, geology, the kind of road network and its maintenance, the remaining

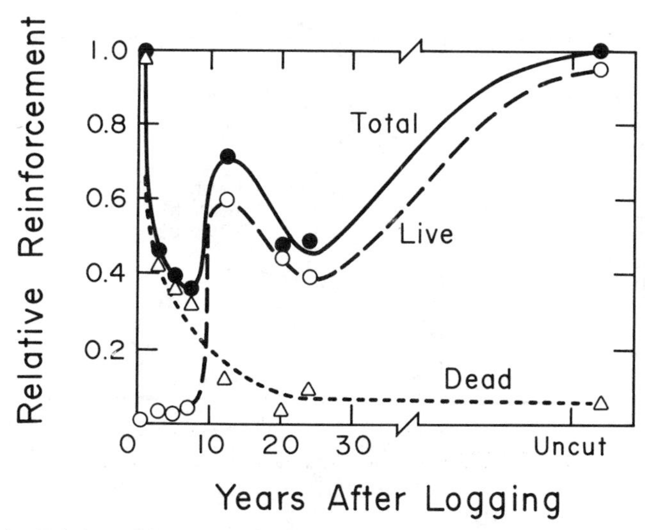

Fig. 10.4. Relative reinforcement of soil by dead roots decreases rapidly with time following logging. The roots of newly established shrub cover with high tensile strength contribute to slope stability. As shrubs die out, however, as they did about 20 yr following logging in this northern California example, the total reinforcement provided by roots again may decrease until a complete cover of trees is finally reestablished. (From Ziemer, 1981.)

vegetative cover, the kind and rate of regrowth, and the type of logging methods utilized. If storms are not severe during the period of initial revegetation, little accelerated erosion may occur following a disturbance. On the other hand, a major storm can trigger massive slope failures if it occurs at a time when root strength is significantly reduced. Similarly, the absence of fallen trees and slash may allow a release of material from hillslopes and the stream channel (Triska and Cromack, 1980; Swanson *et al.*, 1982a).

Following disturbance, different kinds of geomorphic processes can be expected to dominate sequentially. At first, surface erosion is most prevalent due to lack of vegetative cover and other stable barriers. Losses of dissolved nutrient and organic compounds peak at a later time and then decrease as new vegetation becomes well established (Bormann *et al.*, 1969; Chapter 6). Even with complete cover reestablished, the hillslopes may be potentially unstable as root strength may continue to decline for a decade or more following the original disturbance. This projected sequence of events is depicted in Fig. 10.5 following conversion from an old-growth coniferous forest to younger stands of conifers with associated hardwood shrubs and trees.

The location, use, and maintenance of roads are among the factors that affect erosion from a drainage basin. The impact of road building and forest clearing by the Romans 2200 yr ago shows clearly in the amount and kind of lake sediments deposited (Hutchinson *et al.*, 1970). An analysis of a typical mountainous basin

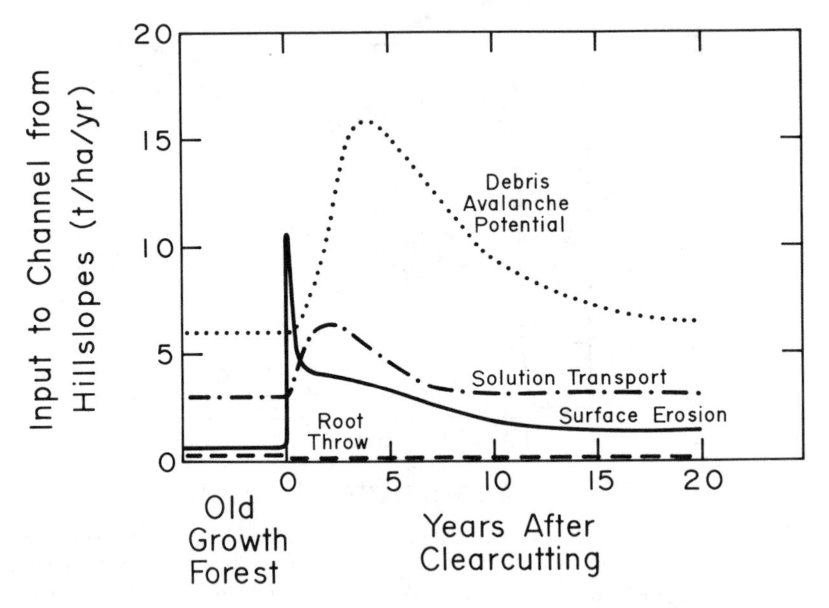

Fig. 10.5. Changing probabilities of various transport mechanisms following logging of an old-growth Douglas fir forest in Oregon, Pacific Northwest, United States. (From Swanson *et al.*, 1982a.)

in western Washington with less than 1% of the area in roads shows that erosion attributed to roads is nearly four times that attributed to natural causes (Table 10.2). Landslides triggered by roads were the most important source of additional sediment, and heavy use on only 6% of the graveled roads accounted for more than 70% of the total surface erosion from this source (Reid and Dunne, 1984).

In less steep areas, such as in the Appalachian Mountains of the southeastern United States, landslides are rare but road density may exceed 10% of the total area because tractor and wheeled vehicles are used in harvesting rather than extraction with elevated cables (Mitchell and Trimble, 1959). Surface erosion from roads is excessive under these circumstances and may average nearly 5000 t/ha/yr on slopes less than 20° (Hoover, 1952).

The choice of logging methods, the location of skid roads, and the selection of sites where logs are temporarily stored greatly influence the degree of erosion associated with harvesting trees (Froehlich, 1976). Use of wheeled or tracked vehicles on slopes greater than about 20° may greatly accelerate erosion. Even on gentle slopes, vehicles may compact the soil so that the infiltration of water is reduced and surface erosion is increased. The effects may persist for decades

TABLE 10.2 Contribution of Various Geomorphic Processes to Erosion in the Clearwater Drainage Basin in Western Washington[a]

Source	Percentage of annual total for basin
Natural	
Landslides	7
Debris flows	3
Bank erosion	7
Treethrow	2
Burrows	1
Subtotal	20
Road-related[b]	
Landslides	50
Debris flows	11
Gullies, rills, and sidecast erosion	2.5
Secondary erosion caused by road drainage	4
Heavily used roads	9
Lightly to moderately used roads	3
Unused roads	0.5
Subtotal	80
Total erosion estimated at 387 t/km/yr	100

[a] From Reid (1981).

[b] Roads represent 1% of the surface in the drainage with 6% receiving heavy use, 44% receiving light to moderate use, and 50% abandoned.

following logging (Froehlich, 1979). The practice of cable logging from a ridge greatly reduces compaction and is advised for most forest harvesting on steep slopes. When care is taken not to move previously fallen trees or to disturb the forest floor, sediment transport downslope and into the stream channel is reduced.

Even with considerable care, cable logging may contribute to increasing erosion from a drainage, sometimes for periods of decades or more if slumps or earthflows are reactivated. Although less erosion is attributed to logging than to road-related activities per unit of area, the fact that logging is conducted over a larger area may make its total contribution of sediment equal or greater than that from roads over an extensive period (Swanson and Fredriksen, 1982; Swanson and Dyrness, 1975).

Herbicides are often applied in many regions to reduce competition between shrub and herbaceous species and tree seedlings and saplings. If these practices are successful in killing shrubs, root strength in the soil profile may be critically reduced on unstable slopes. Sustained treatment with herbicides, fortunately a rare practice in forestry, prevents litter from accumulating and leads to an increase in leaching of dissolved organic compounds (Bormann *et al.*, 1974). Once litter is decomposed, the soil surface bare, and the root network weakened, soil erosion can be expected to increase dramatically. Where herbicides were experimentally applied for 3 consective years on the Hubbard Brook Watershed in the northeastern United States, erosion rates increased by an order of magnitude during the third year of treatment (Bormann *et al.*, 1974).

In this section we have outlined the general hillslope processes involved in transporting material to the stream channel. Models couple climatic conditions to geomorphic processes, but there are few case studies with a complete analysis of storage and sediment transport from one part of a drainage to another. Such studies are arduous but necessary, if hydrologic, geologic, and biological principles are to be combined and predictions of sediment transport applied generally (Swanson *et al.*, 1982b; Dietrich *et al.*, 1982).

STREAM ECOSYSTEMS

Forests strongly influence streams flowing through them. Even in more arid regions, where surface water is not present continually, subsurface water may support a flourishing riparian zone that modifies how sediment and organic matter are transported when conditions permit. Functionally, the riparian zone is the area of direct interactions between aquatic and terrestrial environments. The degree of interaction depends on the type and stature of vegetation, hydrology, and topography. In steep forested topography, the terrestrial vegetation may strongly control stream life, flow rates, and channel profile. In valleys with extensive floodplains and meandering rivers, floodplain forests affect deposition of all transported material and provide a source of food and a habitat for stream life.

Because one stream flows into another, a holistic evaluation requires information for an entire drainage. A forest fire or small logging operation may exert little impact on the transport of sediment or salmon spawning if it is restricted to a small area. On the other hand, mining, pesticide applications, and certain industrial activities, even when restricted to a small portion of a stream, may affect downstream resources over a great distance—including municipal water supplies, beaches, alluvial farmland, and fisheries.

In this section, a continuum viewpoint is emphasized that links transport, storage, and key biological processes together from headwater streams to large rivers and their floodplains (Vannote *et al.*, 1980). We first consider the physical aspects of streams and then the biological.

Stream Mechanics

The sediment load of a stream is determined by bedrock geology, soils, vegetation, precipitation, and land use. Sediment enters the stream by way of geomorphic processes on hillslopes. If relative stability is to be maintained in the stream, a steady state must exist between sediment introduced into a stream, its movement, and subsequent distribution. Changes in sediment load can upset this balance.

Sediment transported in streams may be suspended in the water or carried along the stream bottom as bedload. The rate and extent of movement varies with particle size. In suspension, the larger-sized particles move short distances and settle out quickly; thus, a sequence of sand, silt, and clay soils is developed parallel to rivers with large floodplains. In bedload, the larger-sized particles move slowly but they may also trap some smaller particles, while intermediate-sized particles may be carried along more rapidly (Madej, 1982).

What happens when abnormal amounts of sediment are introduced into a stream? Assuming no change in the volume of streamflow, only the width, depth, or velocity can be altered. The mean velocity of a stream (V) is related to the mean depth (D), gradient (G), and bottom roughness (R) according to (Leopold *et al.*, 1964):

$$V = \frac{D^{2/3}\, G^{1/2}}{R}$$

The stream gradient is likely to remain constant but the average depth will decrease as the added sediment fills in holes on the stream bottom. Roughness will decrease as a result of increasing the sediment load, and decreasing the average roughness forces streams to widen. Streams that have received unusual amounts of sediment can be expected to increase in width and decrease in depth as material is deposited. For example, where logging activities increased sediment loads in western Washington, the width-to-depth ratio for streams increased by nearly twofold while velocity at full flow remained constant (Madej, 1982). As the stream becomes wide and shallow, there is greater effective roughness from the expanded total surface over which water flows (Leopold *et al.*, 1964). Once roughness is greater than that which allows constant velocity, sediment is again deposited, halting further widening of the channel. The ultimate channel configuration is further constrained by bedrock geology and the extent of vegetation along channel edges.

It is not possible to predict the configuration that a stream channel will take without knowing something about peak flow, sediment loads, and erodibility of the banks (Palmer, 1976; Madej, 1982). To classify streams we emphasize (1) linkages among streams; and (2) changes in the relative importance of key variables.

The linkage of streams is defined by the relative number of tributaries that contribute to each stream. This classification defines stream order. At the source, a stream has no tributaries. When two such first-order streams flow together, a second-order stream is created. It may be supplemented by additional first-order streams but only becomes a third-order stream when it is joined by another second-order stream. This classification does not recognize stream size or gradient. In arid regions, first-order streams may be intermittent, in boreal regions they may be frozen part of the year, and in the tropics they may experience frequent cloudburst storms. Within a particular drainage basin, however, flow rates can be expected to increase with stream order unless rivers are heavily exploited for irrigation or drain underground.

Within a drainage, streams may also be classified according to their geologic and hydrologic properties. These include (1) degree of constriction and roughness of channel; (2) valley profile; (3) channel gradient; (4) amount of sediment; and (5) bed material. At least four major river zones may be recognized in most forested regions (Fig. 10.6). Additional refinements in river classification are discussed by Hawkes (1975). The zones relate to different energy levels associated with streamwater velocity and sediment loads (Palmer, 1976). There is a general trend from steep gradient streams that remove coarse sediments to lower gradient streams that deposit fine sediments, but this progression is not continuous. Changes in geology, vegetation, man-made obstructions, and availability of sediment could alter characteristics considerably.

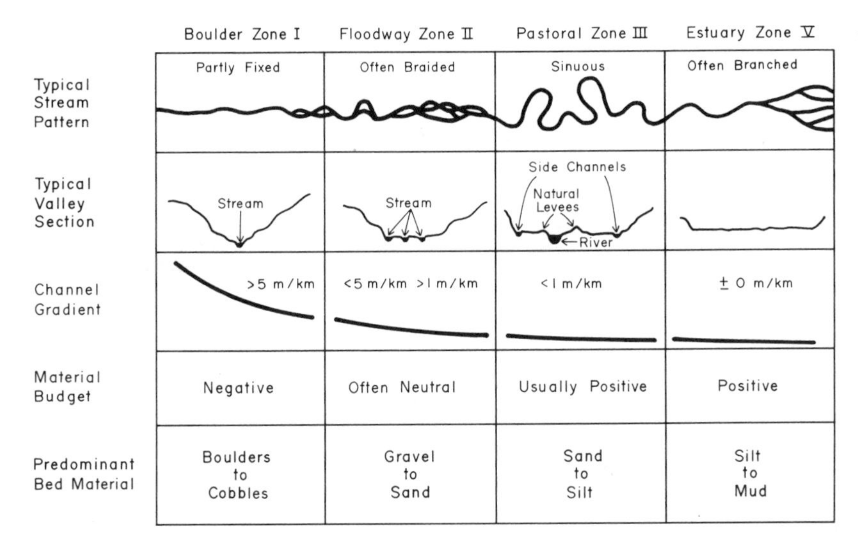

Fig. 10.6. As rivers flow through different topography they change in shape, gradient, and in the amount of sediment transported or deposited. These and other related characteristics may serve to classify rivers as illustrated in the diagrams. (From Palmer, 1976.)

This classification scheme, because it includes the importance of sediment and vegetation, has significant advantages over those concerned only with hydrology. For example, zoning of a floodplain based on the probability of inundation over a decade or century (U.S. Army Corps of Engineers, 1972) takes little account for changing sediment load or how certain structures such as roads, levees, bridges, or dams affect transport mechanics (Palmer, 1976). Maintenance of some of the floodplain forest, the natural levee, and overflow channels can greatly aid in controlling erosion and flood damage.

Certain natural boundaries define an area within which the river channel will remain, although the river may meander (Fig. 10.7). The stream channel might shift over the entire floodplain in a period of centuries. Management policies that attempt to constrain the width of the streamway are less detrimental to stream function and more economical than those that attempt to control the channel position strictly. Attempts to constrain the channel to less than its naturally defined limits results in undercutting the impediments (Palmer, 1976).

Many river systems have been substantially altered by man. In particular, floodplains have been greatly reduced in area. In the United States, the total area of marshes, forested floodplain, and other wetlands has been reduced by nearly 60% in the last 2 centuries, from 87 to 38.5 million ha. Drainage is continuing at a present rate of about 180,000 ha/yr (Shaw and Fredine, 1956; Eberhardt, 1983).

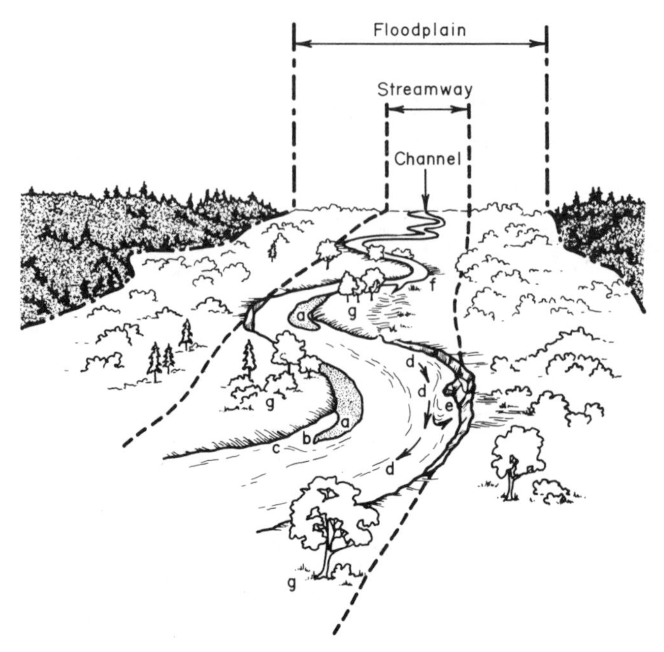

Fig. 10.7. A river in equilibrium with its sediment load creates a streamway on a floodplain within which the channel can be expected to shift. Sand bars (a), still shallows (b), eddies (c), deep, fast current (d), cutbanks (e), annual (f), and perennial vegetation (g) are normal features that aid in dissipating the stream's energy as sediment is transported. (From Palmer, 1976.)

Effects of Terrestrial Vegetation

There are two major biological components that may greatly alter sediment transport, channel width, and stream configuration. One is the presence of large standing or fallen trees in the riparian zone and the other is the presence of large animals. At one time, both of these components affected nearly all permanent streams and large rivers throughout the world. To assess their importance, we can contrast streams with and without these biological features.

Fallen logs form dams across small streams and serve to trap sediment and to dissipate stream energy (Zimmerman *et al.*, 1967; Swanson *et al.*, 1976; Triska *et al.*, 1982; Likens and Bilby, 1982; Megahan, 1982). Fallen logs that are held securely to one bank may force water to scour pools as sediment is washed away below the obstruction. The presence of deep pools and small waterfalls dissipates the force of water by increasing channel roughness.

Standing trees and shrubs in the riparian zone hinder water flow when streams are at flood stage, substantially reducing stream velocity. Reductions in stream velocity, together with increased roughness, cause sediment to be deposited within the riparian zone. Large roots of living trees anchor the soil and reduce bank erosion (D. G. Smith, 1976; Beschta, 1979). Very large boulders also obstruct flow (Megahan, 1982). Moreover, rocks may anchor segments of tree boles and branches that otherwise might wash away (Likens and Bilby, 1982).

In the large tropical river basins, floating islands of vegetation choke backwater lagoons. During floods these islands may come together and restrict flow into the main channel. If these temporary dams are breached increased flow to the channel results (Bonetto, 1975). New islands later form and replace those washed away.

Today, it is easy to overlook the fact that large trees were once prevalent along almost all major rivers from boulder streams to estuaries. Riparian trees are still obvious in relatively undeveloped regions of the tropics and in boreal forest regions, but rare in the temperate zone where human settlements long ago cleared the floodplain forests for agricultural and other uses and opened rivers for navigation. In North America, historical records show the size and number of downed trees removed from rivers in virtually every region of the United States (Sedell *et al.*, 1982). In the Mississippi River, over 800,000 snags were removed from the water in a 50-yr period along the lower 1600 km of river. The snags were primarily cottonwood (*Populus deltoidea*) and sycamore (*Platanus occidentalis*), averaging 1.7 m in diameter at the base with an average length of 35 m.

The Willamette River of western Oregon, the tenth largest river in terms of total discharge from the continental United States, has an excellent record of channel changes since European settlement began in 1830 (Sedell and Froggatt, 1984). The presettlement river had banks 1.5 to 2.6 m above the low water line with a strip of forested floodplain 1.6–3.2 km wide that supported a dense forest of Douglas fir, ash (*Fraxinus oregana*), cottonwood (*Populus trichocarpa*), and other hardwoods. The floodplain was dissected by numerous sloughs and during floods was covered with swiftly running water to a depth of 1.5–3 m. Each year new channels were opened and others closed as the river continually changed course within its streamway (Fig. 10.8). The river often had two to five main channels. A major factor in the changing of channel courses was the presence of fallen trees that closed off or confined water to one channel and then another. From 1870 to 1950, over 65,000 snags and streamside trees were removed, which represented about 550 snags/km of river. As a result of snag removal, drainage, dam formation, and channelizations, the river channel was sequentially confined and its floodplain forest reduced (Fig. 10.8). The shoreline was progressively reduced from 250 km in 1854 to 64 km in 1967. Most rivers in industrialized nations and in developing countries with high human populations have experienced similar reductions in floodplain and habitat diversity. In developing countries where human populations are dependent on annual flooding to

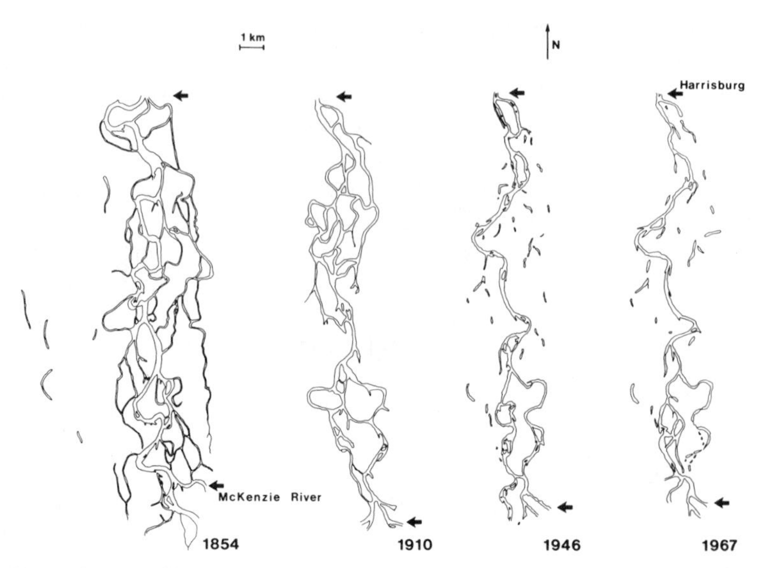

Fig. 10.8. From historical records the Willamette River in Oregon shows a fourfold decrease in surface area from 1854 to 1967 as a result of drainage and removal of much of the original floodplain forest. (From Sedell and Froggatt, 1984.)

renew soil fertility of farm land and on fisheries as a major source of protein, the maintenance of the floodplain forests should be a major priority (Smith, 1981; Goulding, 1980).

Effects of Large Amphibious Animals

At one time, many streams and rivers were inhabitated by large animals that sought refuge in the aquatic environment but obtained some food or other resources from the riparian zone. In the tropics and semitropics, coypu (*Myocastor*), crocodiles (*Crocodylus, Melanosuchus, Caiman,* and *Paleosuchus*), and hippopotami (*Hippopotamus*) are examples; in the temperate and boreal forests, the beaver (*Castor*) was once nearly ubiquitous. These animals manipulate the aquatic environment by (1) removing certain plant species from the riparian zone; (2) transferring organic materials to the aquatic system; (3) modifying the turbidity and nutrient content of the water; and (4) altering the bottom roughness. Animals help determine how the riparian zone responds to water flow, bank erosion, and hydrologic shear stress at flood stage.

The hippopotamus of Africa is the second largest land animal, weighing up to more than 3 t. It spends daylight hours in streams and forages on grass during the night, traveling an average of 5 km away from the channel (Laws, 1981). Normally, population densities are about 8 animals/km^2 but some parks now exceed 60 animals/km^2, which represents a biomass of 200,000 kg/km^2. As the animals move back and forth from the river, they create wide paths through the riparian zone, compact the soil, and make wallows that hold water during dry periods. These changes in the riparian zone may also change water flow during floods. Probably the major influence of large populations of such animals, however, is through the transport of fecal material into the riparian zone and stream. A herd of some 5000 animals grazing along Lake George has been estimated to contribute about 6000–10,000 t of fecal material contributing about 100 t of nitrogen (N) and 200 t of phosphorus (P) each year to the aquatic system (Viner and Smith, 1973). This contribution, if deposited into a stream, is of major significance to aquatic primary production.

Crocodiles in the backwaters of the Amazon basin appear to play a key role in maintaining primary production in otherwise nearly sterile freshwater mouth-lakes (Fittkau, 1970). Following intensive hunting of the beast for hides, the fisheries resource was noted to decrease markedly. Even populations of predatory fish declined. Fittkau studied the exchange of water between the nearly sterile lakes and the Amazon River and concluded that the lakes had insufficient fertility to support algae or almost any other kind of primary production. He speculated that the large numbers of fish that came into the lakes and their tributaries to breed were the main source of nutrients. Only the presence of numerous crocodiles, some that reached more than 4 m in length, assured the retention of the nutrients from the Amazon.

The beaver and coypu once were distributed along nearly all of the forested streams of North and South America, Asia, and Europe. Beaver pelts were items of trade for at least a century during which the beaver populations were drastically reduced. The original population of North American beaver, estimated from fur trade records and the journals of explorers, was in the hundreds of millions (Seton, 1929). For example, nearly every lake, pond, river, and brook in the state of New York was populated with beaver (Ruedemann and Schoonmaker, 1938). Fur trade records indicate similar occupancy from the Oregon territory to the lower Mississippi (Johnson and Chance, 1974; Moore and Martin, 1949).

According to Jenkins and Busher (1979), beaver colonies were once distributed at approximately 0.8/km along streams of all sizes. This density matches estimates made by Smith (1978) along some 1800 km of Wyoming streams. Assuming an average of eight beavers per colony with each beaver requiring about 1000 kg of wood resources per year for growth and survival (Howard, 1982), the annual removal from the riparian zone would be nearly 3000 kg/km of stream channel. As beavers consume less than 25% of the wood cut for food, the rest must decompose in the stream channel. Together with changes in sediment deposition, nutrient enrichment (Naiman and Melillo, 1984), and mixing, beaver profoundly affected all types of stream biological processes from microbial to vertebrate production. The influence of beaver may last centuries after removal. In England, beavers became scarce at the close of the ninth century (Lyell, 1969, p. 149) but their influence extends to the present through the creation of fens and peat areas (Darby, 1956).

In summary, populations of large animals once exerted considerable influence on nearly all forested stream ecosystems by changing the surface area of water, its velocity, nutrient, and sediment load. Such modifications generally helped in maintaining a productive and relatively stable system. We have yet to replace the roles of these animals through intelligent management.

Stream Biology

Stream life is very much dependent on organic resources transferred from the terrestrial system. These inputs usually greatly exceed the organic matter produced within the stream through the photosynthetic activity of algae and other plants. The source of fixed carbon is a basis for stream classification. Small streams dominated by forest cover may derive less than 2% of their organic substrate from primary production within the stream (Fisher and Likens, 1973). Large streams with greater surface area exposed to direct radiation derive more than half their organic material through aquatic primary production (Triska *et al.*, 1982).

When aquatic primary production is evaluated for an entire drainage, the most important variables are water surface area, nutrient availability, and turbidity. Turbidity is important because, as it increases, the penetration of light is reduced and the surfaces supporting photosynthetic organisms may also be increasingly scoured. For streams in Quebec, Canada, where turbidity was of minor importance, primary production was closely correlated with water surface area alone (Table 10.3). The analysis indicates that about 70% of the primary production occurred in seventh- to ninth-order streams, which contained more than 50% of the total stream surface area but represented only 2% of the total length.

The kinds of organisms present in a stream are determined by the availability of various organic substrates. Distinctive groups of invertebrates selectively consume different substrates; their presence and relative abundance is a suitable and easily monitored index (Cummins, 1973, 1974). The interpretation of stream biology in terms of functional groups and processing rates may be of more general use in ecosystem studies than classical taxonomic analysis.

Within most forest streams, four classes of invertebrates, each with distinctive food preferences and associated habitat, may be identified (Fig. 10.9). One group is found on rocks and other surfaces where diatoms, algae, or mosses

TABLE 10.3 Physical Characteristics (%) and Primary Production (%) in Moisie River Basin, Quebec, Canada[a]

Stream order	Total no.	Total length	Total area	Total net primary production
1	78.2	49.9	1.8	0.2
2	16.9	25.5	5.3	1.5
3	3.7	11.9	11.7	5.4
4	0.84	5.8	10.8	6.9
5	0.19	3.3	8.2	6.7
6	0.04	1.5	8.3	8.3
7	0.014	1.1	19.4	22.8
8	0.004	0.9	25.0	33.8
9	0.002	0.2	9.5	14.5

[a] From Naiman and Sedell (1981).

grow. These animals are *grazers* of stream primary production. Another group of invertebrates is described as *gougers* because they bore into wood. A third group is known as *shredders* that eat nearly all but the veins of leaves. Fecal material produced by these or other animals is utilized along with fragments of microbial and plant biomass by *collector* invertebrates that catch specific size classes of

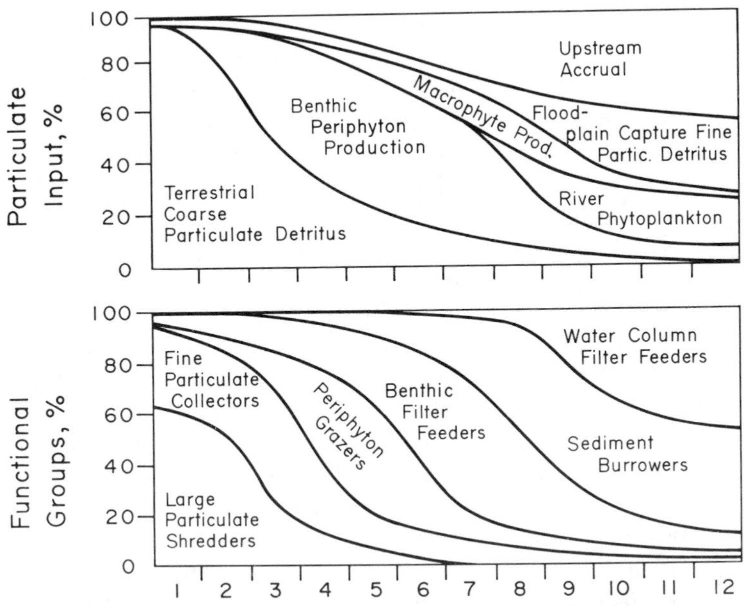

Fig. 10.9. As forested streams increase in size, as indicated here by stream order, associated shifts in the source of particulate material are observed (upper). Specific groups of invertebrates are adapted to utilize various organic substrates and their relative abundance is indicative of major changes in the availability of different substrates. (Vannote *et al.*, 1980; Vannote, 1981.)

particles. Feeding on all of these and other related groups of invertebrates are predators that may include vertebrate as well as invertebrate animals.

The relative abundance of functional groups of invertebrates in a particular segment of stream indicates the relative availability of various organic substrates. An abundance of shredders indicates that leaves are a major energy source, whereas high populations of grazers suggest that photosynthesis in the stream is important. Along small streams, the contribution of riparian leaf fall and woody material may dominate, particularly if the forest is evergreen. On the other hand, larger streams, particularly if not turbid, support large populations of grazers and collectors. For example, the ratio of shredders to grazers changed progressively from more than 100 : 1 in canopy-covered first-order (boulder zone) streams to 1 : 14 in wide eighth-order (pastoral) rivers (Triska *et al.*, 1982). The relative abundance of other functional groups also changes predictably as streams increase in size (Fig. 10.9).

What happens in the small streams has a disproportionate effect upon downstream biological activity. First, because small, steep gradient streams are major sources of sediment, any activity that accelerates erosion from small streams increases turbidity and sedimentation downstream. Second, downstream production processes may be altered substantially if leaf materials and woody debris are not first broken down by organisms in the smaller streams (Triska *et al.*, 1982). This was demonstrated experimentally by applying pesticides to reduce aquatic insect populations by 90% in a small forest stream in the southeastern United States (Wallace *et al.*, 1982). Particulate organic matter resulting from the breakdown of leaves and other organic debris was reduced by more than sevenfold compared to an untreated stream. This indicates the importance of invertebrates in small streams and demonstrates some previously unappreciated effects of local pesticide applications on larger rivers (Wallace *et al.*, 1982).

When organic debris is trapped behind rocks or log dams in streams, there is normally adequate time for microorganisms to colonize the material and for invertebrate populations to develop. For example, in a small mountainous stream in Oregon where obstructions temporarily held 80% of the leaf fall, microbes and invertebrates were able to completely utilize this material over a year. Between 40 and 50% of all organic materials added annually were converted into dissolved products or carbon dioxide (Sedell *et al.*, 1974; Triska *et al.*, 1982).

Fine particulate matter and dissolved compounds are the dominant organic forms normally transported downstream. The fine particulate matter is carried along with inorganic sediment and deposited along the floodplain and channel bottoms. Bottom (benthic) organisms use this material, which also provides nutrients for the growth of macrophytes (rooted aquatic plants) and riparian vegetation (Mills *et al.*, 1966). If sediment loads are high, decomposition may not proceed rapidly because of reduced oxygen levels. Eventually this may lead to the loss of macrophytes and a reduction in the fertility of the floodplain and its river. A good example is provided by the Illinois River, in which the silt load has increased by at least threefold since the turn of the century (Mills *et al.*, 1966). Floodplain lakes have filled with more than 2 m of silt, and the soft unstable bottom now prevents establishment of rooted aquatic plants. Turbidity, even during periods of minimum flow, is sufficient to restrict algae growth (Mills *et al.*, 1966).

The dissolved organic compounds leached from leaves, algae, and other organic materials are rapidly assimilated by microbes in first-order streams with debris dams (Bilby and Likens, 1980; Dahm, 1981). The assimilation by microbes in such situations is twice that recorded in comparable third- and fifth-order streams (Dahm, 1984). In many streams, dissolved organic material is the major form of downstream transport (Wetzel and Manny, 1977; Fisher and Likens, 1973; Moeller *et al.*, 1979; Schlesinger and Melack, 1981).

The importance of large organic debris in streams is related to the stability of these materials. Large pieces of wood become waterlogged and anaerobic, so that decay progresses slowly (Triska and Cromack, 1980). Some large boles submerged in streams may maintain sufficient strength to serve as debris dams for more than a century. In this way, much sediment may be trapped even if hillslope erosion is increased following logging or fire. Removal of large organic debris from streams may release sediment that has accumulated for decades (Likens and Bilby, 1982; Megahan, 1982). Stable debris dams cannot reform until the next forest has matured.

Differences in lignin and nitrogen content help explain variation in decomposition rates of organic detritus from riparian species (Chapter 8). Foliage and woody material from N-fixing plants such as alder are noted for their rapid decay. Most conifer needles and twigs decay more slowly than detritus from deciduous hardwoods (Triska *et al.*, 1982). In general, decay rates of leaves and twigs are two to three times higher in water than on land.

Following mineralization, inorganic nutrients such as N and P usually are rapidly incorporated into microbial biomass in both the riparian and stream systems (Viner, 1975; Minshall *et al.*, 1983). Excess dissolved nutrients pass downstream. The relative retention of labeled P (^{32}P) in streams and riparian zones in relation to flow rate can serve as a measure of biological activity in streams. The rate at which P and other nutrients are removed is often expressed as a cycling distance (Newbold *et al.*, 1981; Elwood *et al.*, 1983; Minshall *et al.*, 1983). An increase in the distance required for a given amount of labeled P to be removed from streamwater is a measure of the impact of disturbance. The time required for stream processes to return to normal following disturbance may also be assessed with the labeling technique.

In general, small streams with forested riparian zones and debris dams retain more nutrients than larger streams (Minshall *et al.*, 1983; Cummins *et al.*, 1984). The smaller streams, once disturbed by fire or forest clearing, may require decades or centuries to recover. On the other hand, midsized streams that constantly undergo large-scale bed movement and lack storage sites may actually be less susceptible to the impact of major disturbance (Minshall *et al.*, 1983). The largest rivers, as long as they have functional floodplain forests, are extremely buffered against major variations in water level and sediment load (Goulding, 1980).

Fish are adapted to life in a wide variety of stream environments. Where fish grow large in size and numbers, they may provide a major source of protein for other animals including humans (Smith, 1981). Anadromous fish such as salmon (*Oncorphynchus*) that spawn in rivers but migrate to large lakes or to the sea have been important in maintenance of large populations of bears (*Ursi*) in Alaska (Holzworth, 1930), eagles in Idaho (McClelland, 1973), and Indian tribes in the Pacific Northwest (Thwaites, 1905). Fish that feed on the bottom or shoreline vegetation, such as carp (*Cyprinus*), catfish (*Ictalurus*), and Amazon *Prochilodus,* may represent more than three-quarters of the total fish biomass harvested by humans from large rivers and associated drainages in various regions of the world (Bonetto, 1975; Mills *et al.*, 1966; Smith, 1981; Goulding, 1980).

Both salmonid and bottom-feeding fish require areas of relatively still water to rest, breed, and forage (Welcomme, 1979). In the case of salmonid species, turbid water limits feeding (Bisson and Bilby, 1982). Large dams or waterfalls may prevent use of potential spawning and rearing habitat by migrating fish such as salmon. Natural dams created by floods or landslides may halt fish migration at low water flow, but these are bypassed by fish during major storms.

Salmonid species are examples of carnivores that feed predominantly on other fish and invertebrates. As predators, such fish are dependent on the maintenance of relatively stable food chains. For this reason, the introduction of an insecticide into the aquatic environment is particularly damaging (Elson and Kerswill, 1966). Peak populations of invertebrates in streams normally correspond with the period of peak growth and development of salmonid fish (Hunt, 1975). Although terrestrial sources of food may only represent 30% of annual consumption, these sources make up more than 80% of the diet for salmonid during critical summer months of maximum growth (Hunt, 1975).

The environment may become temporarily unsuited for fish following fire or logging that destroys the riparian vegetation. Salmonid species are particularly sensitive to water temperatures above 30°C, which may occur in the summer in some small streams. If debris dams and other complex structures remain in streams following the loss of vegetative cover, pools may serve as thermal refuges for fish that require cool water. In streams without pools, the maximum summer temperature of the water may be calculated from knowledge of solar radiation, riparian zone cover, streamflow, and the heat capacity of the streambed (G. W. Brown, 1970). With removal of debris dams, much trapped sediment is released and downstream pools are soon filled. Overzealous removal of stable debris from streams following logging should be prevented. When debris are removed, the gradient of a stream becomes more uniform and may develop into a continuous riffle system (Sedell and Luchessa, 1982; Madej, 1982).

At times, removal of the riparian vegetation may result in warmer water, favoring increased growth rates of salmonid species and associated water column and benthic invertebrates. In nine paired streams scattered throughout Oregon and Washington, Bisson and Sedell (1984) found an average of 50% more fish biomass in streams where much of the riparian tree cover had recently been removed as compared to streams with full cover. Similar changes have been observed following logging along other cold water streams (Burns, 1972; Aho, 1976; Murphy and Hall, 1981). However, these studies also show that the older and larger fish tend to disappear following logging, because the altered environment lacks sufficiently deep pools and cover. The presence of stable debris in streams, on the other hand, maintains habitat that allows the larger fish to remain (Sedell and Swanson, 1984).

In larger streams characteristic of the floodway and pastoral zones, fish are dependent on side channels and pools created by woody debris. In the Amazon basin, with more than 1300 species of fish, a large majority build up fat reserves by eating tree fruits and seeds in shallow water areas (Goulding, 1980; Smith, 1981). The main channel is relatively poor habitat, particularly during flood stage when currents are strong and the water is filled with suspended sediment.

In the Hoh River basin of the Olympic Mountains in Washington, the main river channel contained more than 75% of the total surface water in the basin, yet the density and biomass of resident fish were extremely low (Naiman and Sedell, 1981). The low-gradient tributaries and backwaters harbored up to 75% of the total fisheries resource in the drainage basin (Sedell *et al.*, 1984; Naiman and Sedell, 1981). Goulding (1980) made a similar claim for the importance of floodplain forests in supporting the commercial catch of fish from the Amazon.

SUMMARY

The linkage between terrestrial and aquatic ecosystems is very strong. The input of sediment and organic materials critical for stream life is controlled by streamside vegetation. Certain geomorphic processes and sediment source areas contribute a disproportionate share of sediment. Often, inherent instability of the land can be recognized and measures taken to avoid disturbance or to traverse such areas with special care. Both road construction and logging methods should be adapted to minimize risks of accelerating erosion. In this regard, maintenance of vegetative cover with high root tensile strength is particularly important.

One of the keys to perpetuating a natural complement of stream life and to reducing channel instability is to maintain, as much as possible, the integrity of the riparian zone vegetation. In particular, large fallen trees maintain pools, riffles, and protective cover. Debris dams maintain habitat for invertebrates, fish, and other aquatic organisms. Standing and fallen trees are important along both low- and high-gradient streams. Large animals such as the beaver can be useful in stabilizing degraded streams through their engineering activities, which increase low water volume, reduce sediment transport, and maintain conditions favorable for salmonid and other fish.

Alterations in channel stability or water quality are reflected in reduced utilization of organic compounds. Certain groups of invertebrates reflect the availability of specific substrates. Their absence indicates loss of the substrate or drastic alterations in stream chemistry or sediment load. Another approach to evaluation of the biotic activity in streams involves the introduction of labeled isotopes and monitoring the distance required to remove all radioactivity from the water.

To evaluate the impact of natural disturbances as well as those associated with forest and floodplain management, an entire drainage basin approach is recommended. Four major types of streams may be characterized on the basis of their valley and channel profiles and by the kind of sediment they transport or deposit. This classification is hierarchical, accounting for the number of tributaries that enter a particular channel, starting with first-order streams without any tributaries.

One of the major lessons derived from a basin analysis is that extensive disturbance in mountainous topography will predictably alter valley streams, making them wider, shallower, and more uniform in gradient. Even if streamflow is not increased, the greater amount of sediment transported downstream leads to channel instability and filling of pools and side channels. Floodplain management must, therefore, consider sediment transport as well as water discharge. Maintenance of sloughs, backwaters, and forests along floodplains can help contain the meandering channel to a predictable area. Side channels, natural levees, and flood overflow channels along the valley walls are worth protecting in this regard. Careful location of permanent structures would greatly decrease flood damage to roads, buildings, and bridges.

Flooding and sediment transport should be considered as normal events critical for maintaining the integrity of streams, floodplains, estuaries, and even ocean beaches. Excessive erosion will make a river system unstable. Through an understanding of how river systems operate and their dependence on the adjacent forest, new approaches to river management are possible.

11

Forests and Global Ecology

INTRODUCTION

Buried in the peat bogs of coastal North Carolina are fossils and pollen from forests dominated by jack pine (*Pinus banksiana*) (Watts, 1980). Today, the southernmost occurrence of this species is in Massachusetts, 900 km to the north. Similarly, on desert mountain slopes in west Texas, fossils of pinyon pine (*Pinus cembroides*) and juniper (*Juniperus pinchotii*) are found at elevations 800 m below their present range (Wells, 1966). In both areas the fossils are 15,000–20,000 yr old. These specimens imply that even in areas far south of the glacial ice sheet, the climate of North America was very different during the last ice age. Paleobotanists believe that the distribution of vegetation in North America has changed gradually as glacial conditions have given way to our present climate (Davis, 1981).

Widespread and continuous change in climate and vegetation have important implications for understanding the forest ecosystems of today. For example, it is possible that well-known species assemblages, such as the beech–maple forests of the midwestern United States, are relatively recent associations, rather than the product of long-term coevolution. Even now, forest ecosystems are likely to be slowly changing in composition. When these changes are accompanied by changes in forest biomass, nutrient cycling, and soil development, steady-state models of ecosystem processes (e.g., Chapters 6 and 8) may not be realistic. The response of forest ecosystems to air pollution, acid rain, and species introductions can only be evaluated in the context of slow natural changes in climate that may affect vegetation.

In this chapter we examine the role of forests in global biogeochemical cycles. Does land vegetation in general, and forest in particular, affect the surface environment of the Earth to create conditions that are different from an Earth without vegetation? Are changes in land vegetation by man likely to affect global cycles? Finally, are there indications of past conditions to be found by an examination of plant fossils and tree rings?

FORESTS AND THE GLOBAL ATMOSPHERE

Forests have high rates of net primary production (NPP) compared to other ecosystems (Table 11.1). Aboveground NPP of 400–1000 $g/m^2/yr$ of carbon (C) is over twice that of grasslands and many times greater than the open ocean. Some of this production accumulates as biomass and humus, which constitute net ecosystem production (NEP) (Chapters 3 and 6). The storage of C in forest vegetation and forest soils accounts for about 60% of the organic C stored on the Earth's land surface.

Gross primary production of all land ecosystems removes about 100×10^{15} g C/yr from the atmosphere. This C returns to the atmosphere as carbon dioxide (CO_2) from plant respiration, decomposition, and fires. At these rates, the equivalent of the entire atmospheric content of CO_2 passes through the terrestrial biota every 7 yr, with about 70% of the exchange occurring through the forest ecosystems. The seasonal activity of temperate vegetation, especially forests, is seen in measurements made at the global level. Long-term records of atmospheric CO_2 show an annual decline during the summer months due to photosynthesis. Carbon dioxide increases during the winter as decomposition continues in the absence of C fixation. This annual oscillation of CO_2 (Fig. 11.1) illustrates the effect of temperate vegetation on the Earth's atmosphere. In the southern hemisphere, where land vegetation is less abundant, the oscillation is much less pronounced.

The record of atmospheric CO_2 also shows a long-term increase that has been linked to the burning of fossil fuels. Since CO_2 in the atmosphere reduces the loss of infrared (heat) radiation from the Earth's surface, such an increase is likely to result in a warming of the Earth's climate. The present increase in atmospheric CO_2 appears to be due to man's activities, but there is evidence of large natural changes through geologic time. At the peak of the last ice age,

TABLE 11.1 Net Primary Production (NPP), Biomass, and Soil Organic Matter in World Ecosystems[a]

Ecosystem type	Area (10^6 km^2)	Mean NPP (g C/m^2/yr)	Mean biomass (kg C/m^2)	Total biomass (10^{15} g C)	Mean soil organic (kg C/m^2)	Total soil organic (10^{15} g C)
Tropical forest	24.5	1000	22	460	10.4	255
Temperate forest	12	650	15	175	11.8	142
Boreal forest	12	400	9	108	14.9	179
Woodland and shrubland	8.5	300	3	22	6.9	59
Tropical savanna	15	350	2	27	3.7	56
Temperate grassland	9	250	1	6.3	19.2	173
Tundra and alpine	8	70	0.3	2.4	21.6	173
Desert scrub	18	35	0.3	5.4	5.6	101
Extreme desert	24	1	0.01	0.2	0.1	3
Agricultural	14	325	0.5	7.0	12.7	178
Swamp and marsh	2	1000	7	13.6	68.6	137
Total land	147			827		1456
Open ocean	332	63	0.0015	0.5		
Continental shelf	27	175	0.005	0.15		
Estuaries	2	1000	0.5	1.0		
Lake and stream	2	250	0.01	0.02		
Total aquatic and marine	363			1.67		

[a] From Whittaker (1970) and Schlesinger (1977; reproduced, with permission, from the *Annual Review of Ecology and Systematics*, Vol. 8 © 1977 by Annual Reviews Inc.)

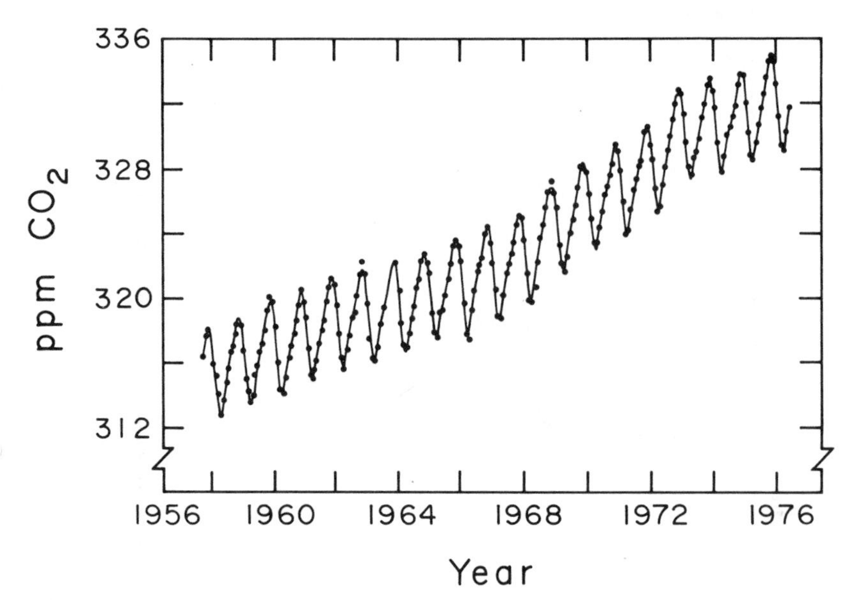

Fig. 11.1. Concentration of atmospheric CO_2 at Mauna Loa Observatory, Hawaii, showing the annual cycle due to the photosynthesis and respiration of terrestrial ecosystems and the long-term increase that is largely attributed to the burning of fossil fuels. Dots indicate the monthly averages based on continuous measurements. [From Bacastow and Keeling, 1981, Atmospheric carbon dioxide concentration and the observed airborne fraction. *In* "Carbon Cycle Modeling" (B. Bolin, ed.), Copyright 1981 by John Wiley & Sons, Ltd. Reprinted by permission of John Wiley & Sons, Ltd.]

atmospheric CO_2 was 200 ppm compared to the present value near 340 ppm (Neftel *et al.*, 1982).

The causes of prehistoric changes in atmospheric CO_2 are highly uncertain. Atmospheric CO_2 is largely determined by equilibrium buffering with the oceans. Nevertheless, changes in the C storage in land vegetation are likely to have occurred during ice age glaciations (Shackleton, 1977). Any shift in the relative land area occupied by forests and grasslands will change the C stored on land, because the C storage per unit area is much larger in forests (Table 11.1). Changes in land vegetation that alter the global balance of photosynthesis and respiration may have played a role in determining the prehistoric record of atmospheric CO_2 (Shackleton, 1977).

Agricultural expansion may be responsible for some of the current increase in atmospheric CO_2, when forests with high biomass and net primary productivity are replaced by croplands (Houghton *et al.*, 1983). When the original vegetation is destroyed, there are also losses of soil C to the atmosphere as CO_2 (Schlesinger, 1986). Conversely, successional forests show a net storage of CO_2 from the atmosphere in areas of agricultural abandonment (Delcourt and Harris, 1980). One may speculate that the relative area of forested and nonforested land through time has affected global climate by affecting the atmospheric level of CO_2.

Increasing levels of atmospheric CO_2 might stimulate the primary productivity of land ecosystems, inasmuch as the level of internal CO_2 in leaves determines the rate of photosynthesis in many instances (Chapter 2). If higher net primary productivity were to lead to greater long-term storage of C in land vegetation and soils, forest growth might buffer the increase in atmospheric CO_2 from fossil fuel combustion. Many scientists believe, however, that this is unlikely in view of the widespread limitation of natural systems by available water, nutrients, and light (Kramer, 1981), especially in forest ecosystems with maximum development of canopy leaf area (Chapter 3). Emissions of nitrogen (N) and sulfur (S) during fossil fuel combustion might "fertilize" terrestrial ecosystems, but the release of these elements is small relative to the release of C during fuel combustion. In fact, the ratios of C to nutrient elements in living biota are much smaller than the ratios in fuel emissions (Table 11.2). Emissions of N increase the ratios of available N/P in soils that receive airborne deposition from industry, shifting the balance of limiting nutrients in these ecosystems (Chapter 9). If CO_2 stimulates a greater storage of organic C in land ecosystems, nutrient limitations and imbalances should become more critical.

Photosynthesis releases O_2 to the atmosphere in a stoichiometric equivalent to the C fixed in carbohydrate. Without photosynthesis, carbon would be bound with oxygen as CO_2, which would be a dominant atmospheric constituent as it is on the nearby planets. Indeed, the presence of O_2 in the Earth's atmosphere appears to be entirely due to the activity of green plants through geologic time; the search for free oxygen as a "signature" of life was a major goal of space explorations to Mars.

We have seen that forest production may affect atmospheric CO_2, so it is tempting to suggest that forest vegetation affects the O_2 content of the atmosphere as well. Some simple calculations suggest that this is not so. If all the organic C in land vegetation and soils (2000×10^{15} g = 167×10^{15} moles; Table 11.1) were oxidized, the oxygen consumed would result in a 0.02%

TABLE 11.2 Elemental Ratios (by Weight) for Major Nutrients[a]

	C	:	N	:	S	:	P
Fossil fuel emissions	9300	:	36	:	130	:	1
Land plants	790	:	7.6	:	3.1	:	1
Ocean plants	129	:	12	:	2.9	:	1
Soil humus	54	:	3	:	1.2	:	1

[a] From Likens *et al.* (1981) Interactions between major biogeochemical cycles in terrestrial ecosystems. *In* "Some Perspectives of the Major Biogeochemical Cycles," (G. E. Likens, ed.)

decline in the atmospheric content. Thus atmospheric O_2 is not balanced by reduced C stored in the present land vegetation. The pool of C in vegetation and soils has a mean residence time (pool/NPP) of about 40 yr. This rapid turnover speaks for the efficiency of decomposers (Chapter 8); there is little reduced C left for storage that would result in free O_2 in the atmosphere.

Atmospheric oxygen is an enormous reservoir that has built up over the last billion years, largely through the storage of reduced C in ocean sediments. The organic C storage in ocean sediments is dominated by the primary production of the marine environment. Transport of organic C in riverflow (0.4×10^{15} g/yr) (Schlesinger and Melack, 1981; Meybeck, 1982) carries about 1% of the terrestrial primary production to the sea, but this is only about 10% of the amount of organic C that is added to the ocean sediments by phytoplankton production (Eppley and Peterson, 1979). Compared to their effect on the global C cycle, forest ecosystems show no obvious or major effect on the global occurrence of free oxygen in the atmosphere.

Anaerobic soils are the source of many reduced trace gases found in the atmosphere (Chapter 6). Emissions from wetlands dominate the annual global production of methane (CH_4) and sulfur gases such as H_2S and dimethyl sulfide $(CH_3)_2S$. Emissions from upland forests are also a significant flux in some global budgets (Adams *et al.*, 1981; Delmas and Servant, 1983). Anaerobic digestion by termites in forest soils may account for 25% of the annual input of methane to the atmosphere (Crutzen, 1983; Zimmerman *et al.*, 1982). As an atmospheric constituent, methane acts like CO_2 to prevent infrared reradiation from the Earth's surface. Forests also release virtually all of the reduced hydrocarbons, such as terpenes and isoprenes, that contribute to summertime atmospheric haze (Peterson and Tingey, 1980).

After N_2, the most important atmospheric N gas is N_2O, which comprises $<0.01\%$ of the atmospheric N content. Denitrification in forest soils may play an important role in the global distribution and occurrence of N_2O (Chapter 6). Recently, Michael McElroy and co-workers have monitored atmospheric concentrations of N_2O in tropical forests (Keller *et al.*, 1983). Their measurements show high concentrations of N_2O emanating from the tropical forests of the Amazon Basin (Fig. 11.2). Atmospheric concentrations of N_2O appear to be increasing at about 0.3% per year (Khalil and Rasmussen, 1983), but the source of this imbalance in the atmospheric budget and the global flux of N_2O from forests are unknown. N_2O also affects the reradiation from the Earth and is consumed by reactions with ozone in the upper atmosphere, producing some of the NO_3^- that is found in rainfall (Chapter 6). Fortunately, the current increase in N_2O concentration appears unlikely to reduce the ozone content and the absorption of ultraviolet light in the upper atmosphere (Crutzen, 1983).

In addition to the production of trace gases in undisturbed forest ecosystems,

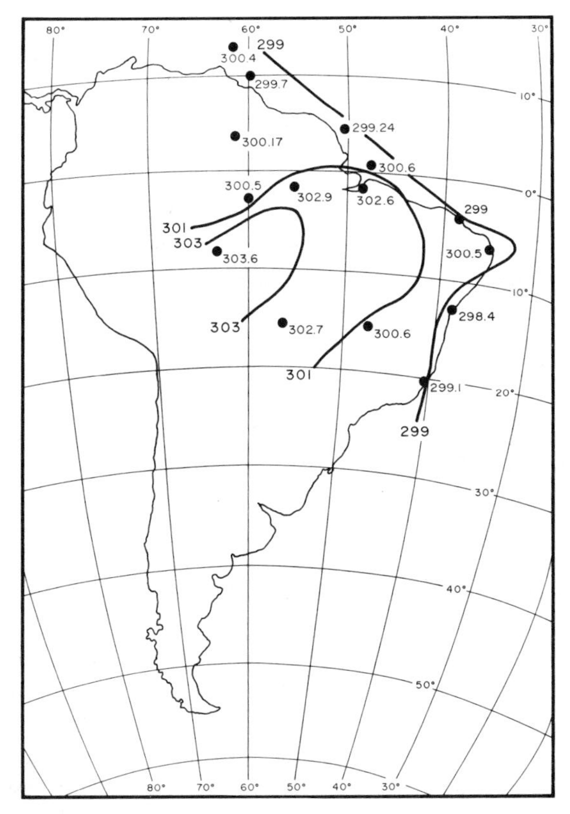

Fig. 11.2. Concentrations of N_2O measured at ground-based stations throughout eastern South America. (Courtesy of and copyright by Steven C. Wofsy and M.B. McElroy, Harvard University.)

there are substantial releases of CH_4, NO, NO_2, N_2O, NH_3, and SO_2 during forest fires. We know little about the role of fire and man's alteration of fire cycles on the global budgets of these gases (Crutzen, 1983; Greenberg *et al.*, 1984). Aerosols are also released from undisturbed forest canopies and during forest fires (Chapter 6). Conversely, interception of fog water, aerosols, and gases by a forest canopy (Chapter 6) enhances the removal of these constituents from the atmosphere in forests compared to the removal by deposition in grasslands or on bare soil. In some cases this deposition increases stream runoff and results in unusual concentrations of ions in forest soils (Potts, 1978; Lovett *et al.*, 1982). While forest canopy processes are unlikely to have a major effect on the global atmospheric burden of aerosols, it is worth noting that the rate of wind erosion of soil is much lower in humid regions, occupied by forests, than in deserts and semiarid grasslands (Marshall, 1973).

FOREST INTERACTIONS WITH THE HYDROSPHERE

Land vegetation affects global biogeochemical cycles by altering the rate of erosion and chemical weathering of the continents and the delivery of dissolved and suspended materials to the sea. Through geologic time, the accumulation of ocean sediments is inversely related to changes in sea level; higher sea level lowers the land/sea ratio of surface area and lowers the total chemical erosion from the continents (Worsley and Davies, 1979). In Chapter 6, we noted that increased streamflow following forest fires or clearcutting results in greater streamwater losses of nutrients from small watersheds. These effects are more difficult to see in major river systems, which integrate a large diversity of land-use practices, watershed characteristics, and rainfall events. Nevertheless, the scale of man's current impact is seen in an increase in riverflow in the Amazon that appears related to forest cutting in the upper basin (Gentry and Lopez-Parodi, 1980), although the interpretation of these data has been questioned (Nordin and Meade, 1982). Based on the increases in erosion following timber harvest (Chapter 10), one can only speculate on the massive rates of erosion that must have occurred before the evolution of land vegetation.

Steamwater losses of nutrients change during the development of forest ecosystems as a result of changes in NEP (Chapter 6). The loss of dissolved ions in streamwater is largely determined by biotic activities in the soil (Chapter 8). Even the loss of nonessential elements such as lead is affected by biotic processes, because some of these elements are retained in humus and accumulate in the ecosystem (Smith and Siccama, 1981). The relative transport of dissolved and suspended sediment in riverflow is also affected by rainfall and topography (Meybeck, 1977); thus, it is difficult to compare the direct influence of vegetation on the global flux to the ocean.

Chemical weathering proceeds most rapidly in warm, humid climates. The removal of dissolved ions per unit land area is greatest with high runoff both in small watersheds (Chapter 6, Fig. 6.9) and in global comparisons of river basins (Van Denburgh and Feth, 1965). Since forest vegetation dominates climatic regions with the greatest excess of precipitation over evapotranspiration, high rates of chemical weathering and removal of dissolved solids are expected. In forests, chemical weathering is driven by carbonation, supplied by CO_2 from root respiration and decomposition (Chapter 6). Presumably carbonation weathering now occurs more rapidly than before the advent of land plants, unless atmospheric CO_2 was much more abundant in earlier geologic epochs (cf. Berner *et al.*, 1983).

The removal of sediment and suspended solids, which are the products of

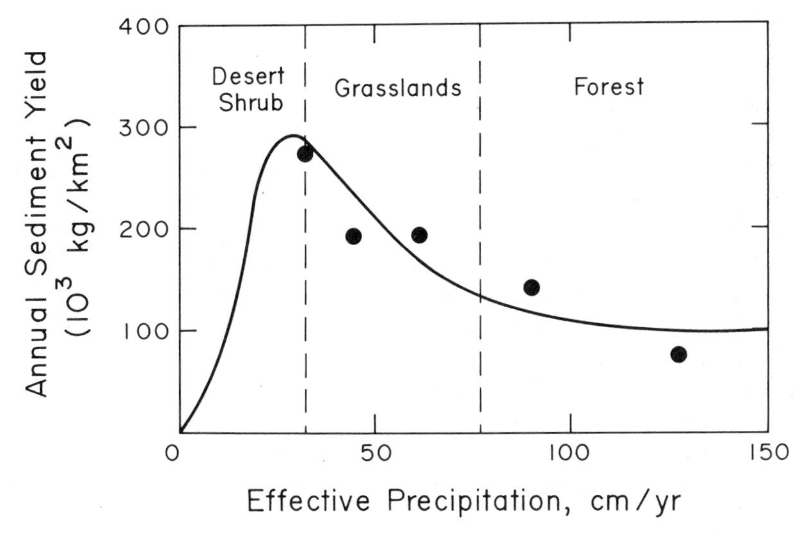

Fig. 11.3. Sediment yield as a function of effective precipitation in terrestrial ecosystems. (From Langbein and Schumm, *Trans. Am. Geophys. Union* **39,** 1076–1084, 1958, Copyright by the American Geophysical Union.)

mechanical weathering, is low in forests relative to grasslands (Fig. 11.3). The greatest rates of sediment transport are found in the semiarid grasslands, where unpredictable torrential rainstorms erode soils with a sparse cover of protective vegetation. During most other periods, there is insufficient water flux for chemical weathering, and in many semiarid and arid soils, salts, such as $CaCO_3$, accumulate in the lower soil profile. High ratios of suspended to dissolved materials characterize the load of rivers draining grasslands compared to those draining forests. This is also reflected in the transport of organic C, which largely consists of dissolved humic substances in forest streams and of particulate matter in grassland streams (Malcolm and Durum, 1976; Schlesinger and Melack, 1981).

On a global basis, one would expect that the relative area of forest versus nonforest has controlled the flux of dissolved and suspended sediment to the oceans through geologic time. Changes in the delivery of dissolved nutrient ions strongly affect the productivity of lakes, estuaries, and the near-shore ocean (Whitehead *et al.*, 1973). Man's influence on riverine nutrient flux does not now affect the total productivity of the sea (Peterson, 1981), but over longer periods large variations in the total NPP of marine environments may have been due to changes in the delivery of dissolved nutrients during glaciations and other events that affect land vegetation (McElroy, 1983).

FORESTS AND GLOBAL CLIMATE

In most regions, there is a greater volume of streamflow after forest cutting. Greater runoff is due to the reduction of leaf area and transpiration from the canopy (Chapter 5). On a global basis, evapotranspiration from land supplies about 20% of the water vapor to the atmosphere annually (Westall and Stumm, 1980). Most of this comes from forest regions that occupy land areas with the greatest mean rainfall. As much as half of the precipitation falling in the Amazon Basin may be derived from evapotranspiration from the rain forest (Dall'Olio *et al.*, 1979; Salati *et al.*, 1979; Salati and Vose, 1984). A reduction in forest cover that lowers global evapotranspiration and increases the percentage of water lost as runoff could affect the distribution and amount of precipitation in adjacent regions (Salati and Vose, 1984) and over the entire Earth (Fig. 11.4). Such a perturbation would change other aspects of climate as well. Since the radiation absorbed by a forest canopy is greater than that absorbed by a grassland, crop-

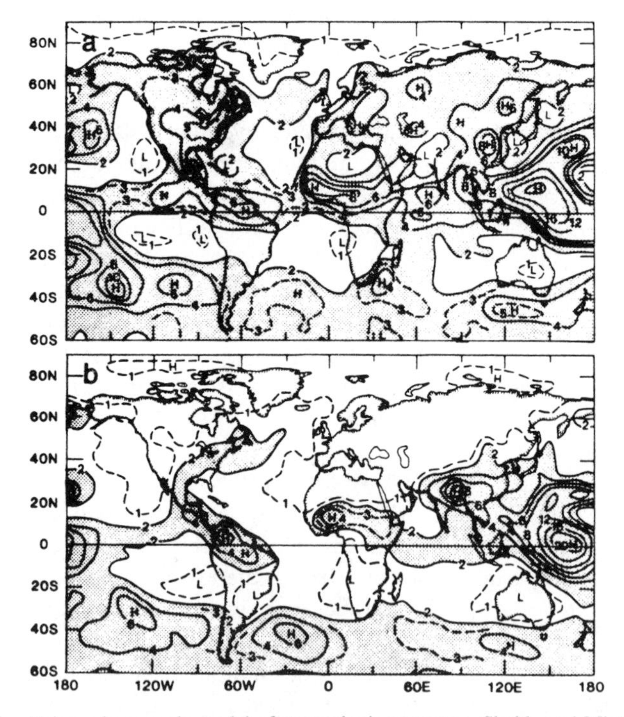

Fig. 11.4. Using a large-scale model of atmospheric processes, Shukla and Mintz (1982; *Science* **215,** 1498–1501, Copyright 1982 by the AAAS) predicted the global distribution of July rainfall in two contrasting conditions: (a) evapotranspiration equal to potential evapotranspiration in all land areas; (b) no evapotranspiration from land.

land, or barren soil, a reduction in forest growth would increase global albedo, or reflectivity (Sagan *et al.*, 1979; Henderson-Sellers and Gornitz, 1984). One result of tropical deforestation could be a cooler, drier global climate (Potter *et al.*, 1975). Such changes may be irreversible. Even today, in areas of steppe woodland in the southern Sahara, the desert area appears to be expanding as overgrazing produces changes in regional albedo and climate. Changes in albedo due to deforestation affect climate in an opposite direction from the effects that are anticipated from increasing atmospheric CO_2. While these effects are speculative, the examples illustrate an effect of forests on global processes such as the circulation of water and climatic patterns.

The presence of coal beneath the Antarctic Ice Cap indicates that large-scale shifts in climate, vegetation, and continental location have occurred through geologic time. Even during the last 2 million yr, as many as 16 continental glaciations have occurred in the northern hemisphere (Davis, 1981). The most recent glacial advance, the Wisconsin, reached a maximum about 18,000 yr ago, covering most of Canada, Europe, and the northern United States. Evidence from fossil pollen leaves little doubt that tundra vegetation occurred along the southern boundary of the glacier. Boreal species, such as spruce and larch, occurred in the central United States (Davis, 1981), and oak hickory forests were found on the Gulf Coastal Plain (Delcourt *et al.*, 1983). Melting of the glacier was relatively rapid, and a northward "migration" of tree species was well advanced by 15,000 yr ago. The present distribution of maple (*Acer* sp.) was established by 6,000 yr ago (Fig. 11.5), based on the occurrence of fossil pollen in dated lake sediments. From migration maps, Davis (1981) calculated rates of northward range extension, ranging from 100 to 400 m/yr for various species. Even some heavy seeded species, such as oaks, moved northward quite rapidly at rates consistent with observation of present-day dispersal of acorns by bluejays (Darley-Hill and Johnson, 1981) and other animals (Chapter 9). Within a span of 5,000 yr, the vegetation of Ohio was dominated sequentially by tundra, spruce forest, and the present forest of hardwoods.

The forest communities of North America now exist in a dynamic equilibrium with climatic conditions. Recent climate has shown short-term variations, such as the "little ice age" from 1450 to 1850, that are related to small-scale changes in forest boundaries (Webb, 1981). These changes are often observed at tree line (e.g., Kearney and Luckman, 1983). For example, the northern range of black spruce (*Picea mariana*) in the boreal forest is apparently limited by temperature at the stage of seedling germination, and the present tree line is 40 km north of present climatic conditions that would allow successful germination (Black and Bliss, 1980).

Postglacial changes in vegetation are documented for other areas of the world, including Europe (Birks and Birks, 1980) and the tropics (Livingstone, 1975; Shackleton, 1977). In the deserts of the southwestern United States, the mid-

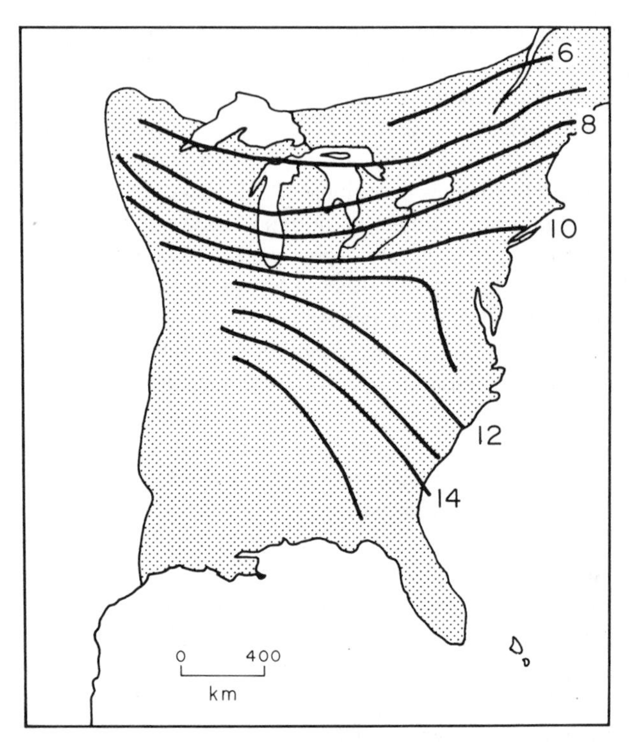

Fig. 11.5. Contour map showing the northward migration of maple (*Acer* sp.), based on dates of occurrence of pollen in dated lake sediments. Data are in 10^3 yr. The present range of the genus is shaded. (From Davis, 1981.)

glacial climate probably included mild, wet winters and cool summers. Semiarid forests of pinyon pine and juniper occurred at lower elevations than today, in unusual floristic associations with present-day desert taxa (Van Devender and Spaulding, 1979; Wells, 1983). Postglacial changes in the distribution of desert vegetation were apparently more rapid than in the eastern United States.

Widespread and long-term shifts in vegetation due to climatic change are incompatible with simple steady-state concepts of ecosystem processes and soil development. Soil horizons and soil chemical properties that affect nutrient cycling are established over thousands of years in association with vegetation (Bockheim, 1980; Jenny, 1980; Chapter 8). For instance, cold temperatures impede decomposition in most of the boreal zone. These ecosystems have shown an accumulation of soil C representing NEP during the entire Holocene period. The boreal forests now contain between 5 and 16% of all soil organic C stored on land (Miller, 1980; Post *et al.*, 1982; Schlesinger, 1984). Future climatic warming in these areas may release CO_2 to the atmosphere (Billings *et al.*, 1982, 1983) and renew the northward extension of the vegetation zones. Since the wintertime albedo of deciduous forest is much greater than that of coniferous forest, changes in regional albedo could also occur with a shift in the northern boundary of temperate forest cover (Bolin, 1982).

In addition to fossil pollen, a record of climate change during more recent years is contained in the annual rings of xylem found in trees. Remembering that the allocation of carbohydrate to stem increment is a sensitive index of environmental stress (Chapter 2), one can extract a record of climate from the variation in the width of the annual increment through the life of a tree. Tree-ring data must be subjected to careful scrutiny in collection and analysis (Fritts, 1976). The most useful trees often grow in stressful environments, where annual variations in environment are likely to cause an immediate response in growth (Fritts *et al.*, 1965). False and missing rings must be recognized by analysis of replicate cores or cross sections. Freestanding trees often produce a clearer record because they have not changed in relative canopy position during growth. Finally, through statistical analysis one must remove the reduction in ring width that is expected as any tree increases in girth. When these corrections are made, the remaining data may produce a strong correlation to recent climatic measurements, and past climatic regimes can be reconstructed by extrapolation (Fig. 11.6).

While some long-lived trees, such as bristlecone pine (*Pinus longaeva*), contain a tree-ring record of 6000 yr or more (La Marche, 1974), the majority of the tree-ring records are 200–400 yr in length. Forty-year records from four species

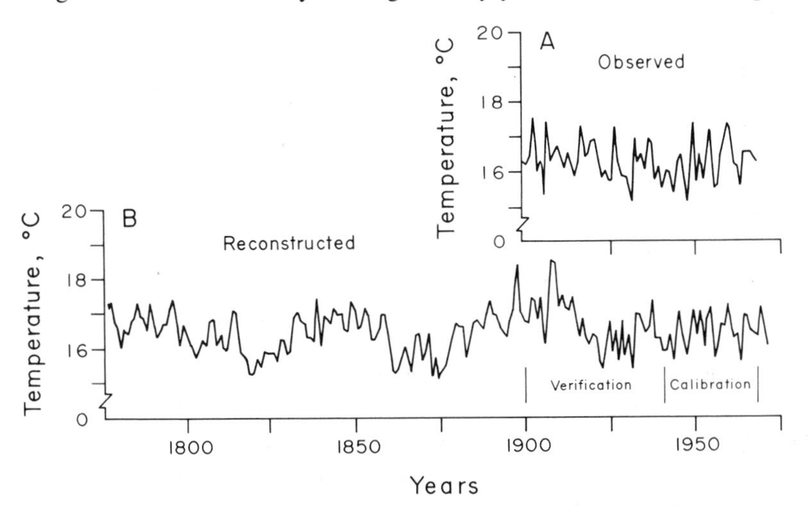

Fig. 11.6. Correlations between tree-ring width and temperature were obtained for the period 1940 to 1970, validated using the data from 1900 to 1940, and used to extrapolate the climate in Tasmania in the interval from 1780 to 1900. (From La Marche and Pittock, 1982, Preliminary temperature reconstructions for Tasmania. *In* "Climate from Tree Rings," Cambridge University Press.)

collected in the Hudson River Valley (New York) show an inverse correlation of ring width to summer drought (Cook and Jacoby, 1977). While industrial pollution and acid rain appear to have reduced forest growth in recent years, the effects of regional drought must also be considered (Cogbill, 1977; A. H. Johnson *et al.*, 1981; Puckett, 1982). La Marche *et al.* (1984) suggest that recent, wide rings in bristlecone pine in the Sierra Nevada may indicate increasing primary production due to higher atmospheric CO_2. If correct, the tree-ring evidence represents the first detection of a biospheric response to changes in CO_2.

Recent work has included measurements of wood density and the isotopic composition of wood cellulose in tree rings. The isotopic ratio of $^{18}O/^{16}O$ in cellulose is closely correlated to that in precipitation, in which it is determined by mean annual temperature. Thus, the oxygen isotopes contain a record of mean annual temperature (Burk and Stuiver, 1981). When corrections for radioactive decay are made, ^{14}C in tree-ring cellulose contains a record of atmospheric ^{14}C levels (Stuiver, 1978a). Both ^{14}C and $^{13}C/^{12}C$ ratios are useful in determining the comparative importance of fossil fuel versus changes in forest biomass as sources of the increase in atmospheric CO_2 (Stuiver, 1978b; Freyer, 1979; Leavitt and Long, 1983). A cooperative, international network of tree-ring studies promises to offer much new insight to global climate fluctuations during the Holocene (Hughes *et al.*, 1982).

SUMMARY

Understanding changes in the world's ecosystems requires monitoring at the global level and a consideration of the globe as a single ecosystem, the biosphere. In many cases, the trace constituents of the atmosphere provide a good index of the activity of the biosphere, since the atmosphere is well mixed and many of these gases are derived only from biotic activity. Satellite monitoring of the Earth's energy budget may allow the detection of climatic warming due to increasing CO_2 (Kiehl, 1983). The extent of the world's ecosystems, including forests, is easily seen by satellite photography (Tucker *et al.*, 1985). Estimates of leaf area and global NPP (e.g., Table 11.1), now derived by averaging individual sites sampled in the field, may be vastly improved and simplified by remote sensing. Already a strong relation has been found between the ratio of reflected infrared and red wavelengths and leaf area index for coniferous forests (Spanner *et al.*, 1984; Botkin *et al.*, 1984). Careful observations of vegetation boundaries, such as tree line and desert–woodland ecotones, can lead to sensitive indices of global climatic change. Comparisons of photographs through time will allow estimates of the area of natural ecosystems that has been affected by man (Woodwell *et al.*, 1983).

Bibliography

Aber, J. D., and Melillo, J. M. (1980). Litter decomposition: Measuring relative contributions of organic matter and nitrogen to forest soils. *Can. J. Bot.* **58,** 416–421.

Aber, J. D., and Melillo, J. M. (1982a). Nitrogen immobilization in decaying hardwood leaf litter as a function of initial nitrogen and lignin content. *Can. J. Bot.* **60,** 2263–2269.

Aber, J. D., and Melillo, J. M. (1982b). ''FORTNITE: A Computer Model of Organic Matter and Nitrogen Dynamics in Forest Ecosystems,'' Agric. Res. Bull. R3130. University of Wisconsin, Madison.

Aber, J. D., Botkin, D. B., and Melillo, J. M. (1979). Predicting the effects of differing harvesting regimes on productivity and yield in northern hardwoods. *Can. J. For. Res.* **9,** 10–14.

Abrahamsen, G. (1980). Acid precipitation, plant nutrients and forest growth. *In* ''Ecological Impact of Acid Precipitation'' (D. Drablos and A. Tollan, eds.), pp. 58–63. Norwegian Interdisciplinary Research Program, SNSF Project, Ås, Norway.

Abrahamsen, G. (1983). Sulphur pollution: Ca, Mg and Al in soil and soil water and possible effects on forest trees. *In* ''Effects of Accumulation of Air Pollutants in Forest Ecosystems'' (B. Ulrich and J. Pankrath, eds.), pp. 207–218. Reidel Publ., Dordrecht, Netherlands.

Abrahamsen, G., and Tveite, B. (1983). Effects of air pollutants on forest and forest growth. *In* ''Ecological Effects of Acid Deposition,'' pp. 199–219. National Swedish Environment Protection Board, Stockholm.

Adams, D. F., Farwell, S. O., Robinson, E., Pack, M. R., and Bamesberger, W. L. (1981). Biogenic sulfur source strengths. *Environ. Sci. Technol.* **15,** 1493–1498.

Adams, M. A., and Attiwill, P. M. (1982). Nitrate reductase activity and growth response of forest species to ammonium and nitrate sources of nitrogen. *Plant Soil* **66,** 373–381.

Aho, P. S. (1976). A population study of the cutthroat trout in an unshaded and shaded section of streams. M.S. Thesis, Oregon State University, Corvallis.

Aina, P. O. (1984). Contribution of earthworms to porosity and water infiltration in a tropical soil under forest and long-term cultivation. *Pedobiologia* **26,** 131–136.

Albrektson, A., Aronsson, A., and Tamm, C. O. (1977). The effect of forest fertilization on primary production and nutrient cycling in the forest ecosystems. *Silva Fenn.* **11,** 233–239.

Aldon, E. F. (1968). Moisture loss and weight of the forest floor under pole-sized ponderosa pine stands. *J. For.* **66,** 70–71.

Allison, F. E., Murphy, R. M., and Klein, C. J. (1963). Nitrogen requirements for the decomposition of various kinds of finely ground woods in soil. *Soil Sci.* **96,** 187–190.

Amman, G. B., and Baker, B. H. (1972). Mountain pine beetle influence on lodgepole pine stand structure. *J. For.* **70,** 204–209.

Amthor, J. S. 1984. The role of maintenance respiration in plant growth. *Plant Cell Environ.* **7,** 561–569.

Anderson, E. A. (1968). Development and testing of snowpack energy balance equations. *Water Resour. Res.* **4,** 19–37.

Anderson, J. E., and McNaughton, S. J. (1973). Effects of low soil temperature on transpiration, photosynthesis, leaf relative water content, and growth among elevationally diverse plant populations. *Ecology* **54,** 1220–1233.

Anderson, J. M. (1973). The breakdown and decomposition of sweet chestnut (*Castanea sativa* Mill.) and beech (*Fagus sylvatica* L.) leaf litter in two deciduous woodland soils. I. Breakdown, leaching and decomposition. *Oecologia* **12,** 251–274.

Anderson, J. M., Proctor, J., and Vallack, H. W. (1983). Ecological studies in four contrasting lowland rain forests in Gunung Mulu National Park, Sarawak. III. Decomposition processes and nutrient losses from leaf litter. *J. Ecol.* **71,** 503–527.

Anderson, J. M., Ineson, P., and Huish, S. A. (1983). Nitrogen and cation mobilization by soil fauna feeding on leaf litter and soil organic matter from deciduous woodlands. *Soil Biol. Biochem.* **15,** 463–467.

Anderson, J. P. E., and Domsch, K. H. (1975). Measurement of bacterial and fungal contributions to respiration of selected agricultural and forest soils. *Can. J. Microbiol.* **21,** 314–322.

Anderson, J. P. E., and Domsch, K. H. (1978). A physiological method for the quantitative measurement of microbial biomass in soils. *Soil Biol. Biochem.* **10,** 215–221.

Anderson, J. P. E., and Domsch, K. H. (1980). Quantities of plant nutrients in the microbial biomass of selected soils. *Soil Sci.* **130,** 211–216.

Appleby, R. F., and Davies, W. J. (1983). A possible evaporation site in the guard cell wall and the influence of leaf structure on the humidity response of stomata by woody plants. *Oecologia* **56,** 30–40.

Arno, S. F. (1976). The historical role of fire on the Bitterroot national forest. *USDA For. Serv. Res. Pap. INT* **INT-187,** 1–29.

Aronsson, A., Elowson, S., and Ingestad, T. (1977). Elimination of water and mineral nutrition as limiting factors in a young pine stand. I. Experimental design and some preliminary results. *Swed. Conifer. For. Proj., Tech. Rep.* **10,** 1–38.

Art, H. W., Bormann, F. H., Voigt, G. K., and Woodwell, G. M. (1974). Barrier island forest ecosystem: Role of meteorologic inputs. *Science* **184,** 60–62.

Arthur, M. A. (1982). The carbon cycle-controls on atmospheric CO_2 and climate in the geologic past. *In* "Climate in Earth History," pp. 55–67. National Academy Press, Washington, D.C.

Ascaso, C., Galvan, J., and Rodriquez-Oascual, C. (1982). The weathering of calcareous rocks by lichens. *Pedobiologia* **24,** 219–229.

Aston, A. R. (1979). Rainfall interception by eight small trees. *J. Hydrol.* **42,** 383–396.

Aweto, A. O. (1981). Organic matter build-up in fallow soil in a part of southwestern Nigeria and its effects on soil properties. *J. Biogeogr.* **8,** 67–74.

Axelsson, B. (1981). "Site Differences in Yield Differences in Biological Production or in Redistribution of Carbon within Trees," Res. Rep. No. 9. Dept. Ecol. Environ. Res., Swed. Univ. Agric. Sci., Uppsala.

Axelsson, B., and Bräkenhielm, S. (1980). Investigation sites of the Swedish coniferous forest project-biological and physiographic features. *Ecol. Bull.* **32,** 25–64.

Azevedo, J., and Morgan, D. L. (1974). Fog precipitation in coastal California forests. *Ecology* **55,** 1135–1141.

Bååth, E., and Söderström, B. (1982). Seasonal and spatial variation in fungal biomass in a forest soil. *Soil Biol. Biochem.* **14,** 353–358.

Bååth, E., Lohm, U., Lundgren, B., Rosswall, T., Söderström, B., Sohlenius, B., and Wirén, A.

(1978). The effect of nitrogen and carbon supply on the development of soil organism populations and pine seedlings: A microcosm experiment. *Oikos* **31,** 153–163.

Bååth, E., Lohm, U., Lundgren, B., Rosswall, T., Söderström, B., and Sohlenius, B. (1981). Impact of microbial-feeding animals on total soil activity and nitrogen dynamics: A soil microcosm experiment. *Oikos* **37,** 257–264.

Bacastow, R. B., and Keeling, C. D. (1981). Atmospheric carbon dioxide concentration and the observed airborne fraction. *In* "Carbon Cycle Modelling" (B. Bolin, ed.), pp. 103–112. Wiley, New York.

Baes, C. F., III, and McLaughlin, S. B. (1984). Trace elements in tree rings: Evidence of recent and historical air pollution. *Science* **224,** 499–496.

Baker, J. P., and Schofield, C. L. (1980). Aluminum toxicity to fish as related to acid precipitation and Adirondack surface water quality. *In* "Ecological Impact of Acid Precipitation" (D. Drabolos and A. Tollan, eds.), pp. 292–293. Norwegian Interdisciplinary Research Program, SNSF Project, Ås, Norway.

Bakke, A. (1983). Host tree and bark beetle interaction during a mass outbreak of Ips typographus in Norway. *Z. Angew. Entomol.* **96,** 118–125.

Bamber, E. K., and Humphreys, F. R. (1965). Variations in sapwood starch level in some Australia forest species. *Aust. For.* **29,** 15–23.

Barber, S. A. (1962). A diffusion and mass-flow concept of soil nutrient availability. *Soil Sci.* **93,** 39–49.

Barklund, P., and Rowe, J. (1981). *Gremeniella abietina (Scleroderris lagerbergii),* a primary parasite in Norway spruce die-back. *Eur. J. For. Pathol.* **11,** 97–108.

Barrett, J. W. (1970). Ponderosa pine saplings respond to control of spacing and understory vegetation. *USDA For. Serv. Res. Pap. PNW* **PNW-106,** 1–16.

Barrett, J. W., and Youngberg, C. T. (1965). Effect of tree spacing and understory vegetation on water use in a pumice soil. *Soil Sci. Soc. Am. Proc.* **29,** 472–475.

Basilier, K. (1979). Moss-associated nitrogen fixation in some mire and coniferous forest environments around Uppsala, Sweden. *Lindbergia* **5,** 84–88.

Bauer, T. (1982). Predation by a carabid bettle specialized for catching Collembola. *Pedobiologia* **24,** 169–179.

Baxter, P., and West, D. (1977). The flow of water into fruit trees. I. Resistances to water flow through roots and stems. *Ann. Appl. Biol.* **87,** 95–101.

Bazzaz, F. A., and Pickett, S. T. A. (1980). The physiological ecology of tropical succession: A comparative review. *Annu. Rev. Ecol. Syst.* **11,** 287–310.

Beadle, N. C. W. (1966). Soil phosphate and its role in molding segments of the Australian flora and vegetation, with special reference to xeromorphy and sclerophylly. *Ecology* **47,** 992–1007.

Bear, J., Zavslavsky, D., and Irmay, S. (1968). "Physical Principles of Water Percolation and Seepage." UNESCO, Paris.

Benecke, U., and Nordmeyer, A. H. (1982). Carbon uptake and allocation by *Nothofagus solandri* var. *cliffortioidies* (Hook.f.) Poole and *Pinus contorta* Dougl. ex. Loundon ssp. *contorta* at montane and subalpine altitudes. *In* "Carbon Uptake and Allocation in Subalpine Ecosystems as a Key to Management" (R. H. Waring, ed.), IUFRO Workshop, pp. 9–21. For. Res. Lab., Oregon State University, Corvallis.

Benoit, R. E., and Starkey, R. L. (1968). Inhibition of decomposition of cellulose and some other carbohydrates by tannin. *Soil Sci.* **105,** 291–296.

Bentley, B. L., and Carpenter, E. J. (1984). Direct transfer of newly-fixed nitrogen from free-living epiphyllous microorganisms to their host plant. *Oecologia* **63,** 52–56.

Berg, B., and Staaf, H. (1981). Leaching, accumulation and release of nitrogen in decomposing forest litter. *In* "Terrestrial Nitrogen Cycles" (F. E. Clark and T. Rosswall, eds.), pp. 163–178. Swed. Nat. Sci. Res. Counc., Stockholm.

Berg, B., Hannus, K., Popoff, T., and Theander, O. (1982). Changes in organic chemical components of needle litter during decomposition. Long-term decomposition in a Scots Pine forest. I. *Can. J. Bot.* **60,** 1310–1319.

Bernays, E. A., and Woodhead, S. (1982). Plant phenols utilized by a phytophagous insect. *Science* **216,** 201–203.

Berndt, H. W., and Fowler, W. B. (1969). Rime and hoarfrost in upper-slope forests of eastern Washington. *J. For.* **67,** 92–95.

Berner, R. A., Lasaga, A. C., and Garrels, R. M. (1983). The carbonate-silicate geochemical cycle and its effect on atmospheric carbon dioxide over the past 100 million years. *Am. J. Sci.* **283,** 641–683.

Berry, J., and Bjorkman, O. (1980). Photosynthetic response and adaptation to temperature in higher plants. *Annu. Rev. Plant Physiol.* **31,** 491–543.

Berry, J. A., and Downton, W. J. S. (1982). Environmental regulation of photosynthesis. *In* "Photosynthesis, Development, Carbon Metabolism, and Plant Productivity" (R. Govindjee, ed.), Vol. 2, pp. 263–343. Academic Press, New York.

Berryman, A. A. (1976). Theoretical explanation of mountain pine beetle dynamics in lodgepole pine forests. *Environ. Entomol.* **5,** 1225–1233.

Bertram, H.-G., and Schleser, G. H. (1982). The $^{13}C/^{12}C$ isotope ratios in a north-German podzol. *In* "Stable Isotopes" (H.-L. Schmidt, H. Forstel, and K. Heinzinger, eds.), pp. 115–120. Elsevier, Amsterdam.

Beschta, R. L. (1979). Debris removal and its effects on sedimentation in an Oregon Coast Range stream. *Northwest Sci.* **53,** 71–77.

Beven, K. (1982). On subsurface stormflow: predictions with simple kinematic theory for saturated and unsaturated flows. *Water Resour. Res.* **18,** 1627–1633.

Beven, K., and Germann, P. (1982). Macropores and water flow in soils. *Water Resour. Res.* **18,** 1311–1325.

Beven, K., and Wood, E. F. (1983). Catchment geomorphology and the dynamics of runoff contributing areas. *J. Hydrol.* **65,** 139–158.

Biederbeck, V. O., and Campbell, C. A. (1973). Soil microbial activity as influenced by temperature trends and fluctuations. *Can. J. Soil Sci.* **53,** 363–376.

Bilby, R. E., and Likens, G. E. (1980). Importance of organic debris dams in the structure and function of stream ecosystems. *Ecology* **61,** 1107–1113.

Billings, W. D. (1978). "Plants and the Ecosystem." Wadsworth Publ. Co., Belmont, California.

Billings, W. D., Luken, J. O., Mortensen, D. A., and Peterson, K. M. (1982). Arctic Tundra: A source or sink for atmospheric carbon dioxide in a changing environment? *Oecologia* **53,** 7–11.

Billings, W. D., Luken, J. O., Mortensen, D. A., and Peterson, K. M. (1983). Increasing atmospheric carbon dioxide: Possible effects on arctic tundra. *Oecologia* **58,** 286–289.

Binkley, D. (1981). Nodule biomass and acetylene reduction rates of red alder and sitka alder on Vancouver Island, B.C. *Can. J. For. Res.* **11,** 281–286.

Binkley, D. (1983). Ecosystem production in Douglas-fir plantations: Interaction of red alder and site fertility. *For. Ecol. Manage.* **5,** 215–227.

Binkley, D., and Matson, P. (1983). Ion exchange resin bag method for assessing forest soil nitrogen availability. *Soil Sci. Soc. Am. J.* **47,** 1050–1052.

Binkley, D., Cromack, K., and Fredriksen, R. L. (1982). Nitrogen accretion and availability in some snowbrush ecosystems. *For. Sci.* **28,** 720–724.

Binns, W. O., and Redfern, D. B. (1983). "Acid Rain and Forest Decline in W. Germany," For. Comm. Res. Dev. Pap. No. 131. For. Comm., Edinburgh, Scotland.

Birk, E. M. (1979). Disappearance of overstory and understory litter in an open eucalypt forest. *Aust. J. Ecol.* **4,** 207–222.

Birk, E. M., and Simpson, R. W. (1980). Steady state and the continuous input model of litter accumulation and decomposition in Australian eucalypt forests. *Ecology* **61,** 481–485.

Birks, H. J. B., and Birks, H. H. (1980). "Quaternary Paleoecology." University Park Press, Baltimore, Maryland.

Bisson, P. A., and Bilby, R. E. (1982). Avoidance of suspended sediment by juvenile coho salmon. *North Am. J. Fish. Manage.* **4,** 371–374.

Bisson, P. A., and Sedell, J. R. (1984). Salmonid populations in logged and unlogged stream sections of western Washington. *In* "Proceedings of a Symposium on Fish and Wildlife Relationships in Old-Growth Forests" (W. R. Meeham, T. R. Merrill, and T. Hanley, eds.). Am. Inst. Fish. Biologist. 121–129.

Biswell, H. H., Kallander, H. R., Komarek, R., Vogl, R. J., and Weaver, H. (1973). "Ponderosa Pine Fire Management. A Task Force Evaluation of Controlled Burning in Ponderosa Pine Forests of Central Arizona," Misc. Publ. No. 2. Tall Timbers Res. Stn., Tallahassee, Florida.

Black, R. A., and Bliss, L. C. (1980). Reproductive ecology of *Picea mariana* (Mill.) BSP at tree line near Inuvik, Northwest Territories, Canada. *Ecol. Monogr.* **50,** 331–354.

Blaise, T., and Garbaye, J. (1983). Effects de la fertilisation minerale sur les ectomycorhizes d'une hetraie. *Acta Oecol. Plant.* **4,** 165–169.

Bockheim, J. G. (1980). Solution and use of chronofunctions in studying soil development. *Geoderma* **24,** 71–85.

Boddy, L., and Rayner, A. D. M. (1983). Ecological roles of Basidiomycetes forming decay communities in attached oak branches. *New Phytol.* **93,** 77–88.

Bolin, B. (1982). "Biogeochemical Processes and Climate Modelling." Dep. Meteorol., University of Stockholm.

Bonetto, A. A. (1975). Hydrologic regime of the Parana River and its influence on ecosystems. *In* "Coupling of Land and Water Systems" (A. D. Hasler, ed.), pp. 175–197. Springer-Verlag, Berlin and New York.

Boring, L. R., Monk, C. D., and Swank, W. T. (1981). Early regeneration of a clear-cut southern Appalachian forest. *Ecology* **62,** 1244–1253.

Bormann, B. T., and Gordon, J. C. (1984). Stand density effects in young red alder plantations: Productivity, photosynthate partitioning, and nitrogen fixation. *Ecology* **65,** 394–402.

Bormann, F. H., and Likens, G. E. (1967). Nutrient cycling. *Science* **155,** 424–429.

Bormann, F. H., and Likens, G. E. (1979a). "Pattern and Process in a Forested Ecosystem." Springer-Verlag, Berlin and New York.

Bormann, F. H., and Likens, G. E. (1979b). Catastrophic disturbance and the steady-state in northern hardwood forests. *Am. Sci.* **67,** 660–669.

Bormann, F. H., Likens, G. E., and Eaton, J. S. (1969). Biotic regulation of particulate and solution losses from a forest ecosystem. *BioScience* **19,** 600–610.

Bormann, F. H., Likens, G. E., Siccama, T. G., Pierce, R. S., and Eaton, J. S. (1974). The export of nutrients and recovery of stable conditions following deforestation at Hubbard Brook. *Ecol. Monogr.* **44,** 255–277.

Bosatta, E., and Staaf, H. (1982). The control of nitrogen turnover in forest litter. *Oikos* **39,** 143–151.

Botkin, D. B., Janak, J. F., and Wallis, J. R. (1972). Some ecological consequences of a computer model of forest growth. *J. Ecol.* **60,** 849–872.

Botkin, D. F., Janak, J. F., and Wallis, J. R. (1973). Estimating the effects of carbon fertilization on forest composition by ecosystem simulation. *In* "Carbon and the Biosphere" (G. M. Woodwell and E. V. Pecan, eds.), CONF-720510, pp. 328–344. U.S. At. Energy Comm., Washington, D.C.

Botkin, D. B., Estes, J. E., MacDonald, R. M., and Wilson, M. V. (1984). Studying the earth's vegetation from space. *BioScience* **34,** 508–514.

Boulding, K. (1956). General systems theory: The skeleton of science. *Gen. Syst.* **1,** 11–17.

Bouma, J., and Wosten, J. H. M. (1984). Characterizing ponded infiltration in a dry cracked clay soil. *J. Hydrol.* **69,** 297–304.

Bowen, G. D. (1969). Nutrient status effects on loss of amides and amino acids from pine roots. *Plant Soil* **30,** 139–142.

Bowen, G. D. (1982). The root-microorganism ecosystem. *In* "Biological and Chemical Interactions in the Rhizosphere," pp. 3–42. Swed. Nat. Sci. Res. Counc., Stockholm.

Bowen, G. D., and Smith, S. E. (1981). The effects of mycorrhizas on nitrogen uptake by plants. Terrestrial nitrogen cycles: Ecosystem strategies and management impacts. *Ecol. Bull.* **33,** 237–247.

Boyle, J. R., and Voigt, G. K. (1973). Biological weathering of silicate minerals. Implications for tree nutrition and soil genesis. *Plant Soil* **38,** 191–201.

Boyle, J. R., Voigt, G. K., and Sawhney, B. L. (1974). Chemical weathering of biotite by organic acids. *Soil Sci.* **117,** 42–45.

Braastad, H. (1975). Yield tables and growth models for *Picea abies*. *Rep. Norw. For. Res. Inst.* **31,** 9.

Bradford, K. J., and Hsiao, T. C. (1982). Physiological responses to moderate water stress. *In* "Encyclopedia of Plant Physiology" (O. L. Lange, P. S. Nobel, C. B. Osmond, and H. Ziegler, eds.), Vol. 12 B, pp. 265–354. Springer-Verlag, Berlin and New York.

Brantseg, A. (1969). Furu sønnafjells. Produksjonstabeller. *Medd. Nor.* Skogfors. Ves. **26** (84).

Braun-Blanquet, J. (1951). "Pflanzensoziologie." Springer-Verlag, Vienna.

Bray, J. R., and Gorham, E. (1964). Litter production in forests of the world. *Adv. Ecol. Res.* **2,** 101–157.

Bremner, J. M., and Blackmer, A. M. (1978). Nitrous oxide: Emission from soils during nitrification of fertilizer nitrogen. *Science* **199,** 295–296.

Breznak, J. A., Brill, W. J., Mertins, J. W., and Coppel, H. C. (1973). Nitrogen fixation in termites. *Nature (London)* **244,** 577–579.

Briggs, G. E., Kidd, F., and West, C. (1920). A quantitative analysis of plant growth. *Ann. Appl. Biol.* **7,** 202–223.

Brix, H. (1971). Effects of nitrogen fertilization on photosynthesis and respiration in Douglas-fir. *For. Sci.* **17,** 407–414.

Brix, H. (1981). Effects of nitrogen fertilizer source and application rates on foliar nitrogen concentration, photosynthesis and growth of Douglas-fir. *Can. J. For. Res.* **11,** 775–780.

Brown, A. K., and Davis, K. P. (1973). "Forest Fire: Control and Use." McGraw-Hill, New York.

Brown, G. W. (1970). Predicting the effect of clearcutting on stream temperature. *J. Soil Water Conserv.* **25,** 11–13.

Brown, H. E. (1970). Status of pilot studies in Arizona. *J. Irrig. Drain. Div., Am. Soc. Civ. Eng.* **96,** 11–23.

Brown, S. (1981). A comparison of the structure, primary productivity, and transpiration of cypress ecosystems in Florida. *Ecol. Monogr.* **51,** 403–427.

Brown, S., and Lugo, A. E. (1982). The storage and production of organic matter in tropical forests and their role in the global carbon cycle. *Biotropica* **14,** 161–187.

Bryant, J. P., Chapin, F. S., III, and Klein, D. R. (1983). Carbon/nutrient balance of boreal plants in relation to vertebrate herbivory. *Oikos* **40,** 357–368.

Buckman, H. O., and Brady, N. C. (1969). "The Nature and Properties of Soils." Macmillan, New York.

Buffo, J., Fritschen, L. J., and Murphy, J. L. (1972). Direct solar radiation on various slopes from 0 to 60 degrees north latitude. *USDA For. Serv. Res. Pap. PNW* **PNW-142,** 74.

Bunnell, F. L., and Tait, D. E. N. (1974). Mathematical simulation models of decomposition processes. *In* "Soil Organisms and Decomposition in Tundra" (A. J. Holding, O. W. Heal, S.

F. MacLean, and P. W. Flanagan, eds.), pp. 207–225. Tundra Biome Steering Comm., Stockholm.

Burger, H. (1929). Holz, Blattmenge und Zuwachs. *Mitt., Schweiz. Zentralanst. Forst. Versuchswes.* **15,** 243–292.

Burk, R. L., and Stuiver, M. (1981). Oxygen isotope ratios in trees reflect mean annual temperature and humidity. *Science* **211,** 1417–1419.

Burke, C. J. (1979). Historic fires in the central western Cascades, Oregon. M.S. Thesis, Oregon State University, Corvallis.

Burke, M. J., Gusta, L. V., Quamme, H. A., Wieser, C. J., and Li, P. H. (1976). Freezing and injury in plants. *Annu. Rev. Plant Physiol.* **27,** 507–528.

Burns, J. W. (1972). Some effects of logging and associated road construction on northern California streams. *Trans. Am. Fish. Soc.* **101,** 1–17.

Burns, R. G. (1982). Enzyme activity in soil: Location and a possible role in microbial ecology. *Soil Biol. Biochem.* **14,** 423–427.

Burton, T. M., and Likens, G. E. (1975). Energy flow and nutrient cycling in salamander populations in the Hubbard Brook Experimental Forest, New Hampshire. *Ecology* **56,** 1068–1080.

Cain, S. A. (1944). "Foundations of Plant Geography." Harper, New York.

Campbell, C. A., Paul, E. A., Rennie, D. A., and MaCallum, K. J. (1967). Applicability of the carbon-dating method of analysis to soil humus studies. *Soil Sci.* **104,** 217–224.

Campbell, G. S. (1977). "An Introduction to Environmental Biophysics." Springer-Verlag, Berlin and New York.

Campbell, R. W., and Sloan, R. J. (1977). Forest stand responses to defoliation by the Gypsy moth. *For. Sci. Monogr.* **19,** 1–34.

Carlisle, A., Brown, A. H. F., and White, E. J. (1966a). Litter fall, leaf production and the effects of defoliation by *Tortrix viridana* in a sessile oak (*Quercus petraea*) woodland. *J. Ecol.* **54,** 65–85.

Carlisle, A., Brown, A. H. F., and White, E. J. (1966b). The organic matter and nutrient elements in the precipitation beneath a sessile oak (*Quercus petraea*) canopy. *J. Ecol.* **54,** 87–98.

Carlquist, S. (1975). "Ecological Strategies of Xylem Evolution." Univ. of California Press, Berkeley.

Cary, A. (1936). White pine and fire. *J. For.* **34,** 62–65.

Cates, R. G., Redak, R., and Henderson, C. B. (1983). Natural product defensive chemistry of Douglas fir, western spruce budworm success, and forest management practices. *Z. Angew. Entomol.* **96,** 173–182.

Chalk, L., and Bigg, J. M. (1956). The distribution of moisture in the living stem in Sitka spruce and Douglas fir. *Forestry* **29,** 5–21.

Chapin, F. S. (1980). The mineral nutrition of wild plants. *Annu. Rev. Ecol. Syst.* **11,** 233–260.

Chapin, F. S., and Kedrowski, R. A. (1983). Seasonal changes in nitrogen and phosphorus fractions and autumn retranslocation in evergreen and deciduous taiga trees. *Ecology* **64,** 376–391.

Chase, E. M., and Sayles, F. L. (1980). Phosphorus in suspended sediments of the Amazon River. *Estuarine Coastal Mar. Sci.* **11,** 383–391.

Chettleburgh, M. R. (1952). Observations on the collection and burial of acorns by jays in Hainault Forest. *Br. Birds* **45,** 359–364.

Chia, L. S., Mayfield, C. I., and Thompson, J. E. (1984). Simulated acid rain induces lipid peroxidation and membrane damage in foliage. *Plant, Cell Environ.* **7,** 333–338.

Childs, E. C. (1969). "An Introduction to the Physical Basis of Soil Water Phenomenon." Wiley, New York.

Christensen, N. L. (1981). Fire regimes in southeastern forests. *Gen. Tech. Rep. WO—U.S., For. Serv.* [*Wash.* Off.] **GTR-WO-26,** 112–136.

Chung, H. H., and Barnes, R. L. (1977). Photosynthate allocation in *Pinus taeda*. I. Substrate requirements for synthesis of shoot biomass. *Can. J. For. Res.* **7,** 106–111.

Clarholm, M. (1981). Protozoan grazing of bacteria in soil-impact and importance. *Microb. Ecol.* **7,** 343–350.

Clark, J., and Gibbs, R. D. (1957). Studies in tree physiology. IV. Further investigations of seasonal changes in moisture content of certain Canadian forest trees. *Can. J. Bot.* **35,** 219–253.

Clary, W. P., and Ffolliott, P. F. (1969). Water holding capacity of ponderosa pine forest floor layers. *J. Soil Water Conserv.* **24,** 22–23.

Clayton, J. L. (1976). Nutrient gains to adjacent ecosystems during a forest fire: an evaluation. *For. Sci.* **22,** 162–166.

Cogbill, C. V. (1977). The effect of acid precipitation on tree growth in Eastern North America. *Water, Air, Soil Pollut.* **8,** 89–93.

Cole, C. V., Elliott, E. T., Hunt, H. W., and Coleman, D. C. (1978). Trophic interactions in soils as they affect energy and nutrient dynamics. V. Phosphorus transformations. *Microb. Ecol.* **4,** 381–387.

Cole, D. W., and Rapp, M. (1981). Elemental cycling in forest ecosystems. *In* "Dynamic Principles of Forest Ecosystems" (D. E. Reichle, ed.), pp. 341–409. Cambridge Univ. Press, London and New York.

Cole, W. E., and Amman, G. D. (1969). Mountain pine beetle infestation in relation to lodgepole pine diameters. *USDA For. Serv. Res. Note INT* **INT-95,** 1–7.

Coleman, D. C., Reid, C. P. P., and Cole, C. V. (1983). Biological strategies of nutrient cycling in soil systems. *Adv. Ecol. Res.* **13,** 1–55.

Coley, P. D. (1982). Rates of hervivory on different tropical trees. *In* "The Ecology of a Tropical Forest" (E. G. Leigh, Jr., A. S. Rand, and D. M. Windsor, eds.), pp. 123–132. Smithson. Inst. Press, Washington, D.C.

Coley, P. D. (1983). Herbivory and defensive characteristics of tree species in a lowland tropical forest. *Ecol. Monogr.* **53,** 209–233.

Connell, J. H. (1971). On the role of natural enemies in preventing competitive exclusion in some marine animals and in rain forest trees. *In* "Dynamic of Populations" (P. J. den Boern and G. Gradwell, eds.), pp. 298–312. Centre for Agricultural Publications and Documentation, Wageningen, Netherlands.

Connell, J. H., and Sousa, W. P. (1983). On the evidence needed to judge ecological stability or persistence. *Am. Nat.* **121,** 789–824.

Conard, S. G., and Radosevich, S. K. (1981). Photosynthesis, xylem pressure potential, and leaf conductance of three montane chaparral species in California. *For. Sci.* **27,** 626–639.

Cook, E. R., and Jacoby, C. C. (1977). Tree-ring-drought relationships in the Hudson Valley, New York. *Science* **198,** 399–401.

Corey, A. T., and Klute, A. (1985). Application of the potential concept to soil water equilibrium and transport. *Soil Sci. Soc. Am. J.* **49,** 3–11.

Coulson, R. N., Hennier, P. B., Flann, R. O., Rykiel, E. J., Hu, L. C., and Payne, T. L. (1983). The role of lightning in the epidemiology of the southern pine beetle. *Z. Angew. Entomol.* **96,** 182–193.

Covington, W. W. (1975). Altitudinal variation of chlorophyll concentration and reflectance of the bark of *Populus tremuloides*. *Ecology* **56,** 715–720.

Covington, W. W. (1981). Changes in forest floor organic matter and nutrient content following clear cutting in northern hardwoods. *Ecology* **62,** 41–48.

Cowan, I. R., and Farquhar, G. D. (1977). Stomatal function in relation to leaf metabolism and environment. *Symp. Soc. Exp. Biol.* **31,** 471–505.

Cox, T. L., Harris, W. F., Ausmus, B. S., and Edwards, N. T. (1978). The role of roots in biogeochemical cycles in an eastern deciduous forest. *Pedobiologia* **18,** 264–271.

Cranswick, A. M. (1979). "Food Reserves of Douglas Fir Trees Defoliated by Lepidopterous Larvae in Kaningaroa SF.,F.R.I.," Intern. Rep. No. 13. Prod. For. Div. Rotorua, New Zealand.

Crawley, M. J. (1983). Herbivory: The dynamics of animal-plant interactions. *Stud. Ecol.* **10,** 1–437.

Cremer, K. W., and Svensson, J. G. P. (1979). Changes in length of *Pinus radiata* shoots reflecting loss and uptake of water through foliage and bark surfaces. *Aust. For. Res.* **9,** 163–172.

Cromack, K., and Monk, C. D. (1975). Litter production, decomposition, and nutrient cycling in a mixed hardwood watershed and a white pine watershed. *In* "Mineral Cycling in Southeastern Ecosystems" (F. G. Howell, J. B. Gentry, and M. H. Smith, eds.), pp. 609–624. Technical Information Center, U.S. Department of Energy, Springfield, Virginia.

Cromack, K., Sollins, P., Graustein, W. C., Speidel, K., Todd, A. W., Spycher, G., Li, C. Y., and Todd, R. L. (1979). Calcium oxalate accumulation and soil weathering in mats of the hypogeous fungus, *Hysterangium crassum. Soil Biol. Biochem.* **11,** 463–468.

Croteau, R., Burbott, A. J., and Loomis, W. D. (1972). Biosynthesis of mono-sesqui-terpenes in peppermint. *Phytochemistry* **11,** 2937–2948.

Crozat, G. (1979). Sur l'émission d'un aérosol riche en potassium par la forêt tropicale. *Tellus* **31,** 52–57.

Crutzen, P. J. (1983). Atmospheric interactions—Homogeneous gas reactions of C, N and S containing compounds. *In* "The Major Biogeochemical Cycles and Their Interactions" (B. Bolin and R. B. Cook, eds.), pp. 67–112. Wiley, New York.

Cummins, K. W. (1973). Trophic relations of aquatic insects. *Annu. Rev. Entomol.* **18,** 183–206.

Cummins, K. W. (1974). Structure and function of stream ecosystems. *BioScience* **24,** 631–641.

Cummins, K. W., Minshall, G. W., Sedell, J. R., Cushing, C. E., and Petersen, R. C. (1984). Stream ecosystem theory. *Verh. Internat. Verein. Limnol.* **22,** 1818–1827.

Dahm, C. N. (1981). Pathways and mechanisms for removal of dissolved organic carbon from leaf leachate in streams. *Can. J. Fish. Aquat. Sci.* **38,** 68–76.

Dahm, C. N. (1983). Uptake of dissolved organic carbon in mountain streams. *Verh. Internat. Verein. Limnol.* **22,** 1842–1846.

Dall'Olio, A., Salati, E., de Azevedo, C. T., and Matsui, E. (1979). Modelo de fracionamento isotópico de áqua na bacia Amazonica (Primeira aproximacao). *Acta Amazonica* **9,** 675–687.

Daniel, T. W., Helm, J. A., and Baker, F. S. (1979). "Principles of Silviculture." McGraw-Hill, New York.

Darby, H. C. (1956). "The Draining of the Fens." Cambridge Univ. Press, London and New York.

Darley-Hill, S., and Johnson, W. C. (1981). Acorn dispersal by the blue jay (*Cyanocitta cristata*). *Oecologia* **50,** 231–232.

Daubenmire, R. (1968). "Plant Communities." Harper & Row, New York.

Davies, R. I. (1971). Relation of polyphenols to decomposition of organic matter and to pedogenetic processes. *Soil Sci.* **111,** 80–85.

Davies, W. J., Wilson, J. A., Sharp, R. E., and Osunubi, D. (1981). Control of stomatal behaviour in water-stressed plants. *In* "Stomatal Physiology" (P. G. Jarvis and T. A. Mansfield, eds.), Soc. Exp. Biol., Semin. Ser. No. 8, pp. 163–185. Cambridge Univ. Press, London and New York.

Davis, L. S., and Cooper, R. W. (1963). How prescribed burning affects wildlife occurrence. *J. For.* **61,** 915–917.

Davis, M. B. (1981). Quaternary history and the stability of forest communities. *In* "Forest Succession" (D. C. West, H. H. Shugart, and D. B. Botkin, eds.), pp. 132–153. Springer-Verlag, Berlin and New York.

Dawson, H. J., Ugolini, F. C., Hrutfiord, B. F., and Zachara, J. (1978). The role of soluble organics in the soil processes of a podzol, central Cascades, Washington. *Soil Sci.* **126,** 290–296.

Day, N. F., and Megaham, W. F. (1975). Landslide occurrence on the Clearwater National Forest. *Geol. Soc. Am., Abstr. Program* **7,** 602–603.

Day, R. J., and MacGillivray, G. R. (1975). Root regeneration of fall-lifted white spruce nursery stock in relation to soil moisture content. *For. Chron.* **51,** 196–199.

DeAngelis, D. L., Gardner, R. H., and Shugart, H. H. (1981). Productivity of forest ecosystems studied during the IBP: The woodland data set. *In* "Dynamic Properties of Forest Ecosystems" (D. E. Reichle, ed.), Int. Biol. Programme No. 23, pp. 567–672. Cambridge Univ. Press, London and New York.

Deans, J. D. (1981). Dynamics of coarse root production in a young plantation of *Picea sitchensis. Forestry* **54,** 139–155.

DeBano, L. F., and Rice, R. M. (1973). Water repellent soils: Their implications in forestry. *J. For.* **71,** 220–223.

DeBell, D. S., and Ralston, C.W. (1970). Release of nitrogen by burning light forest fuels. *Soil Sci. Soc. Am. Proc.* **34,** 936–938.

Delcourt, H. R., and Harris, W. F. (1980). Carbon budget of the southeastern U.S. Biota: Analysis of historical change in trend from source to sink. *Science* **210,** 321–323.

Delcourt, H. R., Delcourt, P. A., and Webb, T. (1983). Dynamic plant ecology: The spectrum of vegetational change in space and time. *Quat. Sci. Rev.* **1,** 153–175.

Delmas, R., and Servant, J. (1983). Atmospheric balance of sulphur above an equatorial forest. *Tellus* **35B,** 110–120.

Delwiche, C. C. (1970). The nitrogen cycle. *Sci. Am.* **223**(3), 136–146.

Dement, W. A., and Mooney, H. A. (1974). Seasonal variation in the production of tannins and cyanogenic glucosides in the chaparral shrub, *Heteromeles arbutifolia. Oecologia* **15,** 65–76.

Denslow, J. S. (1980). Gap partitioning among tropical rain forest trees. *Biotropica* **12,** Suppl., 47–55.

Dickson, B. A., and Crocker, R. L. (1953). A chronosequence of soils and vegetation near Mt. Shasta, California. II. The development of the forest floors and the carbon and nitrogen profiles of the soils. *J. Soil Sci.* **4,** 142–154.

Dickson, R. E., and Broyer, T. C. (1972). Effects of aeration, water supply, and nitrogen source on growth and development of tupelo gum and bald cypress. *Ecology* **53,** 626–634.

Dietrich, W. E., and Dunne, T. (1978). Sediment budget for a small catchment in mountainous terrain. *Z. Geomorphol., Suppl.* **29,** 191–206.

Dietrich, W.E., Dunne, T., Humphrey, N. F., and Reid, L. M. (1982). Construction of sediment budgets for drainage basins. *USDA For. Serv. Gen. Tech. Rep. PNW* **PNW-141,** 5–23.

Distelbarth, H., Kull, U., and Jeremias, K. (1984). Seasonal trends in energy contents of storage substances in evergreen gymnosperms under mild conditions in central Europe. *Flora (Jena)* **175,** 15–30.

Dixon, R. M. (1971). Role of large soil pores in infiltration and interflow. *In* Proceedings of the Third International Symposium for Hydrology Professors, (E. J. Monke, ed.) pp. 136–147. Purdue University, West Lafayette, Indiana.

Dixon, M. A., Grace, J., and Tyree, M. T. (1984). Concurrent measurements of stem density, leaf and stem water potential, stomatal conductance and cavitation on a sapling of *Thuja occidentalis* L. *Plant, Cell Environ.* **7,** 615–618.

Doane, C. C., and McManus, M. L., eds. (1981). The Gypsy moth: Research toward integrated pest management. *U. S., Dep. Agric., Tech. Bull.* **1584,** 65–86.

Doley, D. (1981). Tropical and subtropical forests and woodlands. *In* "Water Deficits and Plant Growth" (T. T. Kozlowski, ed.), Vol. 6, pp. 209–323. Academic Press, New York.

Doyle, T. W. (1981). The role of disturbance in the gap dynamics of a montane rain forest: An application of a tropical forest model. *In* "Forest Succession: Concepts and Applications" (D. C. West, H. H. Shugart, and D. B. Botkin, eds.), pp. 56–73. Springer-Verlag, Berlin and New York.

Drew, A. P., Drew, L. G., and Fritts, H. C. (1972). Environmental control of stomatal activity in mature semiarid site ponderosa pine. *Ariz. Acad. Sci.* **7,** 85–93.

Duckstein, L., Fogel, M. M., and Thames, J. L. (1973). Elevation effects on rainfall: a stochastic model. *J. Hydrol.* **18,** 21–35.

Duddridge, J. A., Malibari, A., and Read, D. J. (1980). Structure and function of mycorrhizal rhizomorphs with special reference to their role in water transport. *Nature (London)* **287,** 834–836.

Dunne, T. (1983). Relation of field studies and modeling in the prediction of storm runoff. *J. Hydrol.* **65,** 25–48.

Dunne, T., and Black, R. D. (1970). Partial area contributions to storm runoff in a small New England watershed. *Water Resour. Res.* **6,** 1296–1311.

Durzan, D. J. (1974). Nutrition and water relations of forest trees: A biochemical approach. *In* "Proceedings of the Third North American Forest Biology Workshop" (P. P. Reid and G. H. Fechner, eds.), Symp. C, pp. 15–63. Colorado State University, Fort Collins.

Duvigneaud, P., and Denaeyer-DeSmet, S. (1970). Biological cycling of minerals in temperate deciduous forests. *In* "Analysis of Temperate Forest Ecosystems" (D. E. Reichle, ed.), pp. 199–225. Springer-Verlag, Berlin and New York.

Dwyer, L. M., and Merriam, G. (1981). Influence of topographic heterogeneity on deciduous litter decomposition. *Oikos* **37,** 228–237.

Eagleson, P. E. (1970). "Dynamic Hydrology." McGraw-Hill, New York.

Eaton, J. S., Likens, G. E., and Bormann, F. H. (1978). The input of gaseous and particulate sulfur to a forest ecosystem. *Tellus* **30,** 546–551.

Eberhardt, R. (1983). "Valuable Wetlands Threatened by Development." Los Angeles Times (published in Gazette-Times, Dec. 4, Sect. B-2, Corvallis, Oregon).

Edmonds, R. L. (1980). Litter decomposition and nutrient release in Douglas fir, red alder, western hemlock and Pacific silver fir ecosystems in western Washington. *Can. J. For. Res.* **10,** 327–337.

Edwards, M., and Meidner, H. (1978). Stomatal responses to humidity and the water potentials of epidermal and mesophyll tissue. *J. Exp. Bot.* **29,** 771–780.

Edwards, N. T. (1975). Effects of temperature and moisture on carbon dioxide evolution in a mixed deciduous forest floor. *Soil Sci. Soc. Am. Proc.* **39,** 361–365.

Edwards, N. T., and Harris, W. F. (1977). Carbon cycling in a mixed deciduous forest floor. *Ecology* **58,** 431–437.

Edwards, N. T., and McLaughlin, S. B. (1978). Temperature-independent diel variations of respiration rates in *Quercus alba* L. and *Liriodendron tulipifera*. *Oikos* **31,** 200–206.

Edwards, N. T., and Sollins, P. (1973). Continuous measurement of carbon dioxide evolution from partitioned forest floor components. *Ecology* **54,** 407–412.

Edwards, N. T., Shugart, H. H., McLaughlin, S. B., Harris, W. F., and Reichle, D. E. (1981). Carbon metabolism in terrestrial ecosystems. *In* InterBiol. Programme No. 23, "Dynamic Properties of Forest Ecosystems" (D. E. Reichle, ed.), pp. 499–536. Cambridge Univ. Press, London and New York.

Edwards, P. J. (1977). Studies of mineral cycling in a montane rain forest in New Guinea. II. The production and disappearance of litter. *J. Ecol.* **65,** 971–992.

Edwards, P. J. (1982). Studies of mineral cycling in a montane rain forest in New Guinea. V. Rates of cycling in throughfall and litterfall. *J. Ecol.* **70,** 807–827.

Edwards, P. J., and Grubb, P. J. (1982). Studies of mineral cycling in a montane rain forest in New Guinea. IV. Soil characteristics and the division of mineral elements between the vegetation and soil. *J. Ecol.* **70,** 649–666.

Edwards, W. R. N., and Jarvis, P. G. (1982). Relations between water content, potential and permeability in stems of conifers. *Plant, Cell Environ.* **5,** 271–277.

Edwards, W. R. N., and Jarvis, P. G. (1983). A method for measuring radial differences in water content of intact tree stems by attenuation of gamma radiation. *Plant, Cell Environ.* **6,** 255–260.

Eggleston, K. O., Israelson, E. K., and Riley, J. P. (1971). "Hybrid Computer Simulation of the Accumulation and Melt Processes in a Snowpack." Utah Water Res. Lab., Coll. Eng., Utah State University, Logan.

Ehleringer, J. R. (1979). Photosynthesis and photorespiration: Biochemistry, physiology, and ecological implications. *HortScience* **14,** 217–222.

Elkins, N. Z., Steinberger, Y., and Whitford, W. G. (1982). Factors affecting the applicability of the AET model for decomposition in arid environments. *Ecology* **63,** 579–580.

Ellenberg, V. H. (1977). Stickstoff als Standortsfaktor, Insbesondere für mitteleuropäische Pflanzengesellschaften. *Oecol. Plant.* **12,** 1–22.

Ellis, R. C. (1969). The respiration of the soil beneath some *Eucalyptus* forest stands as related to the productivity of the stands. *Aust. J. Soil Res.* **7,** 349–357.

Elrick, E. E. (1968). The microhydrological characterization of soils. *IAHS-AISH Pub.* **83,** 311–317.

Elson, P. F., and Kerswill, C. J. (1966). Impact on salmon of spray insecticides over forests. *Adv. Water Pollut. Res.* **1,** 55–69.

Elwood, J. W., Newbold, J. D., O'Neill, R. V., and Van Winkle, W. (1983). Resource spiralling: an operational paradigm for analyzing lotic ecosystems. *In* "The Dynamics of Lotic Ecosystems" (T. D. Fontaine and S. M. Bartell, eds.), pp. 3–27. Ann Arbor Sci. Publ., Ann Arbor, Michigan.

Emmingham, W. H. (1977). Comparison of selected Douglas-fir seed sources for cambial and leader growth patterns in four western environments. *Can. J. For. Res.* **7,** 154–164.

Emmingham, W. H. (1982). Ecological indexes as a means of evaluating climate, species distribution and primary production. *In* "Analysis of Coniferous Forest Ecosystems in the Western United States" (R. L. Edmonds, ed.), US/IBP Ser. No. 14, pp. 45–67. Dowden, Hutchision & Ross, Inc., Stroudsburg, Pennsylvania.

Emmingham, W. H., and Waring, R. H. (1977). An index of photosynthesis for comparing forest sites in western Oregon. *Can. J. For. Res.* **7,** 165–174.

Eppley, R. W., and Peterson, B. J. (1979). Particulate organic matter flux and planktonic new production in the deep ocean. *Nature (London)* **282,** 677–680.

Ericsson, T. (1981). Effects of varied nitrogen stress on growth and nutrition in three *Salix* clones. *Physiol. Plant.* **51,** 423–429.

Eriksson, E. (1955). Air borne salts and the chemical composition of river waters. *Tellus* **7,** 243–250.

Eriksson, E. (1960). The yearly circulation of chloride and sulphur in nature: Meteorological, geochemical and pedological implications. Part II. *Tellus* **12,** 63–109.

Eriksson, H. (1984). Granens produktion i sodra Sverige—trender och fragetecken. *In* "Ekologisk stabilitet och skogsproduktion," Skogsfakta Suppl. No. 3, pp. 26–37. Swed. Univ. Agric. Sci., Uppsala.

Evans, F. C. (1956). Ecosystem as the basic unit in ecology. *Science* **123,** 1127–1128.

Ewel, J., Berish, C., Brown, B., Price, N., and Raich, J. (1981). Slash and burn impacts on a Costa Rican wet forest site. *Ecology* **62,** 816–829.

Ewel, J. J. (1977). Differences between wet and dry successional tropical ecosystems. *Geo-Eco-Trop* **1,** 103–117.

Fahey, T. J. (1979). The effect of night frost on the transpiration of *Pinus contorta* ssp. *latifolia. Oecol. Plant.* **14,** 483–490.

Fahey, T. J. (1983). Nutrient dynamics of aboveground detritus in lodgepole pine (*Pinus contorta* ssp. *latifolia*) ecosystems, Southeastern Wyoming. *Ecol. Monogr.* **53,** 51–72.

Fahey, T. J., and Lang, G. E. (1975). Concrete frost along an elevational gradient in New Hampshire. *Can. J. For. Res.* **5,** 700–705.

Fahey, T. J., and Reiners, W. A. (1981). Fire in the forests of Maine and New Hampshire. *Bull. Torrey Bot. Club* **108,** 362–373.

Fahey, T. J., and Young, D. R. (1984). Soil and xylem water potential and soil water content in contrasting *Pinus contorta* ecosystems, southeastern Wyoming, U.S.A. *Oecologia* **61,** 346–351.

Farquhar, G. D., and Sharkey, T. D. (1982). Stomatal conductance and photosynthesis. *Annu. Rev. Plant Physiol.* **33,** 317–345.

Farquhar, G. D., Wetselaar, R., and Firth, P. M. (1979). Ammonia volatilization from senescing leaves of maize. *Science* **203,** 1257–1258.

Farquhar, G. D., von Caemmerer, S., and Berry, J. A. (1980). A biochemical model of photosynthetic CO_2 assimilation in leaves of C_3 species. *Planta* **149,** 78–90.

Federer, C. A. (1979). A soil-plant atmosphere model for transpiration and availability of soil water. *Water Resour. Res.* **15,** 555–562.

Federer, C. A. (1983). Nitrogen mineralization and nitrification: Depth variation in four New England forest soils. *Soil Sci. Soc. Am. J.* **47,** 1008–1014.

Feller, M. C. (1977). Nutrient movement through western hemlock-western redcedar ecosystems in southwestern British Columbia. *Ecology* **58,** 1269–1283.

Feller, M. C., and Kimmins, J. P. (1979). Chemical characteristics of small streams near Haney in southwestern British Columbia. *Water Resour. Res.* **15,** 247–258.

Feller, M. C., and Kimmins, J. P. (1984). Effects of clearcutting and slash burning on streamwater chemistry and watershed nutrient budgets in southwestern British Columbia. *Water Resour. Res.* **20,** 29–40.

Fellin, D. G. (1980). A review of some interactions between insects and disease. *USDA For. Serv. Gen. Tech. Rep. INT* **INT-90,** 335–414.

Fife, D. N., and Nambiar, E. K. S. (1982). Accumulation and retranslocation of mineral nutrients in developing needles in relation to seasonal growth of young radiata pine trees. *Ann. Bot. (London)* [N.S.] **50,** 817–829.

Firmage, D. H. (1981). Environmental influence on the monoterpene variation in *Hedeoma drummondii. Biochem. Syst. Ecol.* **9,** 53–58.

Fisher, R. F. (1972). Spodosol development and nutrient distribution under *Hydnaceae* fungal mats. *Soil Sci. Soc. Am. Proc.* **36,** 492–495.

Fisher, S. G., and Likens, G. E. (1973). Energy flow in Bear Brook, New Hampshire: An integrative approach to stream ecosystem metabolism. *Ecol. Monogr.* **43,** 421–439.

Fittkau, E.-J. (1970). Role of caimans in the nutrient regime of mouth-lakes of Amazon affluents. (An hypothesis.) *Biotropica* **2,** 138–142.

Flaig, W., Beutelspacher, H., and Rietz, E. (1975). Chemical composition and physical properties of humic substances. *In* ''Soil Components'' (J. E. Gieseking, ed.), Vol. I, pp. 1–211. Springer-Verlag, Berlin and New York.

Flanagan, P. W., and Van Cleve, K. (1983). Nutrient cycling in relation to decomposition and organic-matter quality in taiga ecosystems. *Can. J. For. Res.* **13,** 795–817.

Fogel, R., and Cromack, K. (1977). Effect of habitat and substrate quality on Douglas fir litter decomposition in western Oregon. *Can. J. Bot.* **55,** 1632–1640.

Ford, E. D. (1975). Competition and stand structure in some even-age plant monocultures. *J. Ecol.* **63,** 311–333.

Ford, E. D. (1982). High productivity in a polestage Sitka spruce stand and its relation to canopy structure. *Forestry* **55,** 1–17.

Ford, E. D., and Deans, J. D. (1977). Growth of a sitka spruce plantation: Spatial distribution and seasonal fluctuations of lengths, weights, and carbohydrate concentrations of fine roots. *Plant Soil* **47,** 463–485.

Forman, R. T. T. (1975). Canopy lichens with blue-green algae: A nitrogen source in a Columbian rainforest. *Ecology* **56,** 1176–1184.

Forman, R. T. T., and Boerner, R. E. (1981). Fire frequency and the pine barrens of New Jersey. *Bull. Torrey Bot. Club* **108,** 34–50.

Forrest, W. G., and Ovington, J. D. (1970). Organic matter changes in an age series of *Pinus radiata* plantations. *J. Appl. Ecol.* **7,** 177–186.

Forristall, F. F., and Gessel, S. P. (1955). Soil properties related to forest cover type and productivity on Lee Forest, Snohomish County, Washington. *Soil Sci. Soc. Am. Proc.* **19,** 384–389.

Foster, J. R., and Lang, G. E. (1982). Decomposition of red spruce and balsam fir boles in the White Mountains of New Hampshire. *Can. J. For. Res.* **12,** 617–626.

Fowells, H. A., and Krauss, R. W. (1959). The inorganic nutrition of loblolly pine and Virginia pine with special reference to nitrogen and phosphorus. *For. Sci.* **5,** 95–112.

Fox, L., and Macauley, B. (1977). Insect grazing on *Eucalyptus* in response to variation in leaf tannins and nitrogen. *Oecologia* **29,** 145–162.

Fox, L. R. (1981). Defense and dynamics in plant-herbivore systems. *Am. Zool.* **21,** 853–864.

Franklin, J. F., and Dyrness, C. T. (1973). Natural vegetation of Oregon and Washington. *USDA For. Serv. Gen. Tech. Rep. PNW* **PNW-8.**

Franz, G. (1976). Der Einflub von Niederschlag, Hohenlage und Jahresdurchschnittstemperatur im Untersuchungsgebiet auf Humusgehalt und mikrobielle Aktivitat in Bodenproben aus Nepal. *Pedobiologia* **16,** 136–150.

Freedman, B., Morash, R., and Hanson, A. J. (1981). Biomass and nutrient removals by conventional and whole-tree clear-cutting of a red spruce-balsam fir stand in central Nova Scotia. *Can. J. For. Res.* **11,** 249–257.

Freeland, W. J., and Janzen, D. H. (1974). Strategies in herbivory by mammals: The role of plant secondary compounds. *Am. Nat.* **108,** 269–289.

Freeze, R. A. (1972a). Role of subsurface flow in generating surface runoff. 1. Base flow contributions to channel flow. *Water Resour. Res.* **8,** 609–623.

Freeze, R. A. (1972b). Role of subsurface flow in generating surface runoff. 2. Upstream source areas. *Water Resour. Res.* **8,** 1272–1283.

Freyer, H. D. (1979). On the ^{13}C record in tree rings. Part I. ^{13}C variations in northern hemisphere trees during the last 150 years. *Tellus* **31,** 124–137.

Frissell, S. S., Jr. (1973). The importance of fire as a natural ecological factor in Itasca State Park, Minnesota. *Quat. Res.* **3,** 397–407.

Fritschen, L. J., and Doraiswamy, P. (1973). Dew: An addition to the hydrologic balance of Douglas-fir. *Water Resour. Res.* **9,** 891–894.

Fritts, H. C. (1976). "Tree Rings and Climate." Academic Press, New York.

Fritts, H. C., Smith, D. G., Cardis, J. W., and Budelsky, C. A. (1965). Tree-ring characteristics along a vegetation gradient in northern Arizona. *Ecology* **46,** 393–401.

Froehlich, H. A. (1976). The influence of different thinning systems on damage to soil and trees. *IUFRO World Congr., 16th, 1976,* pp. 333–346.

Froehlich, H. A. (1979). Soil compaction from logging equipment: Effects on growth of young ponderosa pine. *J. Soil Water Conserv.* **34,** 276–278.

Fruchter, J. S., Robertson, D. E., Evans, J. C., Olsen, K. B., Lepel, E. A., Laul, J. C., Abel, K. H., Sanders, R. W., Jackson, P. O., Wogman, N. S., Perkins, R. W., Van Tuyl, H. H., Beauchamp, R. H., Shade, J. W., Daniel, J. L., Erikson, R. L., Schmel, G. A., Lee, R. N., Robinson, A. V., Moss, O. R., Briant, J. K., and Cannon, W. C. (1980). Mount St. Helens ash from the 18 May 1980 eruption: Chemical, physical, mineralogical and biological properties. *Science* **209,** 1116–1125.

Garnier, B. J., and Ohmura, A. (1970). The evaluation of surface variations in shortwave radiation income. *Sol. Energy* **13,** 21–34.

Gartlan, J. S., McKey, D. B., Waterman, P. G., Mbi, C. N., and Struhsaker, T. T. (1980). A comparative study of the phytochemistry of two African Rain Forests. *Biol. Chem. Syst. Ecol.* **8,** 401–422.

Garwood, N. C., Janos, D. P., and Brokaw, N. (1979). Earthquake-caused landslides: A major disturbance to tropical forests. *Science* **205,** 997–999.

Gary, H. L. (1972). Rime contributes to water balance in high-elevation aspen forests. *J. For.* **70,** 93–97.

Gash, J. H. C., and Stewart, J. B. (1977). The evaporation from Thetford forest during 1975. *J. Hydrol.* **35,** 385–396.

Gates, D. M. (1965). Transpiration and leaf temperature. *Annu. Rev. Plant Physiol.* **19,** 211–238.

Gates, D. M. (1980). "Biophysical Ecology." Springer-Verlag, Berlin and New York.

Geiger, R. (1965). "The Climate near the Ground," 4th ed. Harvard Univ. Press, Cambridge, Massachusetts.

Gentry, A. H., and Lopez-Parodi, J. (1980). Deforestration and increased flooding of the upper Amazon. *Science* **210,** 1354–1356.

Gentry, J. B., and Whitford, W. G. (1982). The relationship between wood litter infall and relative abundance and feeding activity of subterranean termites *Reticulitermes* spp. in three southeastern coastal plain habitats. *Oecologia* **54,** 63–67.

Gersper, P. L., and Holowaychuk, N. (1971). Some effects of stem flow from forest canopy trees on chemical properties of soils. *Ecology* **52,** 691–702.

Gessel, S. P., Cole, D. W., Johnson, D., and Turner, J. (1981). The nutrient cycles of two Costa Rican forests. *Prog. Ecol.* **3,** 23–44.

Gholz, H. L. (1982). Environmental limits on aboveground net primary production, leaf area, and biomass in vegetation zones of the Pacific Northwest. *Ecology* **63,** 469–481.

Gholz, H. L., Grier, C. C., Campbell, A. G., and Brown, A. T. (1979). "Equations and Their Use for Estimating Biomass and Leaf Area of Pacific Northwest Plants," Res. Pap. No. 41. For. Res. Lab., Oregon State University, Corvallis.

Gibbs, J. N. (1967). The role of host vigour in the susceptibility of pines to *Fomes annosus. Ann. Bot. (London)* [N.S.] **31,** 803–815.

Gifford, R. M., and Evans, L. T. (1981). Photosynthesis, carbon partitioning, and yield. *Annu. Rev. Plant Physiol.* **32,** 485–509.

Gilbert, L. E. (1979). Development of theory in insect-plant interactions. *In* "Analysis of Ecological Systems" (D. Horn, J. R. Stairs, and R. D. Mitchell, eds.), pp. 117–154. Ohio State University, Columbus.

Gill, R. S., and Lavender, D. P. (1983). Urea fertilization and foliar nutrient composition of western hemlock (*Tsuga heterophylla* (Raf.) Sarc.). *For. Ecol. Manage.* **6,** 333–341.

Gillett, J. B. (1962). Pest pressure, an underestimated factor in evolution. *Syst. Assoc. Publ.* **4,** 37–46.

Goldammer, J. G. (1983). "Sicherung des sudbrasilianischen Kiefernanbaues durch Kontrolliertes Brennen," Forstwirtsch. Vol. 4. Hochschul-Verlag, Freiburg.

Goldberg, D. E. (1982). The distribution of evergreen and deciduous trees relative to soil type: An example from the Sierra Madre, Mexico, and a general model. *Ecology* **63,** 942–951.

Golley, F. B., ed. (1977). "Ecological Succession." Dowden, Hutchinson & Ross, Inc., Stroudsburg, Pennsylvania.

Good, R. (1953). "The Geography of the Flowering Plants," 2nd ed. Longmans, Green, New York.

Goodroad, L. L., and Keeney, D. R. (1984). Nitrous oxide emission from forest, marsh, and prairie ecosystems. *J. Environ. Qual.* **13,** 448–452.

Gorham, E., Vitousek, P. M., and Reiners, W. A. (1979). The regulation of chemical budgets over the course of terrestrial ecosystem succession. *Annu. Rev. Ecol. Syst.* **10,** 53–84.

Gorham, E., Martin, F. B., and Litzau, J. T. (1984). Acid rain: ionic correlations in the Eastern United States, 1980–1981. *Science* **225,** 407–409.

Gosz, J. R. (1981). Nitrogen cycling in coniferous ecosystems. *In* "Terrestrial Nitrogen Cycles" (F. E. Clark and T. Rosswall, eds.), pp. 405–426. Swed. Nat. Sci. Res. Counc., Stockholm.

Gosz, J. R., Likens, G. E., and Bormann, F. H. (1972). Nutrient content of litterfall on the Hubbard Brook Experimental Forest, New Hamphire. *Ecology* **53,** 769–784.

Gosz, J. R., Likens, G. E., and Bormann, F. H. (1973). Nutrient release from decomposing leaf and branch litter in the Hubbard Brook Forest, New Hampshire. *Ecol. Monogr.* **43,** 173–191.

Gosz, J. R., Likens, G. E., and Bormann, F. H. (1976). Organic matter and nutrient dynamics of the forest and forest floor in the Hubbard Brook Forest. *Oecologia* **22,** 305–320.

Gosz, J. R., Brookins, D. G., and Moore, D. I. (1983). Using strontium isotope ratios to estimate inputs to ecosystems. *BioScience* **33,** 23–30.

Goulding, M. (1980). "The Fishes and the Forest." Univ. of California Press, Berkeley.

Grace, J. (1981). Some effects of wind on plants. *In* "Plants and Their Atmospheric Environment" (J. Grace, E. D. Ford, and P. G. Jarvis, eds.), pp. 31–56. Blackwell, Oxford.

Grace, R. A., and Eagleson, P. S. (1966). "The Synthesis of Short-time Increment Rainfall Sequences," Hydrodyn. Lab. Rep. No. 91. Dep. Civ. Eng., MIT, Cambridge, Massachusetts.

Graham, R. L., and Cromack, K. (1982). Mass, nutrient content, and decay rate of dead boles in rain forests of Olympic National Park. *Can. J. For. Res.* **12,** 511–521.

Graustein, W. C., Cromack, K., and Sollins, P. (1977). Calcium oxalate: Occurrence in soils and effect on nutrient and geochemical cycles. *Science* **198,** 1252–1254.

Gray, J. T. (1983). Nutrient use by evergreen and deciduous shrubs of southern California. I. Community nutrient cycling and nutrient-use efficiency. *J. Ecol.* **71,** 21–41.

Greenberg, J. P., Zimmerman, P. R., Heidt, L., and Pollock, W. (1984). Hydrocarbon and carbon monoxide emissions from biomass burning in Brazil. *J. Geophys. Res.* 89D, 1350–1354.

Greenwood, E. A. N., and Beresford, J. D. (1979). Evaporation from vegetation in landscapes developing secondary salinity using the ventilated chamber technique. I. Comparative transpiration from juvenile *Eucalyptus* above saline groundwater seeps. *J. Hydrol.* **42,** 369–382.

Gregory, R. A. (1971). Cambial activity in Alaskan white spruce. *Am. J. Bot.* **58,** 160–171.

Grier, C. C. (1975). Wildfire effects on nutrient distribution and leaching in a coniferous forest ecosystem. *Can. J. For. Res.* **5,** 599–607.

Grier, C. C. (1978). A *Tsuga heterophylla-Picea sitchensis* ecosystem of coastal Oregon: Decomposition and nutrient balances of fallen logs. *Can. J. For. Res.* **8,** 198–206.

Grier, C. C., and Logan, R. S. (1977). Old-growth *Pseudotsuga menziesii* communities of a western Oregon watershed: Biomass distributions and production budgets. *Ecol. Monogr.* **47,** 373–400.

Grier, C. C., and Running, S. W. (1977). Leaf area of mature northwestern coniferous forests: Relation to site water balance. *Ecology* **58,** 893–899.

Grier, C. C., and Waring, R. H. (1974). Conifer foliage mass related to sapwood area. *For. Sci.* **20,** 205–206.

Grier, C. C., Vogt, K. A., Keyes, M. R., and Edmonds, R. L. (1981). Biomass distribution and above- and below-ground production in young and mature *Abies amabilis* zone ecosystems of the Washington Cascades. *Can. J. For. Res.* **11,** 155–167.

Griffin, J. R. (1973). Xylem sap tension in three woodland oaks of central California. *Ecology* **54,** 152–159.

Grubb, P. J. (1977). Control of forest growth and distribution on wet tropical mountains: With special reference to mineral nutrition. *Annu. Rev. Ecol. Syst.* **8,** 83–107.

Gupta, J. P., Aggarwal, R. K., and Raikhy, N. P. (1981). Soil erosion by wind from bare sandy plains in western Rajasthan, India. *J. Arid Environ.* **4,** 15–20.

Hadfield, J. S., and Johnson, D. W. (1977). "Laminated Root Rot. A Guide for Reducing and Preventing Losses in Oregon and Washington Forests," PNW Rep. U.S. For. Serv., Pac. Northwest Reg., Portland, Oregon.

Haines, B. L., Stefani, M., and Hendrix, F. (1980). Acid rain: Threshold of leaf damage in eight species from forest succession. *USDA For. Serv. Gen. Tech. Rep. PSW* **PSW-43,** 235.

Hale, M. G., Moore, L. D., and Griffin, G. J. (1982). Factors affecting root exudation and significance for the rhizophere. *In* "Biological and Chemical Interactions in the Rhizophere," pp. 43–72. Ecol. Res. Comm., Swed. Natl. Sci. Res. Counc., Stockholm.

Hall, A. (1979). A model of leaf photosynthesis and respiration for predicting carbon dioxide assimilation in different environments. *Oecologia* **43,** 299–316.

Halldin, S., Grip, H., Jansson, P.-E., and Lindgren, A. (1980). Micro-meteorology and hydrology of pine forest ecosystems. I. Theory and models. *Ecol. Bull.* **32,** 463–503.

Handley, W. R. C. (1961). Further evidence for the importance of residual leaf protein complexes in litter decomposition and the supply of nitrogen for plant growth. *Plant Soil* **15,** 37–73.

Hanson, A. D., and Hitz, W. D. (1982). Metabolic responses of mesophytes to plant water deficits. *Annu. Rev. Plant Physiol.* **33,** 163–203.

Harborne, J. B. (1982). "Introduction to Ecological Biochemistry." Academic Press, New York.

Harcombe, P. A. (1986). Stand development in a 130-year-old spruce-hemlock forest based on age structure and 50 years of mortality data. *For. Ecol. Manage.* **14,** 41–58.

Harcombe, P. A., and Marks, P. L. (1983). Five years of tree death in a *Fagus-Magnolia* forest, southeast Texas (USA). *Oecologia* **57,** 49–54.

Hare, R. C. (1966). Physiology of resistance to fungal diseases in plants. *Bot. Rev.* **32,** 95–137.

Hari, P., Arovaara, H., Raunemaa, T., and Hautojarvi, A. (1984). Forest growth and the effects of energy production: A method for detecting trends in the growth potential of trees. *Can. J. For. Res.* **14,** 437–440.

Hari, P., Heikinheimo, P., Mäkelä, A., Kaipiainen, L., Korpilahti, E., and Salmela, J. (1985). Pine as a balanced water transport system. *For. Sci.* (in press).

Harley, J. L., and Smith, S. E. (1983). "Mycorrhizal Symbiosis." Academic Press, New York.

Harmon, M. (1985). Succession on fallen logs in the Olympic rain forest. Ph.D. Dissertation, Oregon State University, Corvallis.

Harr, R. D. (1977). Water flux in soil and subsoil on a steep forested slope. *J. Hydrol.* **33,** 37–58.

Harr, R. D. (1982). Fog drip in the Bull Run municipal watershed, Oregon. *Water Resour. Bull.* **18,** 785–789.

Harr, R. D., Fredriksen, R. L., and Rothacher, J. (1979). Changes in streamflow following timber harvest in southwestern Oregon. *USDA For. Serv. Res. Pap. PNW* **PNW-249,** 1–22.

Harris, W. F., Sollins, P., Edwards, N. T., Dinger, B. E., and Shugart, H. H. (1975). Analysis of carbon flow and productivity in a temperate deciduous forest ecosystem. *In* "Productivity of World Ecosystems," pp. 116–122. Special Committee for Int. Biol. Prog., Natl. Acad. Sci., Washington, D.C.

Harrison, A. F. (1971). The inhibitory effect of oak leaf litter tannins on the growth of fungi in relation to litter decomposition. *Soil Biol. Biochem.* **3,** 167–172.

Harrison, A. F. (1982). ^{32}P method to compare rates of mineralization of labile organic phosphorus in woodland soils. *Soil Biol. Biochem.* **14,** 337–341.

Hartshorn, G. S. (1978). Treefalls and tropical forest dynamics. *In* "Tropical Trees as Living Systems" (P. B. Tomlinson and M. H. Zimmerman, eds.), pp. 617–638. Cambridge Univ. Press, London and New York.

Hase, H., and Folster, H. (1982). Bioelement inventory of a tropical (semi-) evergreen seasonal forest on eutrophic alluvial soils, western Llanos, Venezuela. *Acta Oecol.* **3,** 331–346.

Hatch, A. B. (1937). The physical basis of mycotrophy in the genus *Pinus*. *Black Rock For. Bull.* **6,** 1–168.

Hatchell, G. E. (1981). Site preparation and fertilizer increase pine growth on soils compacted in logging. *South. J. Appl. For.* **5,** 79–83.

Haukioja, E., and Niemela, P. (1979). Birch leaves as a resource for herbivores: Seasonal occurrence of increased resistance in foliage after mechanical damage of adjacent leaves. *Oecologia* **39,** 151–159.

Hawkes, H. A. (1975). River zonation and classification. *In* ''River Ecology'' (B. A. Whitton, ed.), pp. 312–374. Blackwell, Oxford.

Haynes, R. J., and Goh, K. M. (1978). Ammonium and nitrate nutrition of plants. *Biol. Rev. Cambridge Philos. Soc.* **53,** 465–510.

Hedley, M. J., Nye, P. J., and White, R. E. (1982). Plant-induced changes in the rhizosphere of rape (*Brassica napus* var. Emerald) seedlings. II. Origin of the pH change. *New Phytol.* **91,** 31–44.

Heichel, G. H., and Turner, N. C. (1983). CO_2 assimilation of primary and regrowth foliage of red maple (*Acer rubrum* L.) and red oak (*Quercus rubra* L.): Response to defoliation. *Oecologia* **57,** 14–19.

Heilman, P. (1981). Root penetration of Douglas-fir seedlings into compacted soil. *For. Sci.* **27,** 660–666.

Heinselman, M. L. (1973). Fire in the virgin forests of the Boundary waters Canoe Area. *Minn. Quat. Rev.* **3,** 329–382.

Heinselman, M. L. (1981). Fire and the disturbance and structure of northern ecosystems. *Gen. Tech. Rep. WO—U.S. For. Serv.* [*Wash.* Off.] **GTR-WO-26,** 7–57.

Helvey, J. D. (1964). Rainfall interception by hardwood forest litter in the southern Appalachians. *U. S., For. Serv., Southeast. For. Exp. Stn., Res. Pap.* **8,** 1–8.

Helvey, J. D. (1971). A summary of rainfall interception by certain conifers of North America. *In* ''Proceedings of the Third International Symposium for Hydrology Professors. Biological Effects in the Hydrological Cycle'' (E. J. Monke, ed.), pp. 103–113. Purdue University, West Lafayette, Indiana.

Helvey, J. D., and Patric, J. H. (1965). Canopy and litter interception by hardwoods of eastern United States. *Water Resour. Res.* **1,** 193–206.

Hemstrom, M., and Adams, V. D. (1982). Modeling long-term forest succession in the Pacific Northwest. *In* ''Forest Succession and Stand Development Research in the Northwest'' (J. E. Means, ed.), pp. 14–23. For. Res. Lab., Oregon State University, Corvallis.

Henderson, G. S., Swank, W. T., Waide, J. B., and Grier, C. C. (1978). Nutrient budgets of Appalachian and Cascade region watersheds: A comparison. *For. Sci.* **24,** 385–397.

Henderson-Sellers, A., and Gornitz, V. (1984). Possible climatic impacts of land cover transformations, with particular emphasis on tropical deforestation. *Clim. Change* **6,** 231–257.

Henry, J. D., and Swan, J. M. A. (1974). Reconstructing forest history from live and dead plant material—an approach to the study of forest succession in southwest New Hampshire. *Ecology* **55,** 772–783.

Hesketh, J., and Baker, D. (1967). Light and carbon assimilation by plant communities. *Crop Sci.* **7,** 285–293.

Hesterberg, G. A., and Jurgensen, M. F. (1972). The relation of forest fertilization to disease incidence. *For. Chron.* **48,** 92–96.

Hewlett, J. D., and Hibbert, A. R. (1967). Factors affecting the response of small watersheds to precipitation in humid areas. *In* ''Forest Hydrology'' (W. E. Sopper and H. W. Lull, eds.), pp. 275–290. Pergamon, Oxford.

Hewlet, J. D., Post, H. E., and Doss, R. (1984). Effect of clear-cut silviculture on dissolved ion export and water yield in the Piedmont. *Water Resour. Res.* **20,** 1030–1038.

Hill, R. D., Rinker, R. G., and Coucouvinos, A. (1984). Nitrous oxide production by lightning. *J. Geophys. Res.* **89**D, 1411–1421.

Hinckley, T. M., Schroeder, M. O., Roberts, J. E., and Bruckerhoff, D. N. (1975). Effect of several environmental variables and xylem pressure potential on leaf surface resistance in white oak. *For. Sci.* **21,** 201–211.

Hinckley, T. M., Lassoie, J. P., and Running, S. W. (1978). Temporal and spatial variations in the water status of forest trees. *For. Sci. Monogr.* **20,** 1–72.

Hinckley, T. M., Teskey, R. O., Duhme, F., and Richter, H. (1981). Temperate hardwood forests. *In* "Water Deficits and Plant Growth" (T. T. Kozlowski, ed.), Vol. 6, pp. 153–208. Academic Press, New York.

Hinckley, T. M., Duhme, F., Hinckley, A. R., and Richter, H. (1983). Drought relations of shrub species: Assessment of the mechanisms of drought resistance. *Oceologia* **59,** 344–350.

Hole, F. D. (1981). Effects of animals on soil. *Geoderma* **25,** 75–112.

Holmes, R. T., and Sturges, F. W. (1973). Annual energy expenditure by the avifauna of a northern hardwoods ecosystem. *Oikos* **24,** 24–29.

Holmes, R. T., Schultz, J. C., and Nothnagle, P. (1979). Bird predation on forest insects: An exclosure experiment. *Science* **206,** 462–463.

Holzworth, J. M. (1930). "The Wild Grizzlies of Alaska." Putnam, New York.

Honda, H., and Fisher, J. B. (1978). Tree branch angle: maximizing effective leaf area. *Science* **199,** 888–890.

Hook, D. D., Langdon, O. G., Stubbs, J., and Brown, C. L. (1970). Effect of water regimes on the survival, growth, and morphology of tupelo seedlings. *For. Sci.* **16,** 304–311.

Hoover, M. D. (1952). Water and timber management. *J. Soil Water Conserv.* **7,** 75–78.

Hoover, M. D., and Leaf, C. F. (1967). Process and significance of interception in Colorado subalpine forest. *In* "Forest Hydrology" (W. E. Sopper and H. W. Lull, eds.), pp. 213–224. Pergamon, Oxford.

Horn, H. S. (1971). "The Adaptive Geometry of Trees." Princeton Univ. Press, Princeton, New Jersey.

Horn, H. S. (1974). The ecology of secondary succession. *Annu. Rev. Ecol. Syst.* **5,** 25–37.

Hornbeck, J. W. (1973). Stormflow from hardwood-forested and cleared watersheds in New Hampshire. *Water Resour. Res.* **9,** 346–354.

Hornbeck, J. W., and Kropelin, W. (1982). Nutrient removal and leaching from a whole-tree harvest of northern hardwoods. *J. Environ. Qual.* **11,** 309–316.

Houghton, R. A., Hobbie, J. E., Melillo, J. M., Moore, B., Peterson, B. J., Shaver, G. R., and Woodwell, G. M. (1983). Changes in the carbon content of terrestrial biota and soils between 1860 and 1980: A net release of CO_2 to the atmosphere. *Ecol. Monogr.* **53,** 235–262.

House, H. L. (1971). Relations between dietary proportions of nutrients, growth rate, and choice of food in the fly larva *Agria affinis*. *J. Insect Physiol.* **17,** 1225–1238.

Hovland, J., Abrahamsen, G., and Ogner, G. (1980). Effects of artificial acid rain on decomposition of spruce needles and mobilisation and leaching of elements. *Plant Soil* **56,** 365–378.

Howard, R. J. (1982). Beaver habitat classification in Massachusetts. M.S. Thesis, University of Massachusetts, Amherst.

Howard, W. E. (1964). Introduced browsing mammals and habitat stability in New Zealand. *J. Wildl. Manage.* **28,** 421–429.

Howard, W. E. (1967). Ecological changes in New Zealand due to introduced mammals. *Proc. Pap. IUCN 10th Tech. Meet.,* IUCN Publ. New Ser. No. 9, Part III, pp. 219–240.

Hsiao, T. C. (1973). Plant responses to water stress. *Annu. Rev. Plant Physiol.* **24,** 519–570.

Hughes, M. K., Kelley, P. M., Pilcher, J. R., and Lamarche, V. C., eds. (1982). "Climate from Tree Rings." Cambridge Univ. Press, London and New York.

Hunt, R. L. (1975). Food relations and behavior of salmonid fishes. *In* "Coupling of Land and Water Systems" (A. D. Hasler, ed.), pp. 137–151. Springer-Verlag, Berlin and New York.

Hutchinson, G. E., Bonatti, E., Cowgill, U. M., Goulden, C. E., Leventhal, E. A., Racek, M. E., Robak, W. A., Stella, E., Wart-Perkins, J. B., and Wellman, T. R. (1970). Ianula: An account of the history and development of the Lago di Monterosi, Latium, Italy. *Trans. Am. Philos. Soc.* **60,** 1–178.

Hütterman, A. (1983). Immissionsschaden im Bereich der Wurzeln von Waldbaumen. *In* "Immissionsbelastungen von Wald okosystemen," Sonderheft Mitt., erweiterte Neuauflage, pp. 10a–14a. Landesanstalt für Ökologie, Landschafts entwicklung und Forstplanung Nordrhein-Westfalen.

Ingestad, T. (1979a). Nitrogen stress in birch seedlings. II. N, K, P, Ca and Mg nutrition. *Physiol. Plant.* **45,** 149–157.

Ingestad, T. (1979b). Mineral nutrient requirements of *Pinus silvestris* and *Picea abies* seedlings. *Physiol. Plant.* **45,** 373–380.

Ingestad, T. (1980). Growth, nutrition, and nitrogen fixation in grey alder at varied rate of nitrogen addition. *Physiol. Plant.* **50,** 353–364.

Ingestad, T. (1982). Relative addition rate and external concentration; driving variables used in plant nutrition research. *Plant, Cell Environ.* **5,** 443–453.

Ino, Y., and Monsi, M. (1969). An experimental approach to the calculation of CO_2 amount evolved from several soils. *Jpn. J. Bot.* **20,** 153–188.

Jackson, T. A., and Voigt, G. K. (1971). Biochemical weathering of calcium-bearing minerals by rhizosphere micro-organisms and its influence on calcium accumulation in trees. *Plant Soil* **35,** 655–658.

Jacobson, J. S. (1984). Effects of acidic aerosol, fog, mist and rain on crops and trees. *Philos. Trans. R. Soc. London, Ser. B* **305,** 327–338.

Janos, D. P. (1980). Vesicular-arbuscular mycorrhizae affect lowland tropical rain forest plant growth. *Ecology* **61,** 151–162.

Jansson, P.-E., and Halldin, S. (1979). Model for annual water and energy flow in a layered soil. *In* "Comparison of Forest Water and Energy Exchange Models" (S. Halldin, ed.), pp. 145–163. Int. Soc. Ecol. Model., Copenhagen.

Janzen, D. H. (1975). "Ecology of Plants in the Tropics." Arnold, London.

Janzen, D. H. (1979). New horizons in the biology of plant defenses. *In* "Herbivores: Their Interaction with Secondary Plant Metabolites" (G. A. Rosenthal and D. H. Janzen, eds.), pp. 331–348. Academic Press, New York.

Janzen, D. H. (1981). Patterns of herbivory in a tropical deciduous forest. *Biotropica* **13,** 271–282.

Janzen, D. H. (1982). How and why horses open: *Crescentia alata* fruits. *Biotropica* **14,** 149–152.

Janzen, D. H., and Martin, P. S. (1982). Neotropical anachronisms: The fruits the Gomphotheres ate. *Science* **215,** 19–27.

Jarvis, P. G. (1975). Water transfer in plants. *In* "Heat and Mass Transfer in the Biosphere" (D. A. deVries and N. H. Afgan, eds.), pp. 369–394. Scripta Book Co., Washington, D.C.

Jarvis, P. G. (1981a). Stomatal conductance, gaseous exchange and transpiration. *In* "Plants and Their Atmospheric Environment" (J. Grace, E. D. Ford, and P. G. Jarvis, eds.), pp. 175–204. Blackwell, Oxford.

Jarvis, P. G. (1981b). Plant water relations in models of tree growth. *Stud. For. Suec.* **160,** 51–60.

Jarvis, P. G., and Stewart, J. B. (1979). Evaporation of water from plantation forests. *In* "The Ecology of Even-aged Forest Plantations" (E. D. Ford, D. C. Malcolm, and J. Atterson, eds.), pp. 327–350. Inst. Terr. Ecol., Nat. Env. Res. Council, Cambridge, England.

Jarvis, P. G., James, G. B., and Landsberg, J. J. (1976). Coniferous forests. *In* "Vegetation and the Atmosphere" (J. L. Monteith, ed.), Vol. 2, pp. 171–240. Academic Press, New York.

Jarvis, P. G., Edwards, W. R. N., and Talbot, H. (1981). Models of plant and crop water use. *In* "Quantitative Aspects of Plant Physiology" (D. A. Rose and D. A. Charles-Edwards, eds.), pp. 151–194. Academic Press, New York.

Jaynes, R. A., and Elliston, J. E. (1982). Hypovirulent isolates of *Endothia parasitica* with large American chestnut trees. *Plant Dis.* **66,** 769–772.

Jenkins, S. H., and Busher, P. E. (1979). *Castor canadensis. Mamm. species.* No. 120, pp. 1–8.

Jenkinson, D. S., and Powlson, D. S. (1976). The effects of biocidal treatments on metabolism in soil. V. A method for measuring soil biomass. *Soil Biol. Biochem.* **8,** 209–213.

Jenkinson, D. S., Davidson, S. A., and Powlson, D. S. (1979). Adenosine triphosphate and microbial biomass in soil. *Soil Biol. Biochem.* **11,** 521–527.

Jenny, H. (1941). "Factors of Soil Formation." McGraw-Hill, New York.

Jenny, H. (1980). "The Soil Resource." Springer-Verlag, Berlin and New York.

Jensen, V. (1974). Decomposition of angiosperm tree leaf litter. *In* "Biology of Plant Litter Decomposition" (C. H. Dickinson and G. J. F. Pugh, eds.), pp. 69–104. Academic Press, New York.

Johnson, A. H., and Siccama, T. G. (1983). Acid deposition and forest decline. *Environ. Sci. Tech.* **17,** 294A–305A.

Johnson, A. H., Siccama, T. G., Wang, D., Turner, R. S., and Barringer, T. H. (1981). Recent changes in patterns of tree growth rate in the New Jersey pinelands: A possible effect of acid rain. *J. Environ. Qual.* **10,** 427–430.

Johnson, D. R., and Chance, D. H. (1974). Presettlement overharvest of upper Columbia River beaver populations. *Can. J. Zool.* **52,** 1519–1521.

Johnson, D. W., and Cole, D. W. (1980). Anion mobility in soils: Relevance to nutrient transport from forest ecosystems. *Environ. Int.* **3,** 79–90.

Johnson, D. W., and Edwards, N. T. (1979). The effects of stem girdling on biogeochemical cycles within a mixed deciduous forest in eastern Tennessee. II. Soil nitrogen mineralization and nitrification rates. *Oecologia* **40,** 259–271.

Johnson, D. W., Henderson, G. S., and Todd, D. E. (1981). Evidence of modern accumulations of adsorbed sulfate in an east Tennessee forested ultisol. *Soil Sci.* **132,** 422–426.

Johnson, D. W., Turner, J., and Kelly, J. M. (1982a). The effects of acid rain on forest nutrient status. *Water Resour. Res.* **18,** 449–461.

Johnson, D. W., West, D. C., Todd, D. E., and Mann, L. K. (1982b). Effects of sawlog vs. whole-tree harvesting on the nitrogen, phosphorus, potassium and calcium budgets of an upland mixed oak forest. *Soil Sci. Soc. Am. J.* **46,** 1304–1309.

Johnson, F. L., and Risser, P. G. (1974). Biomass, annual net primary production, and dynamics of six mineral elements in a post oak-blackjack oak forest. *Ecology* **55,** 1246–1258.

Johnson, N. M. (1971). Mineral equilibria in ecosystem geochemistry. *Ecology* **52,** 529–531.

Johnson, N. M., Likens, G. E., Bormann, F. H., and Pierce, R. S. (1968). Rate of chemical weathering of silicate minerals in New Hampshire. *Geochim. Cosmochim. Acta* **32,** 531–545.

Johnson, N. M., Likens, G. E., Bormann, F. H., Fisher, D. W., and Pierce, R. S. (1969). A working model for the variation in stream water chemistry at the Hubbard Brook Experimental Forest, New Hampshire. *Water Resour. Res.* **5,** 1353–1363.

Johnson, P. C., and Denton, R. E. (1975). Outbreaks of western spruce budworm in the American Northern Rocky Mountain area from 1922 through 1971. *USDA For. Serv. Gen. Tech. Rep. INT* **INT-20,** 1–144.

Jordan, C. F. (1982). Amazon rain forests. *Am. Sci.* **70,** 394–401.

Jorgensen, J. R., and Wells, C. G. (1973). The relationship of respiration in organic and mineral soil layers to soil chemical properties. *Plant Soil* **39,** 373–387.

Jurgensen, M. F. (1973). Relationship between nonsymbiotic nitrogen fixation and soil nutrient status—A review. *J. Soil Sci.* **24,** 512–522.

Kaufmann, M. R. (1977). Soil temperature and drying cycle effects on water relations of *Pinus radiata. Can. J. Bot.* **55,** 2413–2418.

Kaufmann, M. R. (1984). A canopy model (RM-CWU) for determining transpiration of subalpine forests. II. Consumptive water use in two watersheds. *Can. J. For. Res.* **14,** 227–232.

Kaufmann, M. R., and Troendle, C. A. (1981). The relationship of leaf area and foliage biomass to sapwood conducting area in four subalpine forest tree species. *For. Sci.* **27,** 477–482.

Kazimirov, N. I., and Morozova, R. N. (1973). "Biological Cycling of Matter in Spruce Forests of Karelia." Nauka Publ. House, Leningrad.

Kearney, M. S., and Luckman, B. H. (1983). Holocene timberline fluctuations in Jasper National Park, Alberta. *Science* **221,** 261–263.

Keeley, J. E. (1979). Population differentiation along a flood frequency gradient: Physiological adaptations to flooding in *Nyssa sylvantica*. *Ecol. Monogr.* **49,** 89–108.

Keeney, D. R. (1980). Prediction of soil nitrogen availability in forest ecosystems. A literature review. *For. Sci.* **26,** 159–171.

Keller, M., Goreau, T. J., Wofsy, S. C., Kaplan, W. H., and McElroy, M. B. (1983). Production of nitrous oxide and consumption of methane by forest soils. *Geophys. Res. Lett.* **10,** 1156–1159.

Keller, R. A., and Tregunna, E. B. (1976). Effects of exposure on water relations and photosynthesis of western hemlock in habitat forms. *Can. J. For. Res.* **6,** 40–48.

Keller, T., and Wehrmann, J. (1963). CO_2 Assimilation, Wurzelatmung und Ertrag von Fichten- und Kiefersamlingen bei unterschiedlicher Mineralstoffernahrung. *Mitt., Schweiz. Anst. Forstl. Versuchswes.* **39,** 215–242.

Kellman, M., Hudson, J., and Sanmugudas, K. (1982). Temporal variability in atmospheric nutrient influx to a tropical ecosystem. *Biotropica* **14,** 1–9.

Keyes, M. R., and Grier, C. C. (1981). Above- and below-ground net production in 40-year-old Douglas-fir stands on low and high productivity sites. *Can. J. For. Res.* **11,** 599–605.

Khalil, M. A. K., and Rasmussen, R. A. (1983). Increase and seasonal cycles of nitrous oxide in the Earth's atmosphere. *Tellus* **35B,** 161–169.

Khanna, P. K., and Ulrich, B. (1981). Changes in the chemistry of throughfall under stands of beech and spruce following the addition of fertilizers. *Acta Oecol.* **2,** 155–164.

Kiehl, J. T. (1983). Satellite detection of effects due to increased atmospheric carbon dioxide. *Science* **222,** 504–506.

Kilgore, B. M. (1973). The ecological role of fire in Sierra conifer forests: Its application to national park management. *Quat. Res.* **3,** 496–513.

Kilgore, B. M., and Taylor, D. (1979). Fire history of a sequoia-mixed conifer forest. *Ecology* **60,** 129–142.

Kimmins, J. P., Scoullar, K. A., and Feller, M. C. (1981). FORCYTE-9: A computer simulation approach to evaluating the effect of whole tree harvesting on nutrient budgets and future forest productivity. *Mitt. Deutsch. Forstl. Bundesversuchsanst.* **140,** 189–205.

Kimura, M. (1982). Changes in population structure, productivity and dry matter allocation with the progress of wave regeneration of *Abies* stands in Japanese subalpine regions. *In* "Carbon Uptake and Allocation in Subalpine Ecosystems as a Key to Management" (R. H. Waring, ed.), IUFRO Workshop, pp. 57–63. For. Res. Lab., Oregon State University, Corvallis.

Kinerson, R. S. (1975). Relationships between plant surface area and respiration in loblolly pine. *J. Appl. Ecol.* **12,** 965–971.

Kinerson, R. S., Ralston, C. W., and Wells, C. G. (1977). Carbon cycling in a loblolly pine plantation. *Oecologia* **29,** 1–10.

King, D. (1981). Tree dimensions: Maximizing the rate of height growth in dense stands. *Oecologia* **51,** 351–356.

Kira, T., and Ogawa, H. (1971). Assessment of primary production in tropical and equatorial forests. *Ecol. Conserv.* **4,** 309–321.

Kira, T., and Shidei, T. (1967). Primary production and turnover of Organic matter in different forest ecosystems of the western Pacific. *Jpn. J. Ecol.* **17,** 70–87.

Kira, T., Shinozaki, K., and Hozumi, K. (1969). Structure of forest canopies as related to their primary productivity. *Plant Cell Physiol.* **10,** 129–142.

Kittredge, J. (1948). "Forest Influences." McGraw-Hill, New York.

Kloft, W. (1957). Further investigations concerning the inter-relationship between bark conditions of *Abies alba* and infestation by *Adelges piceae typica* and *A. nusslini schneideri*. *Z. Angew. Entomol.* **41,** 438–442.

Knight, D. H., Fahey, T. J., and Running, S. W. (1985). Water and nutrient outflow from contrasting lodgepole pine forests in Wyoming. *Ecol. Monogr.* **55,** 29–48.

Kohyama, T. (1980). Growth pattern of *Abies mariesii* saplings under conditions of open-growth and supression. *Bot. Mag.* **93,** 13–24.

Kohyama, T., and Fujita, N. (1981). Studies on the *Abies* population of Mt. Shimagare. I. Survivorship curve. *Bot. Mag.* **94,** 55–68.

Kozlowski, T. T., and Keller, T. (1966). Food relations of woody plants. *Bot. Rev.* **32,** 293–382.

Kozlowski, T. T., and Pallardy, S. G. (1979). Stomatal responses of *Fraxinus pennsylvanica* seedlings during and after flooding. *Physiol. Plant.* **46,** 155–158.

Kramer, P. J. (1981). Carbon dioxide concentrations, photosynthesis and dry matter production. *BioScience* **31,** 29–33.

Kramer, P. J. (1983). "Water Relations of Plants." Academic Press, New York.

Krueger, K. W. (1967). Nitrogen, phosphorus, and carbohydrate in expanding and year-old Douglas fir shoots. *For. Sci.* **13,** 352–356.

Krygier, J. T. (1971). Comparative water loss of Douglas-fir and Oregon white oak. Ph.D. Thesis, Colorado State University, Fort Collins.

Kulman, H. M. (1971). Effects of insect defoliation on growth and mortality of trees. *Annu. Rev. Entomol.* **16,** 289–324.

Kuzmin, P. P. (1961). "Melting of Snow Cover" (Israel Program for Scientific Translations, Jerusalem, 1972).

Lagenheim, J. H., Arrhenius, S. P., and Nascimento, J. C. (1981). Relationship of light intensity to leaf resin composition and yield in the tropical leguminous genera *Hymenaea* and *Copaifera. Biochem. Syst. Ecol.* **9,** 27–37.

LaMarche, V. C. (1974). Paleoclimatic inferences from long tree-ring records. *Science* **183,** 1043–1048.

LaMarche, V. C., and Pittock, A. B. (1982). Preliminary temperature reconstructions for Tasmania. *In* "Climate from Tree Rings" (M. K. Hughes, P. M. Kelley, J. R. Pilcher, and V. C. Lamarch, eds.), pp. 177–185. Cambridge Univ. Press, London and New York.

LaMarche, V. C., Graybill, D. A., Fritts, H. C., and Rose, M. R. (1984). Increasing atmospheric carbon dioxide: Tree ring evidence for growth enhancement in natural vegetation. *Science* **225,** 1019–1021.

Lambert, M. J., and Turner, J. (1977). Dieback in high site quality *Pinus radiata* stands—the role of sulphur and boron deficiencies. *N. Z. J. For.* **7,** 333–348.

Lambert, R. L., Lang, G. E., and Reiners, W. A. (1980). Loss of mass and chemical change in decaying boles of a subalpine balsam fir forest. *Ecology* **61,** 1460–1473.

Landsberg, J. J., and Fowkes, N. D. (1978). Water movement through plant roots. *Ann. Bot. (London)* [N.S.] **42,** 493–508.

Landsberg, J. J., and Powell, D. B. B. (1973). Surface exchange characteristics of leaves subject to mutural interference. *Agric. Meteorol.* **12,** 169–184.

Lang, G. E., and Forman, R. T. T. (1978). Detrital dynamics in a mature oak forest: Hutcheson Memorial Forest, New Jersey. *Ecology* **59,** 580–595.

Lang, G. E., and Knight, D. H. (1979). Decay rates of boles of tropical trees in Panama. *Biotropica* **11,** 316–317.

Langbein, W. B., and Schumm, S. A. (1958). Yield of sediment in relation to mean annual precipitation. *Trans. Am. Geophys. Union* **39,** 1076–1084.

Larsen, M. J., Jurgensen, M. F., and Harvey, A. E. (1982). N_2 fixation in brown-rotted soil wood in an intermountain cedar-hemlock ecosystem. *For. Sci.* **28,** 292–296.

Larson, M. M. (1980). Effects of atmospheric humidity and zonal soil water stress on initial growth of planted northern red oak seedlings. *Can. J. For. Res.* **10,** 549–554.

Larsson, S., Oren, R., Waring, R. H., and Barrett, J. W. (1983). Attacks of mountain pine beetle as related to tree vigor of ponderosa pine. *For. Sci.* **29,** 395–402.

Larsson, S., Wirén, A., Lundgren, L., and Ericsson, T. (1985). Effects of light and nutrient stress on leaf phenolic chemistry in *Salix dasyclados* and susceptibility to *Galerucella lineola* (Coleoptera) in *Oikos* **47,** 205–210.

Lassoie, J. P. (1979). Stem dimensional fluctuations in Douglas-fir of different crown classes. *For. Sci.* **25,** 132–144.

Lassoie, J. P. (1982). Physiological activity in Douglas-fir. *In* "Analysis of Coniferous Forest Ecosystems in the Western United States" (R. L. Edmonds, ed.), US/IBP Synth. Ser. No. 14, pp. 126–185. Dowden, Hutchinson, & Ross Publ. Co., Stroudsburg, Pennsylvania.

Laws, R. M. (1981). Experience in the study of large mammals. *In* "Dynamics of Large Mammal Populations" (C. W. Fowler and T. D. Smith, eds.), pp. 19–45. Wiley, New York.

Lawson, D. R., and Winchester, J. W. (1979). Sulfur, potassium, and phosphorus associations in aerosols from South American tropical rain forests. *JGR, J. Geophys. Res.* **84,** 3723–3727.

Lea, R., and Ballard, R. (1982a). Relative effectiveness of nutrient concentrations in living foliage and needle fall at predicting response of loblolly pine to N and P fertilization. *Can. J. For. Res.* **12,** 713–717.

Lea, R., and Ballard, R. (1982b). Predicting loblolly pine growth response from N fertilizer using soil-N availability indices. *Soil Sci. Soc. Am. J.* **46,** 1096–1099.

Lea, R., Tierson, W. C., Bickelhaupt, D. H., and Leaf, A. L. (1980). Differential foliar responses of northern hardwoods to fertilization. *Plant Soil* **54,** 419–439.

Leaf, C.F., and Brink, G. E. (1973). Computer simulation of snow-melt within a Colorado subalpine watershed. *U.S., For Serv., Rocky Mt. For. Range Exp. Stn., Res. Pap. RM* **RM-99,** 22.

Leahey, A. (1947). Characteristics of soils adjacent to the MacKenzie River in the Northwest Territories of Canada. *Soil Sci. Soc. Am. Proc.* **12,** 458–461.

Leavitt, S. W., and Long, A. (1983). An atmospheric $^{13}C/^{12}C$ reconstruction generated through removal of climatic effects from tree-ring $^{13}C/^{12}C$ measurements. *Tellus* **35B,** 92–102.

Leigh, E. G., Jr. (1982). Introduction: Why are there so many kinds of tropical trees. *In* "The Ecology of a Tropical Forest" (E. G. Leigh, Jr., A. S. Rand, and D. M. Windsor, eds.), pp. 63–66. Smithson. Inst. Press, Washington, D.C.

Leith, H. (1975). Modeling the primary productivity of the world. *In* "Primary Productivity of the Biosphere" (H. Leith and R. H. Whittaker, eds.), pp. 237–263. Springer-Verlag, Berlin and New York.

Leopold, L. B., Wolman, M. G., and Miller, J. P. (1964). "Fluvial Processes in Geomorphology." Freeman, San Francisco, California.

Leshem, B. (1970). Resting roots of *Pinus halepensis:* Structure, function, and reaction to water stress. *Bot. Gaz. (Chicago)* **131,** 99–104.

Levine, J. S., Augustsson, T. R., Anderson, I. C., and Hoell, J. M. (1984). Tropospheric sources of NO_x: Lightning and biology. *Atmos. Environ.* **18,** 1797–1804.

Levitt, J. (1972). "Responses of Plants to Environmental Stresses." Academic Press, New York.

Levitt, J. (1980). "Responses of Plants to Environmental Stresses," 2nd ed., Vol. 2. Academic Press, New York.

Lewin, R. (1984). Parks: How big is big enough? *Science* **225,** 611–612.

Lewis, W. M. (1981). Precipitation chemistry and nutrient loading by precipitation in a tropical watershed. *Water Resour. Res.* **17,** 169–181.

Lewis, W. M., and Grant, M. C. (1979). Relationships between stream discharge and yield of dissolved substances from a Colorado Mountain watershed. *Soil Sci.* **128,** 353–363.

Li, C. Y., and Bollen, W. B. (1975). Growth of *Phellinus* (Poria) *weirii* on different vitamins and carbon and nitrogen sources. *Am. J. Bot.* **62,** 838–841.

Likens, G. E., and Bilby, R. E. (1982). Development, maintenance, and role of organic-debris dams in New England streams. *USDA For. Serv. Gen. Tech. Rep. PNW* **PNW-141,** 122–128.

Likens, G. E., and Bormann, F. H. (1970). "Chemical Analyses of Plant Tissues from the Hubbard Brook Ecosystem in New Hampshire," Yale Sch. For. Bull. No. 79. Yale University, New Haven, Connecticut.

Likens, G. E., and Bormann, F. H. (1974). Linkages between terrestrial and aquatic ecosystems. *BioScience* **24,** 447–456.

Likens, G. E., and Eaton, J. S. (1970). A polyurethane streamflow collector for trees and shrubs. *Ecology* **51,** 938–939.

Likens, G. E., Bormann, F. H., Johnson, N. M., Fisher, D. W., and Pierce, R. S. (1970). Effects of forest cutting and herbicide treatment on nutrient budgets in the Hubbard Brook Watershed-Ecosystem. *Ecol. Monogr.* **40,** 23–47.

Likens, G. E., Bormann, F. H., Pierce, R. S., Eaton, J. S., and Johnson, N. M. (1977). "Biogeochemistry of a Forested Ecosystem." Springer-Verlag, Berlin and New York.

Likens, G. E., Bormann, F. H., and Johnson, N. M. (1981). Interactions between major biogeochemical cycles in terrestrial ecosystems. *In* "Some Perspectives of the Major Biogeochemical Cycles" (G. E. Likens, ed.), pp. 93–112. Wiley, New York.

Lindberg, S. E., Harriss, R. C., and Turner, R. R. (1982). Atmospheric deposition of metals to forest vegetation. *Science* **215,** 1609–1611.

Linder, S. (1973). The influence of soil temperature upon net photosynthesis and transpiration in seedlings of Scots pine and Norway spruce. Ph.D. Thesis, University of Umeå, Sweden.

Linder, S. (1982). Photosynthetic response of pine, spruce, birch, and willow to light. *In* "Vitality and Quality of Nursery Stock" (P. Puttonen, ed.), Res. Note No. 36, pp. 75–84. Dep. Silvicult., University of Helsinki, Finland.

Linder, S., and Axelsson, B. (1982). Changes in carbon uptake and allocation patterns as a result of irrigation and fertilization in a young *Pinus sylvestris* stand. *In* "Carbon Uptake and Allocation: Key to Management of Subalpine Forest Ecosystems" (R. H. Waring, ed.), IUFRO Workshop, pp. 38–44. For. Res. Lab., Oregon State University, Corvallis.

Linder, S., and Rook, D. A. (1984). Effects of mineral nutrition on carbon dioxide exchange and partitioning of carbon in trees. *In* "Nutrition of Plantation Forests" (G. D. Bowen and E. K. S. Nambiar, eds.), pp. 211–236. Academic Press, London.

Linder, S., and Troeng, E. (1981). The seasonal variation in stem and coarse root respiration of a 20-year-old Scots pine (*Pinus sylvestris* L.). *Mitt. Forstl. Bundesversuchsanst.* **142,** 125–139.

Linder, S., McDonald, J., and Lohammar, T. (1981). "Effect of Nitrogen Status and Irradiance during Cultivation on Photosynthesis and Respiration in Birch Seedlings." Energy For. Proj. (EFP), Swed. Univ. Agric. Sci., Uppsala.

Lindroth, A., and Perttu, K. (1981). Simple calculation of extinction coefficient of forest stands. *Agric. Meteorol.* **25,** 97–110.

Littke, W. R., Bledsoe, C. S., and Edmonds, R. L. (1984). Nitrogen uptake and growth *in vitro* by *Hebeloma crustuliniforme* and other Pacific Northwest mycorrhizal fungi. *Can. J. Bot.* **62,** 647–652.

Livingston, R. B. (1972). Influence of birds, stones and soil on the establishment of pasture Juniper, *Juniperus communis* and red cedar, *J. virginiana* in New England pastures. *Ecology* **53,** 1140–1147.

Livingstone, D. A. (1975). Late quaternary climatic change in Africa. *Annu. Rev. Ecol. Syst.* **6,** 249–280.

Long, J. N. (1982). Productivity of western coniferous forests. *In* "Analysis of Coniferous Forest Ecosystems in the Western United States" (R. L. Edmonds, ed.), pp. 89–125. Dowden, Hutchison and Ross, Stroudsburg, Pennsylvania.

Losch, R., and Tenhunen, J. D. (1981). Stomatal response to humidity: Phenomenon and mechanism. *In* "Stomatal Physiology" (P. G. Jarvis and T. A. Mansfield, eds.), Soc. Exp. Biol., Semin. Ser. No. 8, pp. 137–161. Cambridge Univ. Press, London and New York.

Lousier, J. D., and Parkinson, D. (1976). Litter decomposition in a cool temperate deciduous forest. *Can. J. Bot.* **54,** 419–436.

Lovett, G. M., Reiners, W. A., and Olson, R. K. (1982). Cloud droplet deposition in subalpine balsam fir forests: Hydrologic and chemical inputs. *Science* **218,** 1303–1304.

Lowry, W. P. (1969). "Weather and Life: An Introduction to Biometerology." Academic Press, New York.

Lugo, A. E., Applefield, M., Pool, D. J., and McDonald, R. B. (1983). The impact of Hurricane David on the forest of Dominica. *Can. J. For. Res.* **13,** 201–211.

Luke, R. H., and McArthur, A. G. (1978). "Bushfires in Australia." Wilke & Co., Clayton, Victoria.

Luttge, U., and Higinbotham, H. (1979). "Transport in Plants." Springer-Verlag, Berlin and New York.

Lutz, H. J., and Chandler, R. F. (1946). "Forest Soils." Wiley, New York.

Luxmoore, R. J. (1983). Infiltration and runoff prediction for a grassland watershed. *J. Hydrol.* **65,** 271–278.

Luxmoore, R. J., Gizzard, T., and Strand, R. H. (1981). Nutrient translocation in the outer canopy and understory of an eastern deciduous forest. *For. Sci.* **27,** 505–518.

Lyell, C. (1969). "Principles of Geology," Vol. 2. Johnson Reprint Corp., New York.

Lyons, J. M. (1973). Chilling injury in plants. *Annu. Rev. Plant Physiol.* **24,** 445–466.

McCauley, K. J., and Cook, S. A. (1980). *Phellinus weirii* infestation of two mountain hemlock forests in the Oregon Cascades. *For. Sci.* **26,** 23–29.

MacClaren, P. (1983). Chemical welfare in the forest. A review of allelopathy with regard to New Zealand Forestry. *N. Z. J. For.* **28,** 73–92.

McClaugherty, C. A., Aber, J. D., and Melillo, J. M. (1982). The role of fine roots in the organic matter and nitrogen budgets of two forested ecosystems. *Ecology* **63,** 1481–1490.

McClelland, B. R. (1973). Autumn concentration of bald eagles in Glacier National Park. *Condor* **75,** 121–123.

McColl, J. G., and Grigal, D. F. (1975). Forest fire: Effect on phosphorus movement to lakes. *Science* **188,** 1109–1111.

McCune, B. (1983). Fire frequency reduced by two orders of magnitude in the Bitterroot Canyons, Montana. *Can. J. For. Res.* **13,** 212–218.

McElroy, M. B. (1983). Marine biological controls on atmospheric CO_2 and climate. *Nature (London)* **302,** 328–329.

McGill, W. B., and Cole, C. V. (1981). Comparative aspects of cycling of organic C, N, S and P through soil organic matter. *Geoderma* **26,** 267–286.

McGinty, D. T. (1976). Comparative root and soil dynamics on a white pine watershed and in the hardwood forest in the Coweeta basin. Ph.D. Dissertation, University of Georgia, Athens.

Macias, E. S., Zwicker, J. O., Ouimette, J. R., Hering, S. V., Friedland, S. K., Cahill, T. A., Kuhlmey, G. A., and Richards, L. W. (1981). Regional haze case studies in the southwestern U.S.—I. Aerosol chemical composition. *Atmos. Environ.* **15,** 1971–1986.

McIntosh, R. P., ed. (1978). "Phytosociology." Dowden, Hutchinson & Ross, Inc., Stroudsburg, Pennsylvania.

MacIntyre, F. (1974). The top millimeter of the ocean. *Sci. Am.* **230**(5), 62–77.

McKey, D., Waterman, P. G., Mbi, C. N., Gartlan, J. S., and Struhsaker, T. T. (1978). Phenolic content of vegetation in two African rain forests: Ecological implications. *Science* **202,** 61–64.

MacKney, D. (1961). A podzol development sequence in oakwoods and heath in central England. *J. Soil Sci.* **12,** 22–40.

McLaughlin, S. B., and Shriner, D. S. (1980). Allocation of resources to defense and repair. *Plant Dis.* **5,** 407–431.

McLaughlin, S. B., McConathy, R. K., Barnes, R. L., and Edwards, N. T. (1980). Seasonal changes in energy allocation by white oak (*Quercus alba*). *Can. J. For. Res.* **10,** 379–388.

MacLean, D. A., and Wein, R. W. (1977). Nutrient accumulation for post fire jack pine and hardwood successional patterns in New Brunswick. *Can. J. For. Res.* **7,** 562–578.

McMurray, T. I. (1980). Effects of drought stress induced changes in host tree foliage quality on the growth of two forest insect pests. M.S. Thesis, University of New Mexico, Albuquerque.

McNeil, S., and Southwood, T. R. E. (1978). The role of nitrogen in the development of insect and plant relationships. *In* "Biochemical Aspects of Plant and Animal Coevolution" (J. B. Harborne, ed.), pp. 77–98. Academic Press, London.

Madej, M. A. (1982). Sediment transport and channel changes in an aggrading stream in the Puget Lowland, Washington. *USDA For. Serv. Gen. Tech. Rep. PNW* **PNW-141,** 97–108.

Madgwick, H. A. I., and Olson, D. F. (1974). Leaf area index and volume growth in thinned stands of *Liriodendron tulipifera. J. Appl. Ecol.* **11,** 575–579.

Mahendrappa, M. K., and Ogden, E. D. (1973). Effects of fertilization of a black spruce stand on nitrogen contents of stemflow, throughfall, and litterfall. *Can. J. For. Res.* **3,** 54–60.

Malcolm, R. L., and Durum, W. H. (1976). Organic carbon and nitrogen concentrations and annual organic load of six selected rivers of the United States. *Geol. Surv. Water Supply Pap. (U.S.)* **1817-F,** 1–21.

Manion, P. D. (1981). "Tree Disease Concepts." Prentice-Hall, Englewood Cliffs, New Jersey.

Mann, L. K., McLaughlin, S. B., and Shriner, D. S. (1980). Seasonal physiological responses of white pine under chronic air pollution stress. *Environ. Exp. Bot.* **20,** 99–105.

Mansfield, T. A., Wellburn, A. R., and Moreira, T. J. S. (1978). The role of abscisic acid and farnesol in the alleviation of water stress. *Philos. Trans. R. Soc.* London, Ser. B **284,** 471–482.

Mansfield, T. A., Travis, A. J., and Jarvis, P. G. (1981). Response to light and carbon dioxide. *In* "Stomatal Physiology" (P. G. Jarvis and T. A. Mansfield, eds.), Soc. Exp. Biol., Semin. Ser. No. 8, pp. 119–135. Cambridge Univ. Press, London and New York.

Marchand, P. J. (1984). Dendrochronology of a fir wave. *Can. J. For. Res.* **14,** 51–56.

Marks, P. L. (1974). The role of pin cherry (*Prunus pennsylvanica* L.) in the maintenance of stability in northern hardwood ecosystems. *Ecol. Monogr.* **44,** 73–88.

Marks, P. L., and Bormann, F. H. (1972). Revegetation following forest cutting: Mechanisms for return to steady-state nutrient cycling. *Science* **176,** 914–915.

Marshall, J. D. (1984). Physiological control of fine root turnover in Douglas-fir. Ph.D. Dissertation, Oregon State University, Corvallis.

Marshall, J. D., and Waring, R. H. (1984). Conifers and broadleaf species: Stomatal sensitivity differs in western Oregon. *Can. J. For. Res.* **14,** 905–908.

Marshall, J. D., and Waring, R. H. (1985). Predicting fine root production and turnover by monitoring root starch and soil temperature. *Can. J. For. Res.* **15,** 791–800.

Marshall, J. K. (1973). Drought, land use and soil erosion. *In* "The Environmental, Economic and Social Significance of Drought" (J. V. Lovett, ed.), pp. 55–77. Angus & Robertson Publishers, Sydney, Australia.

Martin, J. M., and Meybeck, M. (1979). Elemental mass-balance of material carried by major world rivers. *Mar. Chem.* **7,** 173–206.

Martin, R. E. (1982). Fire history and its role in succession. *In* "Forest Succession and Stand Development Research in the Northwest" (J. E. Means, ed.), pp. 92–99. Oregon State Univ. Press, Corvallis.

Marumoto, T., Anderson, J. P. E., and Domsch, K. H. (1982). Mineralization of nutrients from soil microbial biomass. *Soil Biol. Biochem.* **14,** 469–475.

Marx, D. H. (1969). The influence of ectotrophic mycorrhizal fungi on the resistance of pine roots to pathogenic infections. I. Antagonism of mycorrhizal fungi to root pathogenic fungi and soil bacteria. *Phytopathology* **59,** 153–163.

Marx, D. H., Hatch, A. B., and Mendicino, J. F. (1977). High soil fertility decreases sucrose content and susceptibility of loblolly pine roots to ectomycorrhizal infection by *Pisolithus tinctorius. Can. J. Bot.* **55,** 1569–1574.

Maser, C., Trappe, J. M., and Ure, D. C. (1978). Implications of small mycophagy to the management of western coniferous forest. *Trans. North Am. Wildl. Nat. Resour. Conf.* **43,** 78–88.

Matson, P. A., and Boone, R. (1984). Natural disturbance and nitrogen mineralization: wave form dieback of mountain hemlock in the Oregon Cascades. *Ecology* **65,** 1511–1516.

Matson, P. A., and Waring, R. H. (1984). Effects of nutrient and light limitation on mountain hemlock: Susceptibility to laminated root-rot. *Ecology* **65,** 1517–1524.

Matteson, P. C., Altieri, M. A., and Gagné, W. C. (1984). Modification of small farmer practices for better pest management. *Annu. Rev. Entomol.* **29,** 383–402.

Mattson, W. H., and Addy, N. D. (1975). Phytophagous insects as regulators of forest primary production. *Science* **190,** 515–522.

Mattson, W. J., Jr. (1980). Herbivory in relation to plant nitrogen content. *Annu. Rev. Ecol. Syst.* **11,** 119–161.

Mattson-Djos, E. (1981). The use of pressure-bomb and porometer for describing plant water status in tree seedlings. *In* "Proceedings of a Nordic Symposium on Vitality and Quality of Nursery Stock" (P. Puttonen, ed.), pp. 45–57. Dep. Silvicult., University of Helsinki, Finland.

May, R. M. (1981). Models for two interacting populations. *In* "Theorectical Ecology" (R. M. May, ed.), pp. 78–104. Blackwell, Oxford.

Meentemeyer, V. (1978a). Macroclimate and lignin control of litter decomposition rates. *Ecology* **59,** 465–472.

Meentemeyer, V. (1978b). Climatic regulation of decomposition rates of organic matter in terrestrial ecosystems. *In* "Environmental Chemistry and Cycling Processes" (D. C. Adriano and I. L. Brisbin, eds.), pp. 779–789. Natl. Tech. Inf. Serv., Springfield, Virginia.

Meentemeyer, V., Box, E. O., and Thompson, R. (1982). World patterns and amounts of terrestrial plant litter production. *BioScience* **32,** 125–128.

Meeuwig, R. O. (1971). Infiltration and water repellency in granitic soils. *USDA For. Serv. Res. Pap. INT* **INT-111.**

Megahan, W. F. (1982). Channel sediment storage behind obstructions in forested drainage basins draining the granitic bedrock of the Idaho batholith. *USDA For. Serv. Gen. Tech. Rep. PNW* **PNW-141,** 114–121.

Melillo, J. M., Aber, J. D., and Muratore, J. F. (1982). Nitrogen and lignin control of hardwood leaf litter decomposition dynamics. *Ecology* **63,** 621–626.

Melillo, J. M., Aber, J. D., Steudler, P. A., and Schimel, J. P. (1983). Denitrification potentials in a successional sequence of northern hardwood forest stands. *In* "Environmental Biogeochemistry" (R. Hallberg, ed.), pp. 217–228. Swed. Nat. Sci. Res. Counc., Stockholm.

Melin, J., Nommik, H., Lohm, U., and Flower-Ellis, J. (1983). Fertilizer nitrogen budget in a Scots pine ecosystem attained by using root-isolated plots and 15_N tracer technique. *Plant Soil* **74,** 249–263.

Metcalf, R. L., and Luckmann, W. H., eds. (1982). "Introduction to Insect Pest Management," 2nd ed. Wiley, New York.

Meybeck, M. (1977). Dissolved and suspended matter carried by rivers: Composition, time and space variations, and world balance. *In* "Interactions Between Sediments and Fresh Water" (H. L. Golterman, ed.), pp. 25–32. Junk, The Hague.

Meybeck, M. (1982). Carbon, nitrogen, and phosphorus transport by world rivers. *Am. J. Sci.* **282,** 401–450.

Middleton, K. R., and Smith, G. S. (1979). A comparison of ammoniacal and nitrate nutrition of perennial ryegrass through a thermodynamic model. *Plant Soil* **53,** 487–504.

Milburn, J. A. (1979). "Water Flow in Plants." Longmans, Green, New York.

Miller, D. E. (1969). Flow and retention of water in layered soils. *U.S., Agric. Res. Serv., Conserv. Res. Rep.* **13.**

Miller, D. H. (1967). Sources of energy for thermodynamically-caused transport of intercepted snow from forest crowns. *In* "Forest Hydrology" (W. E. Sopper and H. W. Lull, eds.), pp. 201–211. Pergamon, Oxford.

Miller, H. G. (1981). Forest fertilization: Some guiding concepts. *Forestry* **54,** 157–167.

Miller, H. G., Cooper, J. M., and Miller, J. D. (1976). Effect of nitrogen supply on nutrients in litterfall and crown leaching in a stand of corsican pine. *J. Appl. Ecol.* **13,** 233–248.

Miller, P. C., ed. (1980). "Carbon Balance in Northern Ecosystems and the Potential Effect of Carbon Dioxide Induced Climate Change." U.S. Dep. Energy, Washington, D.C.

Mills, H. B., Starrett, W. C., and Bellrose, F. C. (1966). Man's effect on the fish and wildlife of the Illinois River. *Biol. Notes, (Ill. Nat. Hist. Surv.)* **57.**

Minderman, G. (1968). Addition, decomposition and accumulation of organic matter in forests. *J. Ecol.* **56,** 355–362.

Minore, D., Smith, C. E., and Woollard, R. F. (1969). Effects of high soil density on seedling root growth of seven northwestern tree species. *USDA For. Serv. Res. Note PNW* **PNW-112,** 1–5.

Minshall, G. W., Petersen, R. G., Cummins, K. W., Bott, T. L., Sedell, J. R., Cushing, C. E., and Vannote, R. L. (1983). Interbiome comparison of stream ecosystem dynamics. *Ecol. Monogr.* **53,** 1–25.

Mitchell, M. J., and Parkinson, D. (1976). Fungal feeding of oribatid mites (Acari: Cryptostigmata) in an aspen woodland soil. *Ecology* **57,** 302–312.

Mitchell, R. G., and Martin, R. E. (1980). Fire and insects in pine culture of the Pacific Northwest. *In* "Fire and Forest Meterology" (R. E. Martin, R. L. Edmonds, D. A. Faulkner, J. B. Harrington, D. M. Fuquay, B. J. Stocks, and S. Barr, eds.), pp. 182–190. Soc. Am. For., Washington, D.C.

Mitchell, R. G., Waring, R. H., and Pitman, G. B. (1983). Thinning lodgepole pine increases tree vigor and resistance to mountain pine beetle. *For. Sci.* **29,** 204–211.

Mitchell, W. C., and Trimble, G. R., Jr. (1959). How much land is needed for the logging transport system? *J. For.* **57,** 10–12.

Mitsch, W. J., and Rust, W. G. (1984). Tree growth responses to flooding in a bottomland forest in northeastern Illinois. *For. Sci.* **30,** 499–510.

Mitsch, W. J., Dorge, C. L., and Wiemhoff, J. R. (1979). Ecosystem dynamics and a phosphorus budget of an alluvial cypress swamp in southern Illinois. *Ecology* **60,** 1116–1124.

Moeller, J. R., Minshall, G. W., Cummins, K. W., Petersen, R. C., Cushing, C. E., Sedell, J. R., Larson, R. A., and Vannote, R. L. (1979). Transport of dissolved organic carbon in streams of differing physiographic characteristics. *Org. Geochem.* **1,** 139–150.

Mohler, C. L., Marks, P. L., and Sprugel, D. G. (1978). Stand structure and allometry of trees during self-thinning of pure stands. *J. Ecol.* **66,** 599–614.

Mohr, H. (1983). Zur Faktorenanalyse des "Baumsterbens"—Bermerkungen eines Pflanzenphysiologen. *Allg. Forst- Jagdztg.* **154,** 105–110.

Monk, C. D. (1966). An ecological significance of evergreenness. *Ecology* **47,** 504–505.

Monteith, J. L. (1965). Evaporation and environment. *Symp. Soc. Exp. Biol.* **19,** 205–236.

Monteith, J. L. (1972). Solar radiation and productivity in tropical ecosystems. *J. Appl. Ecol.* **9,** 747–766.

Monteith, J. L. (1973). "Principles of Environmental Physics." Arnold, London.

Mooney, H. A. (1972). The carbon balance of plants. *Annu. Rev. Ecol. Syst.* **3,** 315–346.

Mooney, H. A., and Chu, C. (1974). Seasonal carbon allocation in *Heteromeles arbutifolia* a California evergreen shrub. *Oecologia* **14,** 295–306.

Mooney, H. A., and Gulmon, S. L. (1982). Constraints on leaf structure and function in reference to herbivory. *BioScience* **32,** 198–206.

Mooney, H. A., Ferrar, P. J., and Slatyer, R. O. (1978). Photosynthetic capacity and carbon allocation patterns in diverse growth forms of *Eucalyptus. Oecologia* **36,** 103–111.

Moore, A., and Swank, W. T. (1975). A model of water content and evaporation for hardwood leaf litter. *In* "Mineral Cycling in Southeastern Ecosystems" (F. G. Howell, J. B. Gentry, and M. H. Smith, eds.), pp. 58–69. U.S. Energy Res. Dev. Admin., Natl. Tech. Inf. Serv., U.S. Dep. Commerce, Springfield, Virginia.

Moore, G. C., and Martin, E. C. (1949). "Status of Beaver in Alabama." Ala. Dep. Conserv., Game, Fish Seafood Div., Walker Printing Co., Montgomery, Alabama.

Morrow, P. A., and LaMarche, V. C. (1978). Tree ring evidence for chronic insect suppression of productivity in subalpine Eucalyptus. *Science* **201,** 1244–1246.

Moulder, B. C., and Reichle, D. E. (1972). Significance of spider predation in the energy dynamics of forest floor arthropod communities. *Ecol. Monogr.* **42,** 473–498.

Moyers, J. L., Ranweiler, L. E., Hopf, S. B., and Korte, N. E. (1977). Evaluation of particulate trace species in southwest desert atmosphere. *Environ. Sci. Tech.* **11,** 789–795.

Mroz, G. D., Jurgensen, M. F., Harvey, A. E., and Larsen, M. J. (1980). Effects of fire on nitrogen in forest floor horizons. *Soil Sci. Soc. Am. J.* **44,** 395–400.

Munger, J. W. (1982). Chemistry of atmospheric precipitation in the north-central United States: Influence of sulfate, nitrate, ammonia and calcareous soil particulates. *Atmos. Environ.* **16,** 1633–1645.

Murphy, M. L., and Hall, J. D. (1981). Varied effects of clear-cut logging on predators and their habitat in small streams of the Cascade Mountains, Oregon. *Can. J. Fish. Aquat. Sci.* **38,** 137–145.

Naiman, R. J., and Melillo, J. M. (1984). Nitrogen budget of a subarctic stream altered by beaver (*Castor canadensis). Oecologia* **62,** 150–155.

Naiman, R. J., and Sedell, J. R. (1981). Stream ecosystem research in a watershed perspective. *Verh.—Int. Ver. Theor. Angew. Limnol.* **21,** 804–811.

Nakano, H. (1971). "Soil and Water Conservation Functions of Forest on Mountainous Lands," For. Influences Div. Rep. Jpn. Govt. For. Exp. Stn. Tsukuba, Japan.

Neftel, A., Oeschger, H., Schwander, J., Stauffer, B., and Zumbrunn, R. (1982). Ice core sample measurements give atmospheric CO_2 content during the past 40,000 yr. *Nature (London)* **295,** 220–223.

Neilson, R. E., Ludlow, M. M., and Jarvis, P. G. (1972). Photosynthesis in sitka spruce (*Picea sitchensis* (Bong.) Carr.). II. Response to temperature. *J. Appl. Ecol.* **9,** 721–745.

Newbold, J. D., Elwood, J. W., O'Neill, R. V., and VanWinkle, W. (1981). Measuring nutrient spiralling in streams. *Can. J. Fish. Aquat. Sci.* **38,** 860–863.

Nihlgard, B. (1971). Pedological influences of spruce planted on former beech forest soils in Scania, South Sweden. *Oikos* **22,** 302–314.

Nilsson, S. I., Miller, H. G., and Miller, J. D. (1982). Forest growth as a possible cause of soil and water acidification: An examination of the concepts. *Oikos* **39,** 40–49.

Nnyamah, J. U., and Black, T. A. (1977). Rates and patterns of water uptake in a Douglas-fir forest. *Soil Sci. Soc. Am. Proc.* **41,** 972–979.

Nobel, P. S. (1983). "Biophysical Plant Physiology and Ecology." Freeman, San Francisco, California.

Nohrstedt, H-O. (1982). Nitrogen fixation by free-living microorganisms in the soil of a mature oak-stand in Upland, Sweden. *Holarctic Ecol.* **5,** 20–26.

Nordin, C. F., and Meade, R. H. (1982). Deforestation and increased flooding of the upper Amazon. *Science* **215,** 426–427.

Norman, J. M., and Jarvis, P. G. (1975). Photosynthesis in Sitka spruce (*Picea sitchensis* (Bong. Carr.). V. Radiation penetration theory and a test case. *J. Appl. Ecol.* **12,** 839–878.

Nye, P. H. (1977). The rate-limiting step in plant nutrient absorption from soil. *Soil Sci.* **123,** 292–297.

Nye, P. H. (1981). Changes of pH across the rhizosphere induced by roots. *Plant Soil* **61,** 7–26.

Nygren, M., and Kellomaki, S. (1983). Effect of shading on leaf structure and photosynthesis in young birches, *Betula pendula* Roth. and *B. pubescens* Ehrh. *For. Ecol. Manage.* **7,** 119–132.

Odum, E. P. (1969). The strategy of ecosystem development. *Science* **164,** 262–270.

O'Loughlin, C. L. (1974). A study of tree root strength deterioration following clearfelling. *Can. J. For. Res.* **4,** 107–113.

O'Loughlin, C. L., and Watson, A. J. (1979). Root wood strength deterioration in radiata pine after clearfelling. *N. Z. J. For. Sci.* **9,** 284–293.

O'Loughlin, C. L., and Watson, A. J. (1981). Root wood strength deterioration in beech (*Nothofagus fusca* and *N. truncata*) after clearfelling. *N. Z. J. For. Sci.* **11,** 183–185.

O'Loughlin, C. L., and Ziemer, R. R. (1982). The importance of root strength and deterioration rates upon edaphic stability in steepland forests. *In* "Carbon Uptake and Allocation in Subalpine Ecosystems as a Key to Management" (R. H. Waring, ed.), IUFRO Workshop, pp. 70–78. For. Res. Lab., Oregon State University, Corvallis.

Olson, J. S. (1963). Energy storage and the balance of producers and decomposers in ecological systems. *Ecology* **44,** 322–331.

O'Neill, R. V., and DeAngelis, D. L. (1981). Comparative productivity and biomass relations of forest ecosystems. *In* "Dynamic Properties of Forest Ecosystems" (D. E. Reichle, ed.), pp. 411–449. Int. Biol. Programme No. 23, Cambridge Univ. Press, London and New York.

O'Neill, R. V., Ausmus, B. S., Jackson, D. R., Van Hook, R. I., Van Voris, P., Washburne, C., and Watson, A. P. (1977). Monitoring terrestrial ecosystems by analysis of nutrient export. *Water, Air, Soil Pollut.* **8,** 271–277.

Onuf, C. P., Teal, J. M., and Valiela, I. (1977). Interactions of nutrients, plant growth, and herbivory in a mangrove ecosystem. *Ecology* **58,** 514–526.

Oren, R., Thies, W. G., and Waring, R. H. (1985). Tree vigor and stand growth of Douglas-fir as influenced by laminated root rot. *Can. J. For. Res.* **15,** 985–988.

Orndorff, K. A., and Lang, G. E. (1981). Leaf litter redistribution in a West Virginia hardwood forest. *J. Ecol.* **69,** 225–235

Ostman, N. L., and Weaver, G. T. (1982). Autumnal nutrient transfer by retranslocation, leaching and litterfall in a chestnut oak forest in southern Illinois. *Can. J. For. Res.* **12,** 40–51.

Ostrofsky, A., and Shigo, A. L. (1984). Relationship between canker size and wood starch in American chestnut. *Eur. J. For. Pathol.* **14,** 65–68.

Outlaw, W. H., Jr. (1983). Current concepts on the role of potassium in stomatal movements. *Physiol. Plant.* **59,** 302–311.

Owen, D. F. (1980). How plants may benefit from the animals that eat them. *Oikos* **35,** 230–235.

Palmer, L. (1976). River management criteria for Oregon and Washington. *In* "Geomorphology and Engineering" (D. R. Coates, ed.), pp. 329–346. Dowden, Hutchinson & Ross, Inc., Stroudsburg, Pennsylvania.

Panshin, A. J., de Zeeuw, C., and Brown, H. P. (1964). "Textbook of Wood Technology," 2nd ed., Vol. 1. McGraw-Hill, New York.

Pardé, J. (1980). Forest biomass. *For. Abstr.* **41,** 343–362.

Parker, G. G. (1983). Throughfall and stemflow in the forest nutrient cycle. *Adv. Ecol. Res.* **13,** 57–133.

Parker, J., and Houston, D. R. (1971). Effects of repeated defoliation on root and root collar extractives of sugar maple trees. *For. Sci.* **17,** 91–95.

Parker, W. C., Pallardy, S. G., Hinckley, T. M., and Teskey, R. O. (1982). Seasonal changes in tissue water relations of three woody species of the *Quercus Carya* forest type. *Ecology* **63,** 1259–1267.

Parkinson, D., Domsch, K. H., and Anderson, J. P. E. (1978). Die Entwicklung mikrobieller Biomassen im organischen Horizont eines Fichtenstandortes. *Oecol. Plant.* **13,** 355–366.

Parmentier, G., and Remacle, J. (1981). Production de litière et dynamisme de retour au sol des éléments minéraux par l'intermédiaire des feuilles de hetre et des aiguilles d'Epicea en Haute Ardenne. *Rev. Ecol. Biol. Sol* **18,** 159–177.

Parsons, D. J., and DeBenedetti, S. H. (1979). Impact of fire suppression on a mixed-conifer forest. *For. Ecol. Manage.* **2,** 21–33.

Pastor, J., and Bockheim, J. G. (1984). Distribution and cycling of nutrients in an aspen-mixed-hardwood-spodosol ecosystem in northern Wisconsin. *Ecology* **65,** 339–353.

Pastor, J., Aber, J. D., McClaugherty, C. A., and Melillo, J. M. (1984). Aboveground production and N and P cycling along a nitrogen mineralization gradient on Blackhawk Island, Wisconsin. *Ecology* **65,** 256–268.

Pate, J. S. (1980). Transport and partitioning of nitrogenous solutes. *Annu. Rev. Plant Physiol.* **31,** 313–340.

Patric, J. H., Douglass, J. E., and Hewlett, J. D. (1965). Soil water absorption by mountain and Piedmont forests. *Soil Sci. Soc. Am. Proc.* **29,** 303–308.

Paul, E. A., and Kucey, R. M. N. (1981). Carbon flow in plant microbial associations. *Science* **213,** 473–474.

Pearson, L. C., and Lawrence, D. B. (1958). Photosynthesis in aspen bark. *Am. J. Bot.* **45,** 383–387.

Pedro, G., Jamagne, M., and Begon, J. C. (1978). Two routes in genesis of strongly differentiated acid soils under humid, cool-temperate conditions. *Geoderma* **20,** 173–189.

Peet, R. K. (1981). Changes in biomass and production during secondary forest succession. *In* "Forest Succession: Concepts and Application" (D. C. West, H. H. Shugart, and D. B. Botkin, eds.), pp. 324–338. Springer-Verlag, Berlin and New York.

Peirson, D. H., Cawse, P. A., Salmon, L., and Cambray, R. S. (1973). Trace elements in the atmospheric environment. *Nature (London)* **241,** 252–256.

Penman, H. L. (1948). Natural evaporation from open water, bare soil and grass. *Proc. R. Soc. Agron.* **199,** 120–145.

Penning de Vries, F. W. T. (1975). The cost of maintenance processes in plant cells. *Ann. Bot. (London)* [N.S.] **39,** 77–92.

Pereira, J. S., and Kozlowski, T. T. (1977). Influence of light intensity, temperature, and leaf area on stomatal aperture and water potential of woody plants. *Can. J. For. Res.* **7,** 145–153.

Persson, H. (1978). Root dynamics in a young Scots pine stand in central Sweden. *Oikos* **30,** 508–519.

Peterken, G. F. (1966). Mortality of holly (*Ilex aquifolium*) seedlings in relation to natural regeneration in the New Forest. *J. Ecol.* **54,** 259–269.

Peterson, B. J. (1981). Perspectives on the importance of the oceanic particulate flux in the global carbon cycle. *Ocean Sci. Eng.* **6,** 71–108.

Peterson, E. W., and Tingey, D. T. (1980). An estimate of the possible contribution of biogenic sources to airborne hydrocarbon concentrations. *Atmos. Environ.* **14,** 79–81.

Petty, J. A., and Worrell, R. (1981). Stability of coniferous tree stems in relation to damage by snow. *Forestry* **54,** 115–128.

Pew, J. C. (1967). Lignin: An inviting storehouse. *Agric. Sci. Rev.* **5,** 1–7.

Pharis, R. P., Hellmers, H., and Shuurmans, E. (1972). The decline and recovery of photosynthesis of ponderosa pine seedlings subjected to low, but above freezing temperatures. *Can. J. Bot.* **50,** 1965–1970.

Philip, J. R. (1969). Theory of infiltration. *In* "Advances in Hydroscience" (V. T. Chow, ed.), pp. 215–305. Academic Press, New York.

Phillipson, J., Abel, R., Steel, J., and Woodell, S. R. J. (1978). Earthworm numbers, biomass and respiratory metabolism in a beech woodland—Wytham Woods, Oxford. *Oecologia* **33,** 291–309.

Phipps, R. L. (1979). Simulation of wetlands forest vegetation dynamics. *Ecol. Model.* **7,** 257–288.

Piene, H. (1980). Effects of insect defoliation on growth and foliar nutrition of young balsam fir. *For. Sci.* **26,** 665–673.

Piene, H., and Percy, K. E. (1984). Changes in needle morphology, anatomy and mineral content during the recovery of protected balsam fir initially defoliated by spruce bud-worm. *Can. J. For. Res.* **14,** 238–245.

Pierson, T. C. (1980). Iezometric response to rainstorms in forested hillslope drainage depressions. *J. Hydrol. (N. Z.)* **19,** 1–10.

Pitman, G. B., Larsson, S., and Tenow, O. (1982). Stem growth efficiency: An index of susceptibility to bark beetle and sawfly attack. *In* "Carbon Uptake and Allocation in Subalpine Ecosystems as a Key to Management" (R. H. Waring, ed.), IUFRO Workshop, pp. 52–56. For. Res. Lab., Oregon State University, Corvallis.

Polster, H., and Fuchs, S. (1963). Winterassimilation und -atmung der Kiefer (*Pinus silvestris* L.) im mitteldeutschen Binnenlandklima. *Arch. Forstwes.* **12,** 1013–1023.

Post, W. H., Emanuel, W. R., Zinke, P. S., and Stangenberger, A. G. (1982). Soil carbon pools and world life zones. *Nature (London)* **298,** 156–159.

Potter, G. L., Ellsaesser, H. W., MacCracken, M. C., and Luther, F. M. (1975). Possible climatic impact of tropical deforestation. *Nature (London)* **258,** 697–698.

Potts, M. J. (1978). Deposition of air-borne salt on *Pinus radiata* and the underlying soil. *J. Appl. Ecol.* **15,** 543–550.

Powers, R. F. (1980). Mineralizable soil nitrogen as an index of nitrogen availability to forest trees. *Soil Sci. Soc. Am. J.* **44,** 1314–1320.

Powles, S. B., and Osmond, C. B. (1978). Inhibition of the capacity and efficiency of photosynthesis in bean leaflets illuminated in a CO_2-free atmosphere at low oxygen: A possible role of photorespiration. *Aust. J. Plant Physiol.* **5,** 619–629.

Prenzel, J. (1979). Mass flow to the root system and mineral uptake of a beech stand calculated from 3-year field data. *Plant Soil* **51,** 39–49.

Price, P. W., Bouton, C. E., Gross, P., McPheron, B. A., Thompson, J. N., and Weis, A. E. (1980). Interactions among three trophic levels: Influence of plants on interactions between insect herbivores and natural enemies. *Annu. Rev. Ecol. Syst.* **11,** 41–65.

Priestley, C. A. (1970). Carbohydrate storage and utilization. *In* "Physiology of Tree Crops" (L. C. Luckwill and C. V. Cutting, eds.), pp. 113–125. Academic Press, New York.

Prinz, B., Krause, G. H. M., and Stratmann, H. (1982). "Waldschaden in der Vorlaufiger Bundesrepublik," LIS Rep. No. 28. Landesanstalt fur Immissionsschutz Nordrhein-Westfalen.

Pritchett, W. L., and Comerford, N. B. (1983). Nutrition and fertilization of slash pine. *In* "The Managed Slash Pine Ecosystem" E. L. Stone, ed.), pp. 69–90. University of Florida, Gainesville.

Proctor, J., and Woodell, S. R. J. (1975). The ecology of serpentine soils. *Adv. Ecol. Res.* **9,** 255–366.

Prudhomme, T. I. (1983). Carbon allocation to antiherbivore compounds in a deciduous and an evergreen subarctic shrub species. *Oikos* **40,** 344–356.

Puckett, L. J. (1982). Acid rain, air pollution and tree growth in southeastern New York. *J. Environ. Qual.* **11,** 376–381.

Puritch, G. S. (1971). Water permeability of the wood of Grand fir (*Abies grandis* (Dougl.) Lindl.) in relation to infestation by the balsam woolly aphid, *Adelges piceae* (Ratz.). *J. Exp. Bot.* **22,** 936–945.

Putz, F. E., and Milton, K. (1982). Tree mortality rates on Barro Colorado Island. *In* "The Ecology of a Tropical Forest" (E. G. Leigh, Jr., A. S. Rand, and D. M. Windsor, eds.), pp. 95–100. Smithson. Inst., Washington, D.C.

Raffa, K. F., and Berryman, A. A. (1982). Accumulation of monoterpenes and associated volatiles following inoculation of grand fir with a fungus transmitted by the fir engraver, *Scolytus ventralis* (Coeoptera:Scolytidae). *Can. J. Entomol.* **114,** 797–810.

Raffa, K. F., and Berryman, A. A. (1983). Physiological differences between lodgepole pines resistant and susceptible to the mountain pine beetle and associated microorganisms. *Environ. Entomol.* **11,** 486–492.

Raison, R. J. (1979). Modification of the soil environment by vegetation fires, with particular reference to nitrogen transformations. A review. *Plant Soil* **51,** 73–108.

Ranney, J. W., and Cushman, J. H. (1982). "Short Rotation Woody Crops Program," Publ. No. 2000. Environ. Sci. Div., Oak Ridge Natl. Lab., Oak Ridge, Tenn.

Ratnayake, M., Leonard, R. T., and Menge, J. A. (1978). Root exudation in relation to supply of phosphorus and its possible relevance to mycorrhizal formation. *New Phytol.* **81,** 543–552.

Raunemaa, T., Hautojarvi, A., Kaisla, K., Gerlander, M., Erkinjuntii, R., Tuomi, T., Hari, P., Kellomaki, S., and Katainen, H.-S. (1982). The effects on forest of air pollution from energy production: Application of the PIXE method to elemental analysis of pine needles from the years 1959–1979. *Can. J. For. Res.* **12,** 384–390.

Raven, P. H., Evert, R. F., and Curtis, H. (1981). "Biology of Plants." Worth Publ., New York.

Reed, K. L. (1982). Simulation of forest stands in a hypothetical watershed using models and cartographic analyses. *In* "Forest Succession and Stand Development Research in the Northwest" (J. E. Means, ed.), pp. 31–43. For. Res. Lab., Oregon State University, Corvallis.

Reed, K. L., Hamerly, E. R., Dinger, B. E., and Jarvis, P. G. (1976). An analytical model for field measurement of photosynthesis. *J. Appl. Ecol.* **13,** 925–942.

Reeves, F. B., Wagner, D., Moorman, T., and Kiel, J. (1979). The role of endomycorrhizae in revegetation practices in the semi-arid west. I. A comparison of incidence of mycorrhizae in severely disturbed vs. natural environments. *Am. J. Bot.* **66,** 6–13.

Reich, P. B., and Hinckley, T. M. (1980). Water-relations, soil fertility and plant nutrient composition of pygmy oak ecosystem. *Ecology* **61,** 400–416.

Reichelt, G. (1983). Untersuchungen zum Nadelbaumsterben in der Region Schwarzwald-Baar-Heuberg. *Allg. Forst- Jagdztg.* **154,** 66–75.

Reichle, D. E., Goldstein, R. A., van Hook, R. I., and Dodson, G. J. (1973). Analysis of insect consumption in a forest canopy. *Ecology* **54,** 1076–1084.

Reid, L. R. (1981). "Sediment Production from Gravel-Surfaced Roads, Clearwater Basin." Washington Fish. Res. Inst., University of Washington, Seattle.

Reid, L. M., and Dunne, T. (1984). Sediment production from forest road surfaces. *Water Resour. Res.* **20,** 1753–1761.

Reiners, W. A. (1972). Structure and energetics of three Minnesota forests. *Ecol. Monogr.* **42,** 71–94.

Reiners, W. A., and Reiners, N. M. (1970). Energy and nutrient dynamics of forest floors in three Minnesota forests. *J. Ecol.* **58,** 497–519.

Reiners, W. A., Worley, I. A., and Lawrence, D. B. (1971). Plant diversity in a chronosequence at Glacier Bay, Alaska. *Ecology* **52,** 55–69.

Reynolds, J. F., and Knight, D. H. (1973). The magnitude of snowmelt and rainfall interception by litter in lodgepole pine and spruce-fir forests in Wyoming. *Northwest Sci.* **47,** 50–60.

Rhoades, D. F., and Cates, R. G. (1976). Toward a general theory of plant antiherbivore chemistry. *Recent Adv. Phytochem.* **10,** 168–213.

Richter, D. D. (1983). Comment on "acid precipitation in historical perspective" and "effects of acid precipitation." *Environ. Sci. Technol.* **17,** 568–570.

Richter, D. D., Ralston, C. W., and Harms, W. R. (1982). Prescribed fire: Effects on water quality and forest nutrient cycling. *Science* **215,** 661–663.

Riekirk, H. (1983). Impacts of silviculture on flatwoods runoff, water quality and nutrient budgets. *Water Resour. Bull.* **19,** 73–79.

Rishbeth, J. (1951). Observations on the biology of *Fomes annosus* with particular reference to East

Anglian pine plantations. III. Natural and experimental infection of pines, and some factors affecting severity of the disease. *Ann. Bot. (London)* [N.S.] **15,** 221–246.

Robertson, G. P., and Tiedje, J. M. (1984). Denitrification and nitrous oxide production in successional and old-growth Michigan forests. *Soil Sci. Soc. Am. J.* **48,** 383–388.

Rodhe, H., Crutzen, P., and Vanderpol, A. (1981). Formation of sulfuric and nitric acid in the atmosphere during long-range transport. *Tellus* **33,** 132–141.

Rogers, J. J. (1973). Design of a system for predicting effects of vegetation manipulation on water yield in the Salt-Verde basin. Ph.D. Thesis, University of Arizonia, Tucson.

Rogers, R., and Hinckley, T. M. (1979). Foliage weight and area related to current sapwood area in oak. *For. Sci.* **25,** 298–303.

Romell, L. G. (1935). Ecological problems of the humus layer in the forest. *Mem.—N.Y., Agric. Exp. Stn. (Ithaca)* **170,** 1–28.

Romme, W. H. (1982). Fire and landscape diversity in subalpine forests of Yellowstone National Park. *Ecol. Monogr.* **52,** 199–221.

Rook, D. A. (1969). The influence of growing temperature on photosynthesis and respiration of *Pinus radiata* seedlings. *N. Z. J. Bot.* **7,** 43–55.

Rose, S. L., and Youngberg, C. T. (1981). Tripartite associations in snowbrush (*Ceanothus velutinus*): Effect of vesicular-arbuscular mycorrhizae on growth, nodulation, and nitrogen fixation. *Can. J. Bot.* **59,** 34–39.

Rosen, K., and Lindberg, T. (1980). Biological nitrogen fixation in coniferous forest watershed areas in central Sweden. *Holarctic Ecol.* **3,** 137–140.

Rosenzweig, M. L. (1968). Net primary productivity of terrestrial communities: Prediction from climatological data. *Am. Nat.* **102,** 67–74.

Roskoski, J. P. (1980). Nitrogen fixation in hardwood forests of the northeastern United States. *Plant Soil* **54,** 33–44.

Ross, B. A., Bray, J. R., and Marshall, W. H. (1970). Effects of long-term deer exclusion on a *Pinus resinosa* forest in north-central Minnesota. *Ecology* **51,** 1088–1093.

Rosswall, T. (1981). The biogeochemical nitrogen cycle. *In* "Some Perspectives of the Major Biogeochemical Cycles" (G. E. Likens, ed.), pp. 25–49. Wiley, New York.

Rosswall, T. (1982). Microbiological regulation of the biogeochemical nitrogen cycle. *Plant Soil* **67,** 15–34.

Rothacher, J. (1963). Net precipitation under a Douglas-fir forest. *For. Sci.* **9,** 423–429.

Rothacher, J. (1971). Regimes of streamflow and their modification by logging. *In* "Forest Land Uses and Stream Environments" (J. T. Krygier and J. D. Hall, eds.), pp. 40–54. Oregon State Univ. Press, Corvallis.

Ruedemann, R., and Schoonmaker, W. J. (1938). Beaver dams as geologic agents. *Science* **88,** 523–525.

Rühling, A., and Skarby, L. (1979). "Landsomfattande kartering av regionala tungmetallhalter i mossa," Statens Naturvårdsverk Paper No. 1191.

Rundel, P. W., and Stecker, R. E. (1977). Morphological adaptations of tracheid structure to water stress gradients in the crown of *Sequoiadendron giganteum*. *Oecologia* **27,** 135–139.

Running, S. W. (1980). Environmental and physiological control of water flux through *Pinus contorta*. *Can. J. For. Res.* **10,** 82–91.

Running, S. W. (1984). Documentation and preliminary validation of H2OTRANS and DAYTRANS, two models for predicting transpiration and water stress in western coniferous forests. *U.S., For. Serv., Rocky Mt. For. Range Exp. Stn., Res. Pap. RM* **RM-252,** 1–45.

Rutter, A. J. (1963). Studies in the water relations of *Pinus sylvestris* in plantation conditions. *J. Ecol.* **51,** 191–203.

Rutter, A. J., Morton, A. J., and Robins, P. C. (1975). A predictive model of rainfall interception in forests. II. Generalization of the model and comparison with observations in some coniferous and hardwood stands. *J. Appl. Ecol.* **12,** 367–380.

Ryan, D. F., and Bormann, F. H. (1982). Nutrient resorption in northern hardwood forests. *Bio-Science* **32,** 29–32.

Sackett, S. S. (1975). Scheduling prescribed burns for hazard reduction in the southeast. *J. For.* **73,** 143–147.

Sagan, C., Toon, O. B., and Pollack, J. B. (1979). Anthropogenic albedo changes and the Earth's climate. *Science* **206,** 1363–1368.

Saggar, S., Bettany, J. R., and Stewart, J. W. B. (1981). Sulfur transformations in relation to carbon and nitrogen in incubated soils. *Soil Biol. Biochem.* **13,** 499–511.

Sakai, A. (1966). Temperature fluctuation in wintering trees. *Physiol. Plant.* **19,** 105–114.

Salati, E., and Vose, P. B. (1984). Amazon Basin: A system in equilibrium. *Science* **225,** 129–138.

Salati, E., Dall'Olio, A., Matsui, E., and Gat, J. R. (1979). Recycling of water in the Amazon Basin: An isotopic study. *Water Resour. Res.* **15,** 1250–1258.

Salati, E., Sylvester-Bradley, R., and Victoria, R. L. (1982). Regional gains and losses of nitrogen in the Amazon Basin. *Plant Soil* **67,** 367–376.

Salisbury, F. B., and Ross, C. W. (1978). "Plant Physiology," 2nd ed. Wadsworth Publ. Co., Belmont, California.

Same, B. I., Robson, A. D., and Abbott, L. K. (1983). Phosphorus, soluble carbohydrates and endomycorrhizal infection. *Soil Biol. Biochem.* **15,** 593–597.

Sanchez, P. A., Bandy, D. E., Villachica, J. H., and Nicholaides, J. J. (1982a). Amazon Basin Soils: Management for continuous crop production. *Science* **216,** 821–827.

Sanchez, P. A., Gichuru, M. P., and Katz, L. B. (1982b). Organic matter in major soils of the tropical and temperate regions. *Int. Congr. Soil Sci., 12th, 1982,* Vol. 1, pp. 99–114.

Sands, R., Greacen, E. L., and Gerard, C. J. (1979). Compaction of sandy soils in radiata pine forests. I. A penetrometer study. *Aust. J. Soil Res.* **17,** 101–113.

Sanford, R. L., Jr., Saldarriaga, J., Clark, K. E., Uhl, C., and Herrera, R. (1985). Amazon rain-forest fires. *Science* **227,** 53–55.

Santantonio, D. (1982). Production and turnover of fine roots of mature Douglas-fir in relation to site. Ph.D. Dissertation, Oregon State University, Corvallis.

Sartwell, C., and Stevens, R. E. (1975). Mountain pine beetle in ponderosa pine. *J. For.* **73,** 136–140.

Schlesinger, W. H. (1977). Carbon balance in terrestrial detritus. *Annu. Rev. Ecol. Syst.* **8,** 51–81.

Schlesinger, W. H. (1978). Community structure, dynamics and nutrient cycling in the Okefenokee cypress swamp-forest. *Ecol. Monogr.* **48,** 43–65.

Schlesinger, W. H. (1984). Soil organic matter: A source of atmospheric CO_2. *In* "The Role of Terrestrial Vegetation in the Global Carbon Cycle" (G. M. Woodwell, ed.), pp. 111–127. Wiley, London.

Schlesinger, W. H. (1986). Changes in soil carbon storage and associated properties with disturbance and recovery. *In* "The Global Carbon Cycle" (J. R. Trabalka and D. E. Reichle, eds.), pp. 194–220. Springer-Verlag, Berlin and New York.

Schlesinger, W. H., and Hasey, M. M. (1980). The nutrient content of precipitation, dry fallout, and intercepted aerosols in the chaparral of southern California. *Am. Midl. Nat.* **103,** 114–122.

Schlesinger, W. H., and Hasey, M. M. (1981). Decomposition of chaparral shrub foliage: Losses or organic and inorganic constituents from deciduous and evergreen leaves. *Ecology* **62,** 762–774.

Schlesinger, W. H., and Marks, P. L. (1977). Mineral cycling and the niche of Spanish moss, *Tillandsia usneoides* L. *Am. J. Bot.* **64,** 1254–1262.

Schlesinger, W. H., and Melack, J. M. (1981). Transport of organic carbon in the world's rivers. *Tellus* **33,** 172–187.

Schlesinger, W. H., Gray, J. T., and Gilliam, F. S. (1982). Atmospheric deposition processes and their importance as sources of nutrients in a chaparral ecosystem of southern California. *Water Resour. Res.* **18,** 623–629.

Schoeneweiss, D. F. (1975). Predisposition, stress, and plant disease. *Annu. Rev. Phytopathol.* **13,** 193–211.

Scholander, P. F., Hammel, H. T., Bradstreet, E. D., and Hemmingsen, E. A. (1965). Sap pressure in vascular plants. *Science* **148,** 339–346.

Schroeder, M. J., and Buck, C. C. (1970). Fireweather. *U.S., Dep. Agric., Agric. Handb.* **360.**

Schroeder, P. E., McCandlish, B., Waring, R. H., and Perry, D. H. (1982). The relationship of maximum canopy leaf area of forest growth in eastern Washington. *Northwest Sci.* **56,** 121–130.

Schroter, H. (1983). Entwicklung des Gesundheitszustandes von Tannen und Fichten auf Beobachtungsflachen des FVA in Badenwurttemberg. *Allg. Forst- Jagdztg.* **154,** 123–131.

Schultz, J. C., and Baldwin, I. T. (1982). Oak leaf quality declines in response to defoliation by gypsy moth larvae. *Science* **217,** 149–151.

Schultz, R. C., and Kormanik, P. P. (1982). Vesicular-arbuscular mycorrhiza and soil fertility influence mineral concentrations in seedlings of eight hardwood species. *Can. J. For. Res.* **12,** 829–834.

Schulze, E. D. (1981). Carbon gain and wood production in trees of deciduous beech (*Fagus silvatica*) and trees of evergreen spruce (*Picea excelsa*). *Mitt. Forstl. Bundesversuchsanst.* **142,** 105–123.

Schulze, E. D., Lange, O. L., Kappen, L., Buschbom, U., and Evenari, M. (1973). Stomatal responses to changes in temperature at increasing water stress. *Planta* **110,** 29–42.

Schulze, E. D., Lange, O. L., Evenari, M., Kappen, L., and Buschbom, U. (1974). The role of air humidity and leaf temperature in controlling stomatal resistance of *Prunus armeniaca* L. under desert conditions. I. A simulation of the daily course of stomatal resistance. *Oecologia* **17,** 159–170.

Schulze, E. D., Fuchs, M. I., and Fuchs, M. (1977). Spacial distribution of photosynthetic capacity and performance in a montane spruce forest of northern Germany. I. Biomass distribution and daily CO_2 uptake in different crown layers. *Oecologia* **29,** 43–61.

Schütt, P. (1981). Ursache und Ablauf des Tannensterbens. Versuch einer Zwischenbilanz. *Forstwiss. Centralbl.* **100,** 286–287.

Sears, S. O., and Langmuir, D. (1982). Sorption and mineral equilibria controls on moisture chemistry in a c-horizon soil. *J. Hydrol.* **56,** 287–308.

Seastedt, T. R., and Crossley, D. A. (1980). Effects of microarthropods on the seasonal dynamics of nutrients in forest litter. *Soil Biol. Biochem.* **12,** 337–342.

Seastedt, T. R., and Tate, C. M. (1981). Decomposition rates and nutrient contents of arthropod remains in forest litter. *Ecology* **62,** 13–19.

Sedell, J. R., and Froggatt, J. L. (1984). Importance of streamside forests to large rivers: The isolation of the Willamette River, Oregon, U.S.A., from its floodplain by snagging and streamside forest removal. Verh. Int. Verein. Limnol. *Int. Assoc. Theor. Appl. Limnol.* **22,** 1828–1834.

Sedell, J. R., and Luchessa, K. J. (1982). Using the historical record as an aid to salmonid habitat enhancement. *In* "Acquisition and Utilization of Aquatic Habitat and Inventory Information" (N. B. Armantrout, ed.), pp. 210–223. Am. Fish. Soc., Bethesda, Maryland.

Sedell, J. R., and Swanson, F. J. (1984). Ecological Characteristics of Streams in Old-growth Forests of the Pacific Northwest. *In* "Proc. Fish and Wildlife Relationships on Old-growth Forests" (W. R. Meeham, T. R. Merrill, and T. Hanley, eds.). Am. Inst. Fish. Res. Biol., pp. 9–16.

Sedell, J. R., Triska, F. J., Hall, J. D., Anderson, N. H., and Lyford, J. H. (1974). Sources and fates of organic inputs in coniferous forest streams. *In* "Integrated Research in the Coniferous Forest Biome" (R. H. Waring and R. L. Edmonds, eds.), Conifer. For. Biome Bull. No. 5, pp. 57–69. Sch. Nat. Resour., University of Washington, Seattle.

Sedell, J. R., Everest, R. H., and Swanson, F. J. (1982). Fish habitat and streamside management:

Past and present. *In* "Proceedings of the Society of American Foresters," pp. 244–255. Soc. Am. For., Bethesda, Maryland.

Sedell, J. R., Yuska, J. E., and Speaker, R. W. (1984). Habitats and salmonid distribution in pristine sediment-rich river valley systems: S. Fork Hoh and Queet River, Olympic National Park. *In* "Proc. Fish and Wildlife Relationships on Old-growth Forests," (W. R. Meeham, T. R. Merrill, and T. Hanley, eds.) Am. Inst. Fish. Res. Biol., pp. 33–46.

Seif el Din, A., and Obeid, M. (1971). Ecological studies of the vegetation of the Sudan IV. The effect of simulated grazing on the growth of *Acacia senegal* (L.) Willd. seedlings. *J. Appl. Ecol.* **8,** 211–216.

Setliff, E. C., Sullivan, J. A., and Thompson, J. H. (1975). *Scleroderris lagerbergii* in large red and Scots pine trees in New York. *Plant Dis. Rep.* **59,** 380–381.

Seton, E. T. (1929). "Lives of Game Animals," Vol. 4, Part 2. Doubleday, Doran & Co., Garden City, New York.

Sexstone, A. J., Parkin, T. B., and Tiedje, J. M. (1985). Temporal response of soil denitrification rates to rainfall and irrigation. *Soil Sci. Soc. Am. J.* **49,** 99–103.

Shackleton, N. J. (1977). Carbon-13 in Uvigerina: Tropical rainforest history and the equatorial Pacific Carbonate dissolution cycles. *In* "The Fate of Fossil Fuel CO_2 in the Oceans" (N. R. Andersen and A. Malahoff, eds.), pp. 401–477. Plenum, New York.

Sharma, M. L., and Luxmoore, R. J. (1979). Soil spatial variability and its consequences on simulated water balance. *Water Resour. Res.* **15,** 1567–1573.

Shaw, S. P., and Fredine, C. G. (1956). Wetlands of the United States. *Circ.—U.S. Fish Wildl. Serv.* **39,** 1–67.

Shear, C. B., Crane, H. L., and Myers, A. T. (1946). Nutrient-element balance: A fundamental concept in plant nutrition. *Proc. Am. Soc. Hortic. Sci.* **47,** 239–248.

Sheriff, D.W., and Whitehead, D. (1984). Photosynthesis and wood structure in *Pinus radiata* D. Don during dehydration and immediately after rewatering. *Plant, Cell Environ.* **7,** 53–62.

Shugart, H. H., and Noble, I. R. (1981). A computer model of succession and fire response of the high altitude *Eucalyptus* forests of the Brindabella Range, Australian Capital Territory. *Aust. J. Ecol.* **6,** 149–164.

Shugart, H. H., and West, D. C. (1977). Development of an Appalanchian deciduous forest succession model and its application to assessment of the impact of the chestnut blight. *J. Environ. Manage.* **5,** 161–179.

Shugart, H. H., and West, D. C. (1980). Forest succession models. *BioScience* **30,** 308–313.

Shugart, H. H., and West, D. C. (1981). Long-term dynamics of forest ecosystems. *Am. Sci.* **66,** 647–652.

Shukla, J., and Mintz, Y. (1982). Influence of land-surface evapotranspiration on the Earth's climate. *Science* **215,** 1498–1501.

Siau, J. F. (1971). "Flow in Wood," Syracuse Wood Sci., Ser. I. Syracuse Univ. Press, Syracuse, New York.

Silkworth, D. R., and Grigal, D. F. (1982). Determining and evaluating nutrient losses following whole-tree harvesting of aspen. *Soil Sci. Soc. Am. J.* **46,** 626–631.

Silvester, W. B., Carter, D. A., and Sprent, J. I. (1979). Nitrogen input by *Lupinus* and *Coriaria* in *Pinus radiata* forest in New Zealand. *In* "Symbiotic Nitrogen Fixation in the Management of Temperate Forests" (J. C. Gordon, C. T. Wheeler, and D. A. Perry, eds.), pp. 253–265. For. Res. Lab., Oregon State University, Corvallis.

Silvester, W. B., Sollins, P., Verhoeven, T., and Cline, S. P. (1982). Nitrogen fixation and acetylene reduction in decaying conifer boles: Effects of incubation time, aeration and moisture content. *Can. J. For. Res.* **12,** 646–652.

Singh, J. S., and Gupta, S. R. (1977). Plant decomposition and soil respiration in terrestrial ecosystems. *Bot. Rev.* **43,** 449–528.

Skene, D. S., and Balodis, V. (1968). A study of vessel length in *Eucalyptus obligua* L'Hérit. *J. Exp. Bot.* **19,** 825–830.

Slatyer, R. O. (1967). "Plant-Water Relationships." Academic Press, New York.

Slatyer, R. O., and Morrow, P. A. (1977). Altitudinal variation in photosynthetic characteristics of snow gum, *Eucalyptus pauciflora* Sieb.ex Spreng. I. Seasonal changes under field conditions in the Snowy Mountains area of south-eastern Australia. *Aust. J. Bot.* **25,** 1–20.

Small, E. (1972). Photosynthetic rates in relation to nitrogen recycling as an adaptation to nutrient deficiency in peat bog plants. *Can. J. Bot.* **50,** 2227–2233.

Smith, B. H. (1978). "Rock Springs District Stream Survey, 1975–1977," Completion Rep. Bur. Land Manage., Rock Springs District, Wyoming.

Smith, D. G. (1976). Effect of vegetation on lateral migration of anastomosed channels of a glacier meltwater river. *Geol. Soc. Am. Bull.* **87,** 857–860.

Smith, D. H. (1946). Storm damage in New England forests. M.S. Thesis, Yale University, New Haven, Connecticut.

Smith, N. J. H. (1981). "Man, Fishes, and the Amazon." Columbia Univ. Press, New York.

Smith, R. B., Waring, R. H., and Perry, D. A. (1981). Interpreting foliar analyses from Douglas-fir as weight per unit of leaf area. *Can. J. For. Res.* **11,** 593–598.

Smith, V. R. (1979). Evaluation of a resin-bag procedure for determining plant-available P in organic, volcanic soils. *Plant Soil* **53,** 245–249.

Smith, W. H. (1976). Character and significance of forest tree root exudates. *Ecology* **57,** 324–331.

Smith, W. H., and Siccama, T. G. (1981). The Hubbard Brook Ecosystem Study: Biogeochemistry of lead in the northern hardwood forest. *J. Environ. Qual.* **10,** 323–333.

Snell, J. K. A., and Brown, J. K. (1978). Comparison of tree biomass estimators-dbh and sapwood area. *For. Sci.* **24,** 455–457.

Söderström, B. E. (1979). Seasonal fluctuations of active fungal biomass in horizons of a podzolized pine forest soil in central Sweden. *Soil Biol. Biochem.* **11,** 149–154.

Sohlenius, B. (1980). Abundance, biomass and contribution to energy flow by soil nematodes in terrestrial ecosystems. *Oikos* **34,** 186–194.

Sollins, P. (1982). Input and decay of coarse woody debris in coniferous stands in western Oregon and Washington. *Can. J. For. Res.* **12,** 18–28.

Sollins, P., Grier, C. C., McCorison, F. M., Cromack, K., Fogel, R., and Fredriksen, R. L. (1980). The internal element cycles of an old growth Douglas-fir ecosystem in western Oregon. *Ecol. Monogr.* **50,** 261–285.

Sollins, P., Goldstein, R. A., Mankin, J. B., Murphy, C. E., and Swartzman, G. L. (1981). Analysis of forest growth and water balance using complex ecosystem models. *In* "Dynamic Properties of Forest Ecosystems" (D. E. Reichle, ed.), Int. Biol. Programme No. 23, pp. 537–565. Cambridge Univ. Press, London and New York.

Sørensen, L. H. (1974). Rate of decomposition of organic matter in soil as influenced by repeated air drying-rewetting and repeated additions of organic material. *Soil Biol. Biochem.* **6,** 287–292.

Soulides, D. A., and Allison, F. E. (1961). Effect of drying and freezing soils on carbon dioxide production, available mineral nutrients, aggregation, and bacterial population. *Soil Sci.* **91,** 291–298.

Spanner, M. A., Teuber, K. B., Acevedo, W., Peterson, D. L., Running, S. W., Card, D. H., and Mouat, D. A. (1984). Remote sensing of the leaf area index of temperate coniferous forests. 1984 Symposium on Machine Processing of Remotely Sensed Data.

Specht, R. L. (1981). Nutrient release from decomposing leaf litter of *Banksia ornata,* Dark Island heathland, South Australia. *Aust. J. Ecol.* **6,** 59–63.

Sprugel, D. G. (1976). Dynamic structure of wave-regenerated *Abies balsamea* forests in the north-eastern United States. *J. Ecol.* **64,** 889–911.

Staaf, H. (1982). Plant nutrient changes in beech leaves during senescence as influenced by site characteristics. *Oecol. Plant.* **17,** 161–170.

Staaf, H., and Berg, B. (1982). Accumulation and release of plant nutrients in decomposing Scots pine needle litter. Long-term decomposition in a Scots Pine forest. II. *Can. J. Bot.* **60,** 1561–1568.

Stachurski, A., and Zimka, J. R. (1975). Methods of studying forest ecosystems: Leaf area, leaf production, and withdrawal of nutrients from leaves of trees. *Ekol. Pol.* **23,** 637–648.

Stanhill, G. (1981). Radiation balance of plants. *In* "Plants and Their Atmospheric Environment" (J. Grace, E. D. Ford, and P. G. Jarvis, eds.), pp. 57–73. Blackwell, Oxford.

Stanhill, G. L. (1970). Some results of helicopter measurements of albedo of different land surfaces. *Sol. Energy* **11,** 59–66.

Stanley, S. R., and Ciolkosz, E. J. (1981). Classification and genesis of spodosols in the central Appalachians. *Soil Sci. Soc. Am. J.* **45,** 912–917.

Steingraeber, D. A., Kascht, L. J., and Franck, D. H. (1979). Variation of shoot morphology and bifurcation ratio in sugar maple (*Acer saccharum*) saplings. *Am. J. Bot.* **66,** 441–445.

Stephens, G. R., and Waggoner, P. E. (1970). The forests anticipated from 40 years of natural transitions in mixed hardwoods. *Bull.—Conn. Agric. Exp. Stn., New Haven* **707,** 1–58.

Stephenson, G. R., and Freeze, R. A. (1974). Mathematical simulation of subsurface flow contributions to snowmelt runoff. Reynolds Creek watershed, Idaho. *Water Resour. Res.* **10,** 284–294.

Stocker, O. (1960). Die photosynthethetischen Leistungen der Steppen und Wüstenpflanzen. *In* "Handbuch der Pflanzenphysiologie" (W. Ruhland, ed.), Vol. 5, pp. 460–491. Springer-Verlag, Berlin and New York.

Stoiber, R. E., and Jepsen, A. (1973). Sulfur dioxide contributions to the atmosphere by volcanoes. *Science* **182,** 577–578.

Stoker, H. S., and Seager, S. L. (1976). "Environmental Chemistry: Air and Water Pollution," 2nd ed. Scott, Foresman & Co., Glenview, Illinois.

Stone, E. L., and Boonkird, S. (1963). Calcium accumulation in the bark of *Terminalia* spp. in Thailand. *Ecology* **44,** 586–588.

Stone, E. L., and Kszystyniak, R. (1977). Conservation of potassium in the *Pinus resinosa* ecosystem. *Science* **198,** 192–194.

Strain, B. R., and Johnson, P. L. (1963). Corticular photosynthesis and growth in *Populus tremuloides. Ecology* **44,** 581–584.

Strain, B. R., Higginbotham, K. O., and Mulroy, J. C. (1976). Temperature preconditioning and photosynthetic capacity of *Pinus taeda* L. *Photosynthetica* **10,** 47–53.

Stuiver, M. (1978). Radiocarbon timescale tested against magnetic and other dating methods. *Nature (London)* **273,** 271–274.

Stuiver, M. (1978). Atmospheric carbon dioxide and carbon reservoir changes. *Science* **199,** 253–258.

Suberkropp, K., Godshalk, G. L., and Klug, M. J. (1976). Changes in the chemical composition of leaves during processing in a woodland stream. *Ecology* **57,** 720–727.

Sucoff, E. (1972). Water potential in red pine: Soil moisture, evapotranspiration, crown position. *Ecology* **53,** 681–686.

Sukwong, S., Frayer, W. E., and Mogren, E. W. (1971). Generalized comparisons of the precision of fixed-radius and variable-radius plots for basal-area estimates. *For. Sci.* **17,** 263–271.

Swain, T. (1977). Secondary compounds as protective agents. *Annu. Rev. Plant Physiol.* **28,** 479–501.

Swank, W. T. (1972). Water balance, interception, and transpiration studies on a watershed in the Pudget Lowland region of western Washington. Ph.D. Dissertation, University of Washington, Seattle.

Swank, W. T., and Caskey, W. H. (1982). Nitrate depletion in a second-order mountain stream. *J. Environ. Qual.* **11,** 581–584.

Swank, W. T., and Douglas, J. E. (1974). Streamflow greatly reduced by converting deciduous hardwood stands to pine. *Science* **185,** 857–859.

Swank, W. T., and Henderson, G. S. (1976). Atmospheric input of some cations and anions to forest ecosystems in North Carolina and Tennessee. *Water Resour. Res.* **12,** 541–546.

Swank, W. T., and Schreuder, H. T. (1973). Temporal changes in biomass, surface area and net production for a *Pinus strobus* L. forest. *In* "IUFRO Biomass Studies. Working Party on the Mensuration of the Forest Biomass" (H. E. Young, ed.), pp. 171–182. Coll. Life Sci. Agric., University of Maine, Orono.

Swank, W. T., and Schreuder, H. T. (1974). Comparison of three methods of estimating surface area and biomass for a forest of young eastern pine. *For. Sci.* **20,** 91–100.

Swanson, F. J., and Dyrness, C. T. (1975). Impact of clear-cutting and road construction on soil erosion by landslides in the western Cascade Range, Oregon. *Geology* **3,** 393–396.

Swanson, F. J., and Fredriksen, R. L. (1982). Sediment routing and budgets: Implications for judging impacts of forestry practices. *USDA For. Serv. Gen. Tech. Rep. PNW* **PNW-141,** 129–137.

Swanson, F. J., and James, M. E. (1975). "Geology and Geomorphology of the H. J. Andrews Experimental Forest, Western Cascades, Oregon," PNW-188. U. S. Dep. Agric., For. Serv., Pac. Northwest For. Range Exp. Stn., Portland, Oregon.

Swanson, F. J., Lienkaemper, G. W., and Sedell, J. R. (1976). History, physical effects, and management implications of large organic debris in western Oregon streams. *USDA For. Serv. Gen. Tech. Rep. PNW* **PNW-56,** 1–15.

Swanson, F. J., Fredriksen, R. L., and McCorison, F. M. (1982). Material transfer in a western Oregon forested watershed. *In* "Analysis of Coniferous Forest Ecosystems in the Western United States" (R. L. Edmonds, ed.), US/IBP Synth. Ser. No. 14, pp. 233–266. Dowden, Hutchinson & Ross, Inc., Stroudsburg, Pennsylvania.

Swanson, F. J., Gregory, S. V., Sedell, J. R., and Campbell, A. G. (1982). Land-water interactions: The riparian zone. *In* "Analysis of Coniferous Forest Ecosystems in the Western United States" (R. L. Edmonds, ed.), US/IBP Synth. Ser. No. 14, pp. 267–291. Dowden, Hutchinson & Ross, Inc., Stroudsburg, Pennsylvania.

Swanston, D. N. (1970). Mechanics of debris avalanching in shallow till soils of southeast Alaska. *USDA For. Serv. Res. Pap. PNW* **PNW-103,** 1–17.

Swanston, D. N., and Swanson, F. J. (1976). Timber harvesting, mass erosion, and steepland forest geomorphology in the Pacific Northwest. *In* "Geomorphology and Engineering" (D. R. Coates, ed.), pp. 199–221. Dowden, Hutchinson & Ross, Inc., Stroudsburg, Pennsylvania.

Swedish Ministry of Agriculture (1982). "Acidification Today and Tomorrow." Risbergs Tryckeri AB, Uddevalla, Sweden.

Swift, L. W., Swank, W. T., Mankin, J. B., Luxmoore, R. J., and Goldstein, R. A. (1975). Simulation of evapotranspiration and drainage from mature and clearcut deciduous forests and young pine plantation. *Water Resour. Res.* **11,** 667–673.

Swift, M. J., Heal, O. W., and Anderson, J. M. (1979). "Decomposition in Terrestrial Ecosystems." Univ. of California Press, Berkeley.

Swift, M. J., Russell-Smith, A., and Perfect, T. J. (1981). Decomposition and mineral-nutrient dynamics of plant litter in a regenerating bush-fallow in sub-humid tropical Nigeria. *J. Ecol.* **69,** 981–995.

Switzer, G. L., and Nelson, L. E. (1972). Nutrient accumulation and cycling in loblolly pine (*Pinus taeda* L.). Plantation ecosystems: The first twenty years. *Soil Sci. Soc. Am. Proc.* **36,** 143–147.

Tadaki, Y., Sato, A., Sakurai, S., Takeuchi, I., and Kawahara, T. (1977). Studies on the production structure of forest. XVII. Structure and primary production in subalpine "dead tree strips" *Abies* forest near Mt. A. Sahi. *Jpn. J. Ecol.* **27,** 83–90.

Tan, K. H., and Troth, P. S. (1982). Silica-sesquioxide ratios as aids in characterization of some temperate region and tropical soil clays. *Soil Sci. Soc. Am. J.* **46,** 1109–1114.

Tansey, M. R. (1977). Microbial facilitation of plant mineral nutrition. *In* "Microorganisms and Minerals" (E. D. Weinberg, ed.), pp. 343–385. Dekker, New York.

Tappeiner, J. C., and Alm, A. A. (1975). Undergrowth vegetation effects on the nutrient content of litterfall and soils in red pine and birch stands in northern Minnesota. *Ecology* **56,** 1193–1200.

Temple, S. A. (1977). Plant-animal mutualism: Coevolution with dodo leads to near extinction of plant. *Science* **197,** 885–886.

Teskey, R. O. (1982). Acclimation of *Abies amabilis* to water and temperature in a natural environment. Ph.D. Thesis, University of Washington, Seattle.

Tharp, M. L. (1979). Modeling major perturbations of a forest ecosystem. M.S. Thesis, University of Tennessee, Knoxville.

Thomas, W. A. (1969). Accumulation and cycling of calcium by dogwood trees. *Ecol. Monogr.* **39,** 101–120.

Thwaites, R. G. (1905). "Original Journals of the Lewis and Clark Expedition 1804–1808." Dodd, Mead & Co., New York.

Tilman, D. (1978). Cherries, ants and tent caterpillars: Timing of nectar production in relation to susceptibility of caterpillars to ant predation. *Ecology* **59,** 686–692.

Tilton, D. L. (1978). Comparative growth and foliar element concentrations of *Larix laricina* over a range of wetland types in Minnesota. *J. Ecol.* **66,** 499–512.

Timmer, V. R., and Stone, E. L. (1978). Comparative foliar analysis of young balsam fir fertilized with nitrogen, phosphorus, potassium, and lime. *Soil Sci. Soc. Am. J.* **42,** 125–130.

Titus, J. S., and Kang, S.-M. (1982). Nitrogen metabolism, translocation, and recycling in apple trees. *Hortic. Rev.* **4,** 204–246.

Tjepkema, J. D. (1979). Nitrogen fixation in forests of central Massachusetts. *Can. J. Bot.* **57,** 11–16.

Tjepkema, J. D., Cartica, R. J., and Hemond, H. F. (1981). Atmospheric concentration of ammonia in Massachusetts and deposition on vegetation. *Nature (London)* **294,** 445–446.

Tomanek, G. W. (1969). Dynamics of mulch layer in grassland ecosystems. *In* "The Grassland Ecosystem: A Preliminary Synthesis" (R. L. Dix and R. G. Beidleman, eds.), Range Sci. Dep., Sci. Ser. No. 2, pp. 225–240. Colorado State University, Fort Collins.

Tranquillini, W. (1979). Physiological ecology of the alpine timberline. *Ecol. Stud.* **31,** Springer-Verlag, New York.

Trewavas, A. J. (1982). Growth substance sensitivity: The limiting factor in plant development. *Physiol. Plant* **55,** 60–72.

Triska, F. J., and Cromack, K., Jr. (1980). The role of wood debris in forests and streams. *In* "Forests: Fresh Perspectives from Ecosystem Analysis" (R. H. Waring, ed.), Proc. 40th Annu. Biol. Colloq., pp. 171–190. Oregon State Univ. Press, Corvallis.

Triska, F. J., Sedell, J. R., and Gregory, S. V. (1982). Coniferous forest streams. *In* "Analysis of Coniferous Forest Ecosystems in the western United States" (R. L. Edmonds, ed.), US/IBP Synth. Ser. No. 14, pp. 292–332. Dowden, Hutchinson & Ross, Inc., Stroudsburg, Pennsylvania.

Troendle, C. A. (1979). A variable source area model for stormflow prediction on first order forested watersheds. Ph.D. Dissertation, University of Georgia, Athens.

Troeng, E., and Linder, S. (1982). Gas exchange in a 20-year-old stand of Scots pine. I. Net photosynthesis of current and one-year-old shoots within and between seasons. *Physiol. Plant.* **54,** 7–14.

Tucker, G. F., and Emmingham, W. H. (1977). Morphological changes in leaves of residual western hemlock after clear and shelterwood cutting. *For. Sci.* **23,** 195–203.

Tucker, C. J., Townshed, J. R. G., and Goff, T. E. (1985). Africal land-cover classification using satellite data. *Science* **227,** 369–375.

Tukey, H. B., Jr. (1970). The leaching of substances from plants. *Annu. Rev. Plant Physiol.* **21,** 305–324.

Turner, J. (1982). The mass flow component of nutrient supply in three western Washington forest types. *Acta Oecol.* **3,** 323–329.

Turner, J., and Olson, P. R. (1976). Nitrogen relations in a Douglas-fir plantation. *Ann. Bot. (London)* [N.S.] **40,** 1185–1193.

Turner, N. C., and Jarvis, P. G. (1975). Photosynthesis in Sitka spruce (Picea sitchensis (Bong.) Carr.). IV. Responses to soil temperature. *J. Appl. Ecol.* **12,** 561–576.

Tworkoski, T. J., Burger, J. A., and Smith, D. W. (1983). Soil texture and bulk density affect early growth of white oak seedlings. *Tree Plant. Notes* **34,** 22–25.

Tyler, G. (1972). Heavy metals pollute nature—may reduce productivity. *Ambio* **1,** 52–59.

Ugolini, F. C., Dawson, H., and Zachara, J. (1977). Direct evidence of particle migration in the soil solution of a podzol. *Science* **198,** 603–605.

Ugolini, F. C., Minden, R., Dawson, H., and Zachara, J. (1977). An example of soil processes in the *Abies amabilis* zone of central Cascades, Washington. *Soil Sci.* **124,** 291–302.

Ulrich, B., Mayer, R., and Khanna, P. K. (1980). Chemical changes due to acid precipitation in a loess-derived soil in central Europe. *Soil Sci.* **130,** 193–199.

Ulrich, B., Benecke, P., Harris, W. F., Khanna, P. K., and Mayer, R. (1981). Soil processes. *In* "Dynamic Properties of Forest Ecosystems" (D. E. Reichle, ed.), Int. Biol. Programme No. 23, pp. 265–339. Cambridge Univ. Press, London and New York.

U.S. Army Corps of Engineers (1956). "Snow Hydrology." North Pac. Div., USACE, Portland, Oregon.

U.S. Army Corps of Engineers (1972). "Flood Proofing Regulations," Doc. No. EP 1165 2314. USACE, Washington, D.C.

U.S. Department of Agriculture (1965). "Silvics of Forest Trees of the United States," Agric. Handb. No. 271. USDA For. Serv., Washington, D.C.

U.S. Forest Service (1972). "The Resources Capability System. A User's Guide." Watershed Syst. Dev. Unit, U.S. For. Serv., Berkeley, California.

Van Cleve, K. (1971). Energy and weight loss functions for decomposing foliage in birch and aspen forests in interior Alaska. *Ecology* **52,** 720–723.

Van Cleve, K., and Alexander, V. (1981). Nitrogen cycling in tundra and boreal ecosystems. *In* "Terrestrial Nitrogen Cycles" (F. E. Clark and T. Rosswall, eds.), pp. 375–404. Swed. Nat. Sci. Res. Counc., Stockholm.

Van Cleve, K., and Viereck, L. A. (1981). Forest succession in relation to nutrient cycling in the boreal forest of Alaska. *In* "Forest Succession: Concepts and Application" (D. C. West, H. H. Shugart, and D. B. Botkin, eds.), pp. 185–211. Springer-Verlag, Berlin and New York.

Van Cleve, K., and White, R. (1980). Forest-floor nitrogen dynamics in a 60-year-old paper birch ecosystem in interior Alaska. *Plant Soil* **54,** 359–381.

Van Cleve, K., Barner, R., and Schlentner, R. (1981). Evidence of temperature control of production and nutrient cycling in two interior Alaska black spruce ecosystems. *Can. J. For. Res.* **11,** 258–273.

Van Cleve, K., Oliver, L., Schlentnee, R., Viereck, L. A., and Dyrñess, C. T. (1983). Production and nutrient cycling in taiga forest ecosystems. *Can. J. For. Res.* **13,** 747–766.

Van Denburgh, A. S., and Feth, J. H. (1965). Solute erosion and chloride balance in selected river basins of the western conterminous United States. *Water Resour. Res.* **1,** 537–541.

van den Driessche, R. (1974). Prediction of mineral nutrient status of trees by foliar analysis. *Bot. Rev.* **40,** 347–394.

Van Devender, T. R., and Spaulding, W. G. (1979). Development of vegetation and climate in the southwestern United States. *Science* **204,** 701–710.

Van Emden, H. F. (1966). Plant insect relationships and pest control. *World Rev. Pest Control* **5,** 115–123.

Van Emden, H. F., and Bashford, M. A. (1971). The performance of Breviocoryne brassicae and Myzus persicae in relation to plant age and leaf amino acids. *Entomol. Exp. Appl.* **14,** 349–360.

Van Hook, R. I., Johnson, D. W., West, D. C., and Mann, L. K. (1982). Environmental effects of harvesting forests for energy. *For. Ecol. Manage.* **4,** 79–94.

Vannote, R. L. (1981). The river continuum: A theoretical construct for analysis of river ecosystems. *In* "Proceedings of the National Symposium on Freshwater Inflow to Estuaries," Vol. 2, pp. 289–304. U.S. Dep. Int. Fish Wildl. Serv. Office of Biol. Serv. 81-04.

Vannote, R. L., Minshall, G. W., Cummins, K. W., Sedell, J. R., and Cushing, C. E. (1980). The river continuum concept. *Can. J. Fish. Aquat. Sci.* **37,** 130–137.

Van Soest, P. J. (1967). Development of a comprehensive system of feed analysis and its application to forages. *J. Anim. Sci.* **26,** 119–128.

Van Wagner, C. E. (1978). Age-class distribution and the forest fire cycle. *Can. J. For. Res.* **8,** 220–227.

Viner, A. B. (1975). The supply of minerals to tropical rivers and lakes (Uganda). *In* "Coupling of Land and Water Systems" (A. A. Hasler, ed.), pp. 227–261. Springer-Verlag, Berlin and New York.

Viner, A. B., and Smith, I. R. (1973). Geographical, historical and physical aspects of Lake George. *Proc. R. Soc. London, Ser. B* **184,** 235–270.

Vitousek, P. (1982). Nutrient cycling and nutrient use efficiency. *Am. Nat.* **119,** 553–572.

Vitousek, P. M. (1984). Litterfall, nutrient cycling, and nutrient limitation in tropical forests. *Ecology* **65,** 285–298.

Vitousek, P. M., and Matson, P. A. (1984). Mechanisms of nitrogen retention in forest ecosystems: A field experiment. *Science* **225,** 51–52.

Vitousek, P. M., and Melillo, J. M. (1979). Nitrate losses from disturbed forests: Patterns and mechanisms. *For. Sci.* **25,** 605–619.

Vitousek, P. M., and Reiners, W. A. (1975). Ecosystem succession and nutrient retention: A hypothesis. *BioScience* **25,** 376–381.

Vitousek, P. M., Gosz, J. R., Grier, C. C., Melillo, J. M., and Reiners, W. A. (1982). A comparative analysis of potential nitrification and nitrate mobility in forest ecosystems. *Ecol. Monogr.* **52,** 155–177.

Vogt, K. A., Grier, C. C., Meier, C. E., and Edmonds, R. L. (1982). Mycorrhizal role in net primary production and nutrient cycling in *Abies amabilis* ecosystems in western Washington. *Ecology* **63,** 370–380.

Vogt, K. A., Grier, C. C., Meier, C. E., and Keyes, M. R. (1983). Organic matter and nutrient dynamics in forest floors of young and mature *Abies amabilis* stands in western Washington, as affected by fine-root input. *Ecol. Monogr.* **53,** 139–157.

Voroney, R. P., and Paul, E. A. (1984). Determination of K_c and K_n *in situ* for calibration of the chloroform fumigation-incubation method. *Soil Biol. Biochem.* **16,** 9–14.

Wagener, W. W. (1961). Past fire incidence in Sierra Nevada Forest. *J. For.* **59,** 739–748.

Waggoner, P. E., and Turner, N. C. (1971). Transpiration and its control by stomata in a pine forest. *Bull.—Conn. Agric. Exp. Stn., New Haven,* **726,** 1–87.

Walda, H., and Foster, R. (1978). *Zunacetha annulata* (Lepidotera: Dipotidae), an outbreak insect in a neotropical forest. *Geo-Eco-Trop* **2,** 443–454.

Walker, J. C. G. (1977). "Evolution of the Atmosphere." Macmillan, New York.

Walker, T. W., and Syers, J. K. (1976). The fate of phosphorus during pedogenesis. *Geoderma* **15,** 1–19.

Wallace, J. B., Webster, J. R., and Cuffney, T. F. (1982). Stream detritus dynamics: regulation by invertebrate consumers. *Oecologia* **53,** 197–200.

Walter, H. (1954). "Grundlangen der Pflanzenverbreitung. III. Einfuhrung in die Pflanzengeographie." Ulmer, Stuttgart.

Wargo, P. M. (1972). Defoliation-induced chemical changes in sugar maple roots stimulate growth of *Armillaria mellea*. *Phytopathology* **62,** 1278–1283.

Wargo, P. M. (1979). Starch storage and radial growth in woody roots of sugar maple. *Can. J. For. Res.* **9,** 49–56.

Wargo, P. M., Parker, J., and Houston, D. R. (1972). Starch content in roots of defoliated sugar maple. *For. Sci.* **18,** 203–204.

Waring, R. H. (1970). Matching species to site. *In* "Regeneration of Ponderosa Pine" (R. K. Hermann, ed.), pp. 54–61. For. Res. Lab., Oregon State University, Corvallis.

Waring, R. H. (1983). Estimating forest growth and efficiency in relation to canopy leaf area. *Adv. Ecol. Res.* **13,** 327–354.

Waring, R. H. (1985). Imbalanced ecosystems: Assessments and consequences. *For. Ecol. Manage.* **12,** 93–112.

Waring, R. H., and Cleary, B. D. (1967). Plant moisture stress: evaluation by pressure bomb. *Science* **155,** 1248–1254.

Waring, R. H., and Franklin, J. F. (1979). The evergreen coniferous forests of the Pacific Northwest. *Science* **204,** 1380–1386.

Waring, R. H., and Major, J. (1964). Some vegetation of the California a coastal redwood region in relation to gradients of moisture, nutrients, light, and temperature. *Ecol. Monogr.* **34,** 167–215.

Waring, R. H., and Pitman, G. B. (1983). Physiological stress in lodgepole pine as a precursor for mountain pine beetle attack. *Z. Angew. Entomol.* **96,** 265–270.

Waring, R. H., and Pitman, G. B. (1985). Modifying lodgepole pine stands to change susceptibility to mountain pine beetle attack. *Ecology* **66,** 889–897.

Waring, R. H., and Running, S. W. (1978). Sapwood water storage: Its contribution to transpiration and effect upon water conductance through the stems of old-growth Douglas-fir. *Plant, Cell, Environ.* **1,** 131–140.

Waring, R. H., and Youngberg, C. T. (1972). Evaluating forest sites for potential growth responses of trees to fertilizer. *Northwest Sci.* **46,** 67–75.

Waring, R. H., Gholz, H. L., Grier, C. C., and Plummer, M. L. (1977). Evaluating stem conducting tissue as an estimator of leaf area in four woody angiosperms. *Can. J. Bot.* **55,** 1474–1477.

Waring, R. H., Whitehead, D., and Jarvis, P. G. (1979). The contribution of stored water to transpiration in Scots pine. *Plant, Cell, Environ.* **2,** 309–317.

Waring, R. H., Whitehead, D., and Jarvis, P. G. (1980). Comparison of an isotopic method and the Penman-Monteith equation for estimating transpiration from Scots pine. *Can. J. For. Res.* **10,** 555–558.

Waring, R. H., Newman, K., and Bell, J. (1981a). Efficiency of tree crowns and stemwood production at different canopy leaf densities. *Forestry* **54,** 15–23.

Waring, R. H., Rogers, J. J., and Swank, W. T. (1981b). Water relations and hydrologic cycles. *In* "Dynamic Properties of Forest Ecosystems" (D. E. Reichle, ed.), Int. Biol. Programme No. 23, pp. 205–264. Cambridge Univ. Press, London and New York.

Waring, R. H., Schroeder, P. E., and Oren, R. (1982). Application of the pipe model theory to predict canopy leaf area. *Can. J. For. Res.* **12,** 556–560.

Waring, R. H., McDonald, A. J. S., Larsson, S., Ericsson, T., Wirén, A., Arwidsson, E., Ericsson, A., and Lohammar, T. (1985). Differences in chemical composition of plants grown at constant relative growth rates with stable mineral nutrition. *Oecologia* **66,** 157–160.

Watts, W. A. (1980). The late Quaternary vegetation history of the southeastern United States. *Annu. Rev. Ecol. Syst.* **11,** 387–409.

Watts, W. R., Neilson, R. E., and Jarvis, P. G. (1976). Photosynthesis in sitka spruce (*Picea sitchensis* (Bong.) Carr.). VII. Measurements of stomatal conductance and $^{14}CO_2$ uptake in a forest canopy. *J. Appl. Ecol.* **13,** 623–638.

Weaver, H. (1951). Fire as an ecological factor in the southwestern ponderosa pine forests. *J. For.* **49,** 93–98.

Webb, T. (1981). The past 11,000 years of vegetational change in eastern North America. *BioScience* **31,** 501–506.

Webb, W. L. (1981). Relation of starch content to conifer mortality and growth loss after defoliation by the Douglas-fir tussock moth. *For. Sci.* **27,** 224–232.

Webb, W. L., Szarek, S., Lauenroth, W., Kinerson, R., and Smith, M. (1978). Primary productivity and water use in native forest, grassland, and desert ecosystems. *Ecology* **59,** 1239–1247.

Webb, W. L., Lauenroth, W. K., Szarek, S. R., and Kinerson, R. S. (1983). Primary production and abiotic controls in forests, grasslands, and desert ecosystems in the United States. *Ecology* **64,** 134–151.

Weier, T. E., Stocking, C. R., and Barbour, M. G. (1974). "Botany, An Introduction to Plant Biology." Wiley, New York.

Wein, R. W., and Moore, J. M. (1977). Fire history and rotaations in the New Brunswick Acadian Forest. *Can. J. For. Res.* **7,** 285–294.

Welbourn, M. L., Stone, E. L., and Lassoie, J. P. (1981). Distribution of net litter inputs, with respect to slope position and wind direction. *For. Sci.* **27,** 651–659.

Welcomme, R. L. (1979). "Fisheries Ecology of Floodplain Rivers." Longmans, Green, New York.

Wells, P. V. (1966). Late Pleistocene vegetation and degree of pluvial climatic change in the Chihuahuan Desert. *Science* **153,** 970–975.

Wells, P. V. (1983). Paleobiogeography of montane islands in the Great Basin since the last glaciopluvial. *Ecol. Monogr.* **53,** 341–382.

West, D. C., McLaughlin, S. B., and Shugart, H. H. (1980). Simulated forest response to chronic air pollution stress. *J. Environ. Qual.* **9,** 43–49.

West, D. C., Shugart, H. H., and Botkin, D. B., eds. (1981). "Forest Succession: Concepts and Application." Springer-Verlag, Berlin and New York.

Westall, J., and Stumm, W. (1980). The Hydrosphere. *In* "The Handbook of Environmental Chemistry" (O. Hutzinger, ed.), Vol. 1, Part A, pp. 17–49. Springer-Verlag, Berlin and New York.

Wetzel, R. G., and Manny, B. A. (1977). Seasonal changes in particulate and dissolved organic carbon and nitrogen in a hardwater stream. *Arch. Hydrobiol.* **80,** 20–39.

Whisler, F. D., and Bouwer, H. (1970). Comparison of methods for calculating vertical drainage and infiltration for soils. *J. Hydrol.* **10,** 1–19.

Whisler, F. D., Watson, K. K., and Perrens, S. J. (1972). The numerical analysis of infiltration into heterogeneous porous media. *Soil Sci. Soc. Am. Proc.* **36,** 868–874.

White, E. J., and Turner, F. (1970). A method of estimating income of nutrients in a catch of airborne particles by a woodland canopy. *J. Appl. Ecol.* **7,** 441–451.

White, J. (1981). The allometric interpretation of the self-thinning rule. *J. Theor. Biol.* **89,** 475–500.

White, T. C. R. (1969). An index to measure weather-induced stress of trees associated with outbreaks of psyllids in Australia. *Ecology* **50,** 905–909.

White, T. C. R. (1974). A hypothesis to explain outbreaks of looper caterpillars, with special reference to populations of *Selidosema suavis* in a plantation of *Pinus radiata* in New Zealand. *Oecologia* **16,** 279–301.

Whitehead, D. (1978). The estimation of foliage area from sapwood basal area in Scots pine. *Forestry* **51,** 35–47.

Whitehead, D., and Jarvis, P. G. (1981). Coniferous forests and plantations. *In* "Water Deficits and Plant Growth" (T. T. Kozlowski, ed.), Vol. 6, pp. 49–152. Academic Press, New York.

Whitehead, D., Okali, D. U. U., and Fasehun, F. E. (1981). Stomatal response to environmental variables in two tropical forest species during the dry season in Nigeria. *J. Appl. Ecol.* **18,** 571–587.

Whitehead, D., Edwards, W. R. N., and Jarvis, P. G. (1984). Conducting sapwood area, foliage area and permeability in mature trees of *Picea sitchensis* and *Pinus contorta*. *Can. J. For. Res.* **14,** 940–947.

Whitehead, D. R., Rochester, H., Rissing, S. W., Douglass, C. B., and Sheehan, M. C. (1973). Late glacial and postglacial productivity changes in a New England pond. *Science* **181,** 744–747.

Whitford, W. G., Meentemeyer, V., Seastedt, T. R., Cromack, K., Crossley, D. A., Santos, P., Todd, R. L., and Waide, J. B. (1981). Exceptions to the AET model: Deserts and clear-cut forest. *Ecology* **62,** 275–277.

Whittaker, R. H. (1970). "Communities and Ecosystems." Macmillan, New York.

Whittaker, R. H. (1975). "Communities and Ecosystems." 2nd ed. Macmillan, New York.

Whittaker, R. H., and Woodwell, G. M. (1967). Surface area relations of woody plants and forest communities. *Am. J. Bot.* **54,** 931–939.

Whittaker, R. H., and Woodwell, G. M. (1968). Dimension and production relations of trees and shrubs in the Brookhaven forest, New York. *J. Ecol.* **56,** 1–25.

Whittaker, R. H., Walker, R. B., and Kruckeberg, A. R. (1954). The ecology of serpentine soils. *Ecology* **35,** 258–288.

Whittaker, R. H., Bormann, F. H., Likens, G.E., and Siccama, T. G. (1974). The Hubbard Brook ecosystem study: Forest biomass and production. *Ecol. Monogr.* **44,** 233–254.

Wicker, E. F., and Leaphart, C. E. (1976). Fire and dwarf mistletoe (*Arceuthbium spp.*) relationships in the northern Rocky Mountains. *Proc. Tall Timbers Fire Ecol. Conf.* **14,** 279–298.

Wickman, B. E. (1980). Increased growth of white fir after a Douglas-fir tussock moth outbreak. *J.For.* **78,** 31–33.

Wieder, R. K., and Lang, G. E. (1982). A critique of the analytical methods used in examining decomposition data obtained from litter bags. *Ecology* **63,** 1636–1642.

Williams, R. F. (1946). The physiology of plant growth with special reference to the concept of net assimilation rate. *Ann. Bot. (London)* [N.S.] **10,** 41–72.

Wilson, C. L. (1952). "Botany." Dryden Press, New York.

Winner, W. E., Smith, C. L., Koch, G. W., Mooney, H. A., Bewley, J. D., and Krouse, H. R. (1981). Rates of emission of H_2S from plants and patterns of stable sulphur isotope fractionation. *Nature (London)* **289,** 672–673.

Witkamp, M. (1971). Soils as components of ecosystems. *Annu. Rev. Ecol. Syst.* **2,** 85–110.

Witkamp, M. (1974). Direct and indirect counts of fungi and bacteria as indexes of microbial mass and productivity. *Soil Sci.* **118,** 150–155.

Witkamp, M., and Ausmus, B. S. (1976). Processes in decomposition and nutrient transfer in forest systems. *In* "The Role of Terrestrial and Aquatic Organisms in Decomposition Processes" (J. M. Anderson and A. MacFadyen, eds.), pp. 375–396. Blackwell, Oxford.

Witkamp, M., and Crossley, D. A. (1966). The role of arthropods and microflora in breakdown of white oak litter. *Pedobiologia* **6,** 293–303.

Wodzicki, T. J., and Brown, C. L. (1970). Role of xylem parenchyma in maintaining the water balance of trees. *Acta Soc. Bot. Pol.* **39,** 617–621.

Wong, S. C., Cowan, I. R., and Farquhar, G. D. (1978). Leaf conductance in relation to assimilation in *Eucalyptus pauciflora* Sieb. ex Spreng. Influence of irradiance and partial pressure of carbon dioxide. *Plant Physiol.* **62,** 670–674.

Wong, S. C., Cowan, I. R., and Farquhar, G.D. (1979). Stomatal conductance correlates with photosynthetic capacity. *Nature (London)* **282,** 424–426.

Wood, T., and Bormann, F. H. (1975). Increases in foliar leaching caused by acidification of an artificial mist. *Ambio* **4,** 169–171.

Wood, T., Bormann, F. H., and Voigt, G. K. (1984). Phosphorus cycling in a northern hardwood forest: Biological and chemical control. *Science* **223,** 391–393.

Wood, T. G. (1976). The role of termite (Isoptera) in decomposition processes. *In* "The Role of Terrestrial and Aquatic Organisms in Decomposition Processes" (J. M. Anderson and A. MacFadyen, eds.), pp. 145–168. Blackwell, Oxford.

Woodmansee, R. G. (1978). Additions and losses of nitrogen in grassland ecosystems. *BioScience* **28,** 448–453.

Woods, D. B., and Turner, N. C. (1971). Stomatal response to changing light by four tree species of varying shade tolerance. *New Phytol.* **70,** 77–84.

Woods, F. W., and O'Neal, D. (1965). Tritiated water as a tool for ecological field studies. *Science* **147,** 148–149.

Woodwell, G. M. (1974). Variation in the nutrient content of leaves of *Quercus alba, Quercus coccinea,* and *Pinus rigida* in the Brookhaven forest from bud-break to abscission. *Am. J. Bot.* **61,** 749–753.

Woodwell, G. M., Hobbie, J. E., Houghton, R. A., Melillo, J. M., Peterson, B. J., Shaver, G. R., Stone, T. A., Moore, B., and Park, A. B. (1983). "Deforestation Measured by LANDSAT: Steps Toward a Method." U.S. Dep. Energy, Washington, D.C.

Worrell, R. (1983). Damage by the spruce bark beetle in South Norway, 1970–1980: A survey, and factors affecting its occurrence. *Rep. Norw. For. Sci. Res. Inst.* **38,** 1–34.

Worsley, T. R., and Davies, T. A. (1979). Sea-level fluctuations and deep-sea sedimentation rates. *Science* **203,** 455–456.

Wright, H. A., and Bailey, A. W. (1982). "Fire Ecology, United States and Southern Canada." Wiley, New York.

Wright, R. F. (1976). The impact of forest fire on the nutrient influxes to small lakes in northeastern Minnesota. *Ecology* **57,** 649–663.

Yarie, J. (1980). The role of understory vegetation in the nutrient cycle of forested ecosystems in the mountain hemlock biogeoclimatic zone. *Ecology* **61,** 1498–1514.

Yarie, J. (1981). Forest fire cycles and life tables: A case study from interior Alaska. *Can. J. For. Res.* **11,** 554–562.

Yawney, H. W., Leaf, A. L., and Leonard, R. E. (1978). Nutrient content of throughfall and stemflow in fertilized and irrigated *Pinus resinosa* Alt stands. *Plant Soil* **50,** 433–445.

Yoda, K., Kira, T., Ogawa, H., and Hozumi, K. (1963). Self-thinning in overcrowded stands under cultivated and natural conditions. *J. Biol., Osaka City Univ.* **14,** 107–129.

Youngberg, C. T. (1959). The influence of soil conditions, following tractor logging, on the growth of planted Douglas-fir seedlings. *Soil Sci. Soc. Am. Proc.* **23,** 76–78.

Youngberg, C. T., and Wollum, A. G. (1976). Nitrogen accretion in developing *Ceanothus velutinus* stands. *Soil Sci. Soc. Am. Proc.* **40,** 109–112.

Zackrisson, O. (1977). Influence of forest fires on the North Swedish boreal forest. *Oikos* **29,** 22–32.

Zech, W., and Popp, E. (1983). Magnesiummangel, einer der Grunde fur das Fichten- und Tannensterben in NO-Bayern. *Forstwiss. Centralbl.* **102,** 50–55.

Ziemer, R. R. (1981). Roots and the stability of forested slopes. *IAHS-AISH Publ.* **132,** 343–357.

Zimka, J. R., and Stachurski, A. (1976). Vegetation as a modifier of carbon and nitrogen transfer to soil in various types of forest ecosystems. *Ekol. Pol.* **24,** 493–514.

Zimmerman, P. R., Greenberg, J. P., Wandiga, S. O., and Crutzen, P. J. (1982). Termites: A potentially large source of atmospheric methane, carbon dioxide, and molecular hydrogen. *Science* **218,** 563–565.

Zimmerman, R. C., Goodlett, J. C., and Comer, G. H. (1967). The influence of vegetation on channel form of small streams. *IASH-AISH Publ.* **75,** 255–275.

Zimmermann, M. H. (1978). Hydraulic architecture of some diffuse-porous trees. *Can. J. Bot.* **56,** 2286–2295.

Zimmermann, M. H., and Brown, C. L. (1971). "Trees, Structure and Function." Springer-Verlag, Berlin and New York.

Zinke, P. J. (1967). Forest interception studies in the United States. *In* "Forest Hydrology" (W. E. Sopper and W. Lull, eds.), pp. 137–160. Pergamon, Oxford.

Zinke, P. J. (1980). Influence of chronic air pollution on mineral cycling in forests. *U.S., For. Serv., Pac. Southwest For. Range Exp. Stn., Gen. Tech. Rep.* **43,** 88–99.

Zisa, R. P., Halverson, H. G., and Stoul, B. B. (1980). Establishment and early growth of conifers on compacted soils. *USDA For. Serv. Res. Pap. NE* **NE-451,** 1–8.

Zöttl, H. W., and Mies, E. (1983). Die Fichtenerkrankung in Hochlagen des Südschwarzwaldes. *Allg. Forst- Jagdztg.* **154,** 110–114.

Zucker, W. V. (1983). Tannins: Does structure determine function? An ecological perspective. *Am. Nat.* **121,** 335–365.

Index

A

D

X

Z